# Theory and Applications of Natural Language Processing

Series Editors:
Graeme Hirst (Textbooks)
Eduard Hovy (Edited volumes)
Mark Johnson (Monographs)

## Aims and Scope

The field of Natural Language Processing (NLP) has expanded explosively over the past decade: growing bodies of available data, novel fields of applications, emerging areas and new connections to neighboring fields have all led to increasing output and to diversification of research.

"Theory and Applications of Natural Language Processing" is a series of volumes dedicated to selected topics in NLP and Language Technology. It focuses on the most recent advances in all areas of the computational modeling and processing of speech and text across languages and domains. Due to the rapid pace of development, the diversity of approaches and application scenarios are scattered in an ever-growing mass of conference proceedings, making entry into the field difficult for both students and potential users. Volumes in the series facilitate this first step and can be used as a teaching aid, advanced-level information resource or a point of reference.

The series encourages the submission of research monographs, contributed volumes and surveys, lecture notes and textbooks covering research frontiers on all relevant topics, offering a platform for the rapid publication of cutting-edge research as well as for comprehensive monographs that cover the full range of research on specific problem areas.

The topics include applications of NLP techniques to gain insights into the use and functioning of language, as well as the use of language technology in applications that enable communication, knowledge management and discovery such as natural language generation, information retrieval, question-answering, machine translation, localization and related fields.

The books are available in printed and electronic (e-book) form:

* Downloadable on your PC, e-reader or iPad
* Enhanced by Electronic Supplementary Material, such as algorithms, demonstrations, software, images and videos
* Available online within an extensive network of academic and corporate R&D libraries worldwide
* Never out of print thanks to innovative print-on-demand services
* Competitively priced print editions for eBook customers thanks to MyCopy service http://www.springer.com/librarians/e-content/mycopy

For other titles published in this series, go to
www.springer.com/series/8899

Iryna Gurevych • Jungi Kim
Editors

# The People's Web Meets NLP

## Collaboratively Constructed Language Resources

Foreword by Nicoletta Calzolari

Springer

*Editors*
Iryna Gurevych
Jungi Kim
Department of Computer Science
Ubiquitous Knowledge Processing (UKP) Lab
Technische Universität Darmstadt
Darmstadt
Germany

*Foreword by*
Nicoletta Calzolari
Istituto Linguistica Computazionale "Antonio Zampolli"
Pisa
Italy

ISSN 2192-032X ISSN 2192-0338 (electronic)
ISBN 978-3-642-35084-9 ISBN 978-3-642-35085-6 (eBook)
DOI 10.1007/978-3-642-35085-6
Springer Heidelberg New York Dordrecht London

Library of Congress Control Number: 2013934709



Printed on acid-free paper

Springer is part of Springer Science+Business Media (www.springer.com)

*Firstly, we would like to thank our families for their love and support while working on this project. Secondly, we would like to dedicate this book to the people who have shaped our professional life the way it is. It's great luck to have met you, and thanks for the inspiration we draw from the contact with you. Thanks to all of you for being on our side. Without this, the present book would not have happened. We also extend our warmest thanks to the Volkswagen Foundation, which has funded the project as part of the Lichtenberg-Professorship Program under grant No. I/82806.*

*Iryna Gurevych and Jungi Kim*
*September 2012, Darmstadt, Germany*

# Foreword

It's a pleasure to write the Foreword for the book on Collaboratively Constructed Language Resources. I believe that the trend of collaborative construction of Language Resources (LRs) represents both a "natural" evolution of computerised resource building (I'll try to give few historical hints) and a "critical" evolution for the future of the field of language resources.

## 1 Some Historical Hints

Where does collaborative resource construction position itself in the language resource field? I'll just give a glimpse here at some historical antecedents of the current collaborative methodology, without mentioning the obvious ones, like Wikipedia or Wiktionary.

### *1.1 A Nineteenth Century Lexicographic Enterprise*

We have not invented collaborative construction of language resources, or even crowdsourcing, just recently.

George P. Marsh used it already in 1859 for the Philological Society of London for "the preparation of a complete lexicon or thesaurus of the English language", the New English Dictionary (now known as the Oxford English Dictionary). Acting as Secretary in America he decided to "adopt this method of bringing the subject to the notice of persons in this country who may be disposed to contribute to the accomplishment of the object, by reading English books and noting words ...". Moreover: "... the labors of the English contributors are wholly gratuitous".

Given that not much material was collected after this appeal, a similar appeal[1] was re-launched 20 years later, in 1879, by the dictionary's editor James Murray when "volunteer readers were recruited to contribute words and illustrative quotations": "... the Committee want help from readers in Great Britain, America, and the British Colonies, to finish the volunteer work so enthusiastically commenced 20 years ago ...", and "A thousand readers are wanted, and confidently asked for, to complete the work as far as possible within the next 3 years, so that the preparation of the Dictionary may proceed upon full and complete materials."

We can't deny that this is a clear example of collaborative construction of a language resource! It could even be defined as an early example of crowdsourcing.

## *1.2 More Recent Examples: Some EC Resource Projects of the Twentieth Century*

Other – more recent – examples could be found in the policy adopted in projects funded by the European Commission, in the 1990s, where many language resources had to be collaboratively built inside a consortium of many partners. Also because of this "enforced" collaboration, some features and trends presenting clear connections with the current notion of "collaborative building" emerged in the first half of the 1990s:

1. The need to build a core set of LRs, designed in a harmonised way, for all the EU languages
2. The need to base LR building on commonly accepted standards
3. The need to make the LRs that are created available to the community by large, i.e. the need for a distribution policy (at that time we introduced the notion of distributing resources, not yet sharing them!).

By the way, these requirements are strictly implied by and related to the emerging notion in the 1990s of the "infrastructural role" of LRs.

I just mention two types of collaborative resource building in EC projects, representing two partially different building models.

One method could be represented by the EuroWordNet projects [9]: each partner was building the WordNet for her/his language, all modelled on – and linked to – the original Princeton WordNet, and altogether constituting a homogeneous and interrelated set of lexicons.

Another method is represented by projects like PAROLE [11] and SIMPLE [1,7], for the construction and acquisition of harmonised resources. They were, to my knowledge, the first attempt at developing together medium-size coverage lexicons for so many languages (12 European languages), with a harmonised common model,

[1] http://public.oed.com/history-of-the-oed/archived-documents/april-1879-appeal/

and with encoding of structured semantic types and syntactic (subcategorisation) and semantic frames on a large scale. Reaching a common agreed model grounded on sound theoretical approaches within a very large consortium, and for so many languages, was in itself a challenging task. The availability of these uniformly structured lexical and textual resources, based on agreed models and standards, in so many EU languages, offered the benefits of a standardised base, creating an infrastructure of harmonised LRs throughout Europe.

What was interesting was that these projects positioned themselves inside the strategic policy – supported by the EC – aiming at providing a core set of language resources for the EU languages based on the principle of "subsidiarity". According to the subsidiarity concept, the process started at the EU level continued at the national level, extending to real-size the core sets of resources in the framework of a number of National Projects.

This achievement was of major importance in a multilingual space like Europe, where all the difficulties connected with the task of LR building are multiplied by the language factor. All the various language resource projects determined also the beginning of the interest in standardisation in Europe. It was seen as a waste of money, effort and time the fact that every new project was redoing from scratch the same type of (fragments of) LRs, without reusing what was already available, while LRs produced by the projects were usually forgotten and left unused. From here, the notion of "reusability" arose [2]. As a remedy, a clear demand for interoperability standards and for common terms of reference emerged.

## *1.3 Reusability and Integration of Language Resources*

Other requirements with connections to collaborative construction of LRs are the possibility to reuse and integrate different language resources.

LRs (i.e. data) started to be understood as critical to make steps forward in NLP already in the 1980s, marking a sort of revolution with respect to times and approaches in which they were even despised as an uninteresting burden. The 1986 Grosseto (Tuscany) Workshop "On automating the lexicon" [10] was the event marking this inversion of tendency and the starting point of the process which gradually brought the major actors of the NLP sector to pay more and more attention to so-called "reusable" language resources.

In 1998, in a keynote talk at the 1st LREC in Granada, I could state that "Integration of different types of LRs, approaches, techniques and tools must be enforced" as a compelling requirement for our field: "The integration aspect is becoming – fortunately – a key aspect for the field to grow. This is in fact a sign of maturity: today various types of data, techniques, and components are available and waiting to be integrated with not too great an effort. I believe that this integration task is an essential step towards ameliorating the situation, both in view of new applicative goals and also in view of new research dimensions. The integration of many existing components gives in fact more than the sum of the parts, because their combination adds a different quality."

Among the combinations to be explored I mentioned: interaction between lexicon and corpus, integration of different types of lexicons, of various components in a chain (what we call today workflows), of Written and Spoken LRs towards multimedia and multimodal LRs, and also integration of symbolic and statistical approaches. I observed that "a single group simply does not have the means, or the interest, to carry them out. . . . everything is tied together, which makes our overall task so interesting – and difficult. What we must have is the ability to combine the overall view with its decomposition into manageable pieces. No one perspective – the global and the sectorial – is really fruitful if taken in isolation. A strategic and visionary policy has to be debated, designed and adopted for the next few years, if we hope to be successful." [3].

Collaborative construction of LRs is linked to and is an evolution of both the reusability notion and the integration requirement.

## 2 Language Technology As a Data Intensive Field: The Data-Driven Approach

LRs were not conceived as an end in themselves, but as an essential component to develop robust systems and applications. They were the obvious prerequisite and the critical factor in the emergence and the consolidation of the data-driven approach in human language technology. Today we recognise that Language Technology is a data-intensive field and that major breakthroughs have stemmed from a better use of more and more Language Resources.

### *2.1 From Murray Appeal, Through Corpus-Based Lexicography, Back to Collaborative Work!*

In the 1990s computer-aided corpus-based lexicography became the "normal" lexicographic practice for the identification and selection of documentation – through text-processing methods, frequency lists, patterns spotting, context analysis, and so on. No need to ask for 10,000 contributors!

Data-driven methods and automatic acquisition of linguistic information started in the late 1980s with the ACQUILEX project [4], aiming at acquiring lexical information from so-called machine-readable dictionaries. The needs of "language industry" applications compelled to rely on actual usage of languages, as attested in large corpora, for acquiring linguistic information, instead of relying on human introspection as the source of linguistic information and of testing linguistic hypotheses with small amounts of data. This meant developing statistical techniques, machine learning, text mining, and so on.

All this was/is very successful, but all these techniques on one side rely on bigger and bigger collections of data (LRs), possibly annotated in many ways and often with human intervention, and on the other side they are never 100 % correct, thus requiring again human intervention. Therefore, if more and bigger (processed) LRs are needed, if statistical techniques arrive at a certain limit, new ways to cope with this need of "Big Data" must be found and explored. Natural ways of coping with the big data paradigm and the need of accumulation of extremely large (linguistic) knowledge bases are:

(i) Collaborative building of resources on one side, and
(ii) Putting again human intelligence in the loop on the other side, recognising that some tasks are better performed by humans: crowdsourcing as a form of global human-based computation.

Collaborative building vs. crowdsourcing can be paralleled to the difference between involvement and contribution of colleagues (as in the EC projects above) vs. involvement of the layman/everyone (as in Murray appeal). Even if both can be said to rely on collective intelligence or on the "wisdom of the crowd", they clearly represent quite different approaches and methodologies and require different organisations.

## 3 Language Resources and the Collaborative Framework: To Achieve the Status of a Mature Science

The traditional LR production process is too costly. A new paradigm is pushing towards open, distributed language infrastructures based on sharing LRs, services and tools. Joining forces and working together on big experiments that collect thousands of researchers is – since many years – my dream, what I think is the only way for our field to achieve the status of a mature science.

It is urgent to create a framework enabling effective cooperation of many groups on common tasks, adopting the paradigm of accumulation of knowledge so successful in more mature disciplines, such as biology, astronomy, physics. This requires enabling the development of web-based environments for collaborative annotation and enhancement of LRs, but also the design of a new generation of multilingual LRs, based on open content interoperability standards. The rationale behind the need of open LR repositories is that accumulation of massive amounts of (high-quality) multi-dimensional data about many languages is the key to foster advancement in our knowledge about language and its mechanisms. We must finally be coherent and take concrete actions leading to the coordinated gathering – in a shared effort – of as many (processed/annotated) language data as we are collectively able to produce. This initiative compares to the astronomers/astrophysics' accumulation of huge amounts of observation data for a better understanding of the universe.

Consistently with the vision of an open distributed space of sharable knowledge available on the web for processing, the "multilingual Semantic Web" may help in determining the shape of the LRs of the future and may be crucial to the success of an infrastructure – critically based on interoperability – aimed at enabling/improving sharing and collaborative building of LRs for a better accessibility to multilingual content. This will serve better the needs of language applications, enabling building on each other achievements, integrating results, and having them accessible to various systems, thus coping with the need of more and more 'knowledge intensive' large-size LRs for effective multilingual content processing. This is the only way to make a giant leap forward.

## 3.1 Relations with Other Dimensions Relevant to the LR Field

In the "FLaReNet Final Blueprint" [6,8], the actions recommended for a strategy for the future of the LR field are organised around nine dimensions: (a) Infrastructure, (b) Documentation, (c) Development, (d) Interoperability, (e) Coverage, Quality and Adequacy, (f) Availability, Sharing and Distribution, (g) Sustainability, (h) Recognition, (i) International Cooperation. Taken together, as a coherent system, these directions contribute to a sustainable LR ecosystem.

Let's not forget that the same requirements apply whatever the method of LR building: collaboratively built resources undergo the same rules/recommendations. An implication of collaboration is that interoperability acquires even more value. The same is true for sustainability, for data infrastructure enabling international collaboration, and also for notions such as authority and trust. Moreover, when collaborative building is explicitly performed, there is the need to better define all the small steps inside an overall methodology. These recommendations could be taken as a framework in which to insert our future work strategy also in the collaborative paradigm.

## 3.2 Let's Organise Our Future!

One of the challenges for the collaborative model to succeed will be to ensure that the community is engaged at large! This can also be seen as an effort to push towards a culture of "service to the community" where everyone has to contribute. This "cultural change" is not a minor issue. This requirement was for example at the basis of the LRE Map idea, a collaborative bottom-up means of collecting metadata on LRs from conference authors, contributing to the promotion of a large movement towards an accurate and massive bottom-up documentation of LRs [5].[2]

My final remark is that, as with any new development, it is important on one side to leave space to the free rise of new ideas and methods inside the collaborative paradigm, but is also important to start organising its future. There must be a bold vision and an international group able to push for it (with both researchers and policy makers involved) and to organise some grand challenge that, via a distribution of efforts and exploiting the sharing trend, involves the collaboration of a consistent portion of our community. Could we envision a large "Language Library" as the beginning of a big Genome project for languages, where the community collectively deposits/creates increasingly rich and multi-layered LRs, enabling a deeper understanding of the complex relations between different annotation layers/language phenomena?

Pisa, Italy Nicoletta Calzolari

## References

1. Busa F, Calzolari N, Lenci A (2001) Generative lexicon and the SIMPLE model: developing semantic resources for NLP. In: Bouillon P, Busa F (eds) The language of word meaning. Cambridge University Press, Cambridge, pp 333–349
2. Calzolari N (1991) Lexical databases and textual corpora: perspectives of integration for a lexical knowledge base. In: Zernik U (ed) Lexical acquisition: exploiting on-line resources to build a lexicon. Lawrence Erlbaum Associates, Hillsdale, pp 191–208
3. Calzolari N (1998) An overview of written language resources in Europe: a few reflection, facts, and a vision. In: Proceedings of the first international conference on language resources and evaluation (LREC), vol I, Granada, Spain, May. European Language Resources Association (ELRA), Paris, pp 217–224
4. Calzolari N, Briscoe T (1995) ACQUILEX I and II. Acquisition of lexical knowledge from machine-readable dictionaries and text corpora. Cah Lexicol 67(2):95–114
5. Calzolari N, Soria C, Del Gratta R, D'Onofrio L, Goggi S, Quochi V, Russo I, Choukri K, Mariani J, Piperidis S (2010) The LREC2010 resource map. In: Proceedings of the seventh international conference on language resources and evaluation (LREC'10), Valletta, Malta, May. European Language Resources Association (ELRA), pp 19–21
6. Calzolari N, Quochi V, Soria C (2012) The strategic language resource agenda. FLaReNet final deliverable. CNR-ILC, Pisa. http://www.flarenet.eu/
7. Lenci A, Bel N, Busa F, Calzolari N, Gola E, Monachini M, Ogonowsky A, Peters I, Peters W, Ruimy N, Villegas M, Zampolli A (2000) SIMPLE: a general framework for the development of multilingual lexicons. Int J Lexicogr 13(4):249–263
8. Soria C, Bel N, Choukri K, Mariani J, Monachini M, Odijk JEJM, Piperidis S, Quochi V, Calzolari N (2012) The FLaReNet strategic language resource agenda. In: Proceedings of the eight international conference on language resources and evaluation (LREC'12). ELRA, Istanbul, pp 1379–1386

[2] see also http://www.resourcebook.eu/ with metadata for about 4,000 LRs from many conferences

9. Vossen P (ed) (1998) EuroWordNet: a multilingual database with lexical semantic networks. Kluwer, Dordrecht, p 179
10. Walker D, Zampolli A, Calzolari N (eds) (1995) Automating the lexicon: research and practice in a multilingual environment. Clarendon Press/OUP, Oxford, p 413
11. Zampolli A, Calzolari N (1996) LE-PAROLE: its history and scope. ELRA Newsl 1(4):5–6

# Preface

In the last years, researchers from a variety of computer science fields including computer vision, language processing and distributed computing have begun to investigate how collaborative approaches to the construction of information resources can improve the state-of-the-art. Collaboratively constructed language resources (CCLRs) have been recognized as a topic of its own in the field of Natural Language Processing (NLP) and Computational Linguistics (CL). In this area, the application of collective intelligence has yielded CCLRs such as Wikipedia, Wiktionary, and other language resources constructed through crowdsourcing approaches, such as Games with a Purpose and Mechanical Turk.

The emergence of CCLRs generated new challenges to the research field. Collaborative construction approaches yield new, previously unknown levels of coverage, while also bringing along new research issues related to the quality and the consistency of representations across domains and languages. Rather than a small group of experts, the data prepared by volunteers for knowledge construction comes from multiple sources, experts or non-experts with all gradations in-between in a crowdsourcing manner. The resulting data can be employed to address questions that were not previously feasible due to the lack of the respective large-scale resources for many languages, such as lexical-semantic knowledge bases or linguistically annotated corpora, including differences between languages and domains, or certain seldom occurring phenomena.

The research on CCLRs has focused on studying the nature of resources, extracting valuable knowledge from them, and developing algorithms to apply the extracted knowledge in various NLP tasks. Because the CCLRs themselves present interesting characteristics that distinguish them from conventional language resources, it is important to study and understand their nature. The knowledge extracted from CCLRs can substitute for or supplement customarily utilized resources such as WordNet or linguistically annotated corpora in different NLP tasks. Other important research directions include interconnecting and managing CCLRs and utilizing NLP techniques to enhance the collaboration processes while constructing the resources.

CCLRs contribute to NLP and CL research in many different ways, as demonstrated by the diversity and significance of the topics and resources addressed in the chapters of this volume. They promote the improvement of the respective methodologies, software, and resources to achieve deeper understanding of the language, at the larger scale and more in-depth. As the topic of CCLRs matures as a research area, it has been consolidated in a series of workshops in the major CL and artificial intelligence conferences,[3] and a special of issue of the Language Resources and Evaluation journal [1]. Besides, the community produced a number of widely used tools and resources. Examples of them include word sense alignments between WordNet, Wikipedia, and Wiktionary [2–4],[4] folksonomy and named entity ontologies [5, 6], multiword terms [7],[5] ontological resources [8, 9],[6] annotated corpora [10],[7] and Wikipedia and Wiktionary APIs.[8]

## *Purpose of This Book*

The present volume provides an overview of the research involving CCLRs and their applications in NLP. It draws upon the current great interest in collective intelligence for information processing in general. Several meetings have taken place at the leading conferences in the field, and the corresponding conference tracks, e.g. "NLP for Web, Wikipedia, Social Media" have been established. The editors of this volume, thus, recognized the need to summarize the achieved results in a contributed book to advance and focus the further research effort. In this regard, the subject of the book "The People's Web Meets NLP: Collaboratively Constructed Language Resources" is very timely. There is no monograph, textbook or a contributed book on this topic to comprehensively cover the state-of-the-art on CCLRs in a single volume yet. Thus, we very much hope that such a book will become a major point of reference for researchers, students and practitioners in this field.

## *Book Organization*

The chapters in the present volume cover the three main aspects of CCLRs, namely construction approaches to CCLRs, mining knowledge from and using CCLRs in NLP, and interconnecting and managing CCLRs.

---

[3]People's Web Meets NLP workshop series at ACL-IJCNLP 2009, COLING 2010, and ACL 2012

[4]http://www.ukp.tu-darmstadt.de/data/sense-alignment/, http://lcl.uniroma1.it/babelnet/

[5]http://www.ukp.tu-darmstadt.de/data/multiwords/

[6]http://www.ukp.tu-darmstadt.de/data/lexical-resources, http://www.h-its.org/english/research/nlp/download/wikinet.php

[7]http://anawiki.essex.ac.uk/

[8]JWPL (http://www.ukp.tu-darmstadt.de/software/jwpl/), wikixmlj (http://code.google.com/p/wikixmlj/), JWKTL (http://www.ukp.tu-darmstadt.de/software/jwktl/)

## Part 1: Approaches to Collaboratively Constructed Language Resources

Collaboratively constructed resources have different forms and are created by means of different approaches, such as collaborative writing tools, human computation platforms, games with a purpose, or collecting user feedback on the Web.

Some of them are constructed by applying Social Web tools, such as wikis, to existing forms of knowledge production. For example, Wikipedia was created through the use of wikis to construct an electronic encyclopedia. In a similar way, Wiktionary was created through the use of wikis to construct a user-generated dictionary. Major research questions in this area of research are: how to utilize a Social Web tool to come up with a useful resource, motivating users to contribute, how to extract the knowledge, quality issues, varied coverage, or incompleteness of the resulting resources.

Further CCLRs result from the purposeful use of human computation platforms on the Web, such as Amazon Mechanical Turk, to perform expert-like or highly subjective tasks by a large number of non-expert volunteers paid for their work. Thereby, a complex task is typically modeled as a set of simpler tasks solved by means of a web-based interface. In other settings, platforms for collaborative annotation by non-paid peers may be used to construct language resources collaboratively. Major research questions in this context are, for example, how to model a complex task in such a way that it is feasible to be solved by non-experts, how to prevent spam, or monetary, quality and labor management issues.

The third approach to the construction of CCLRs by means of crowdsourcing is modeling the data management tasks, such as data collection or data validation as a game. The players of such a game contribute their knowledge collectively either for fun, or for learning purposes. These works address research questions such as how to convert the task into a game, how to motivate players for continuous participation, and how to manage the quality of the resulting data.

## Part 2: Mining Knowledge from and Using Collaboratively Constructed Language Resources

Much effort have been put into utilizing CCLRs in various NLP tasks and demonstrating their effectiveness. The present volume includes a number of examples for research works in this area, specifically, construction of semantic networks, word sense disambiguation, computational analysis of writing, or sentiment analysis.

The first approach to mining knowledge from CCLRs is to construct or improve semantic networks. There exist manually constructed semantic resources such as WordNet and FrameNet. Resources constructed through collective intelligence such as Wikipedia, Wiktionary, and Open Mind Common Sense[9] can provide rich and real-world knowledge at large scale that may be missing in manually constructed

[9]http://openmind.media.mit.edu

resources. In addition, combining resources that are complementary in coverage and granularity can yield a higher quality resource.

The second approach to utilizing CCLRs is mining the vast amount of user-generated content in the Web to create specific corpora which can be used as resources in computational intelligence tasks. Much of this data implicitly carries semantic annotations by users, as the corpora typically evolve around a certain domain of discourse and therefore represent its inherent knowledge structure. NLP applications exemplified in this book include the computational analysis of writing using Wikipedia revision history, organizing and analyzing consumer reviews, and word sense disambiguation utilizing Wikipedia articles as concepts.

The applications of CCLRs in NLP are certainly not limited to the example topics explained in this book; one can find a large number of research works with similar goals and approaches in the literature.

### Part 3: Interconnecting and Managing Collaboratively Constructed Language Resources

Readily available technology and resources such as Amazon Mechanical Turk and Wikipedia have lowered the barriers to collaborative resource construction and its enhancements. They also have led to a large number of sporadic efforts creating resources in different domains and with different coverage and purposes. This often results in resources that are disparate, poorly documented and supported, with unknown reliability. That is why the resources run the risk of not extensively being used by the community and can therefore disappear very quickly.

The research question is then how to create linguistic resources, expert-built and collaboratively constructed alike, more sustainable, such that the resources are more usable, accessible, and also easily maintained, managed, and improved.

In this part of the book, a number of ongoing community efforts to link and maintain multiple linguistic resources are presented. Considered resources range from lexical resources to annotated corpora. The chapters of the volume also introduce special interest groups, frameworks, and ISO standards for linking and maintaining such resources.

## *Target Audience*

The book is intended for advanced undergraduate and graduate students, as well as professionals and scholars interested in various aspects of research on CCLRs.

**Acknowledgements** We thank all program committee members who generously invested their expertise and time for providing constructive reviews. This book would not have been possible without their support, especially considering the tight schedule and multiple review rounds. We also thank Nicoletta Calzolari for her insightful and inspiring foreword.

Darmstadt, Germany Iryna Gurevych and Jungi Kim

## *Program Committee*

Iñaki Alegria
Chris Biemann
Erik Cambria
Jon Chamberlain
Christian Chiarcos
Johannes Daxenberger
Ernesto William De Luca
Gianluca Demartini
Judith Eckle-Kohler
Nicolai Erbs
Oliver Ferschke
Bilel Gargouri
Catherine Havasi
Sebastian Hellmann
Johannes Hoffart
Inas Mahfouz
Michael Matuschek
Gerard de Melo
Christian M. Meyer
Rada Mihalcea
Tristan Miller
Günter Neumann
Alessandro Oltramari
Simone Paolo Ponzetto
Michal Ptaszynski
Martin Puttkammer
Ruwan Wasala
Magdalena Wolska
Jianxing Yu
Torsten Zesch

## References

1. Gurevych I, Zesch T (2012) Special issue on collaboratively constructed language resources. Language resources and evaluation. Springer, Netherlands
2. Niemann E, Gurevych I (2011) The people's web meets linguistic knowledge: automatic sense alignment of Wikipedia and WordNet. In: Proceedings of the international conference on computational semantics (IWCS), Oxford, UK, pp 205–214
3. Meyer CM, Gurevych I (2011) What psycholinguists know about chemistry: aligning Wiktionary and WordNet for increased domain coverage. In: Proceedings of the 5th international joint conference on natural language processing (IJCNLP), Chiang Mai, Thailand, Nov, pp 883–892
4. Navigli R, Ponzetto SP (2010) BabelNet: building a very large multilingual semantic network. In: Proceedings of the 48th annual meeting of the association for computational linguistics (ACL), Uppsala, Sweden, July, pp 216–225
5. Tomuro N, Shepitsen A (2009) Construction of disambiguated folksonomy ontologies using Wikipedia. In: Proceedings of the 2009 workshop on the people's web meets NLP: collaboratively constructed semantic resources, Suntec, Singapore, August, pp 42–50
6. Shibaki Y, Nagata M, Yamamoto K (2010) Constructing large-scale person ontology from Wikipedia. In: Proceedings of the 2nd workshop on the people's web meets NLP: collaboratively constructed semantic resources, Beijing, China, August, pp 1–9
7. Hartmann S, Szarvas G, Gurevych I (2011) Mining multiword terms from Wikipedia. In: Pazienza MT, Stellato A (eds) Semi-automatic ontology development: processes and resources. IGI Global, Hershey, pp 226–258
8. Meyer CM, Gurevych I (2011) OntoWiktionary – constructing an ontology from the collaborative online dictionary Wiktionary. In: Pazienza MT, Stellato A (eds) Semi-automatic ontology development: processes and resources. IGI Global, Hershey, pp 131–161

9. Nastase V, Strube M, Börschinger B, Zirn C, Elghafari A (2010) WikiNet: a very large scale multi-lingual concept network. In: Proceedings of the 7th international conference on language resources and evaluation (LREC), Valletta, Malta, May, pp 19–21
10. Chamberlain J, Kruschwitz U, Poesio M (2009) Constructing an anaphorically annotated corpus with non-experts: assessing the quality of collaborative annotations. In: Proceedings of the 2009 workshop on the people's web meets NLP: collaboratively constructed semantic resources, Suntec, Singapore, August, pp 57–62

# Contents

# Part I
# Approaches to Collaboratively Constructed Language Resources

# Chapter 1
# Using Games to Create Language Resources: Successes and Limitations of the Approach

**Jon Chamberlain, Karën Fort, Udo Kruschwitz, Mathieu Lafourcade, and Massimo Poesio**

**Abstract** One of the more novel approaches to collaboratively creating language resources in recent years is to use online games to collect and validate data. The most significant challenges collaborative systems face are how to train users with the necessary expertise and how to encourage participation on a scale required to produce high quality data comparable with data produced by "traditional" experts. In this chapter we provide a brief overview of collaborative creation and the different approaches that have been used to create language resources, before analysing games used for this purpose. We discuss some key issues in using a gaming approach, including task design, player motivation and data quality, and compare the costs of each approach in terms of development, distribution and ongoing administration. In conclusion, we summarise the benefits and limitations of using a gaming approach to resource creation and suggest key considerations for evaluating its utility in different research scenarios.

## 1.1 Introduction

Recent advances in human language technology have been made possible by groups of people collaborating over the Internet to create large-scale language resources. This approach is motivated by the observation that a group of individuals can

J. Chamberlain (✉) · U. Kruschwitz · M. Poesio
University of Essex, Wivenhoe Park, Colchester CO4 3SQ, England
e-mail: jchamb@essex.ac.uk; udo@essex.ac.uk; poesio@essex.ac.uk

K. Fort
INIST-CNRS/LIPN, 2, allée de Brabois, 54500 Vandoeuvre-lès-Nancy, France
e-mail: karen.fort@inist.fr

M. Lafourcade
LIRMM, UMR 5506 – CC 477, 161 rue Ada, 34392 Montpellier Cedex 5, France
e-mail: mathieu.lafourcade@lirmm.fr

I. Gurevych and J. Kim (eds.), *The People's Web Meets NLP*, Theory and Applications of Natural Language Processing, DOI 10.1007/978-3-642-35085-6_1,

contribute to a collective solution, which has a better performance and is more robust than an individual's solution. This is demonstrated in simulations of collective behaviour in self-organising systems [34].

Web-based systems such as Wikipedia[1] and similar large initiatives have shown that a surprising number of individuals can be willing to participate in projects.

One of the more novel approaches to collaboratively creating language resources in recent years is to use online games to collect and validate data. The ESP Game,[2] the first mass market online *game-with-a-purpose (GWAP)*, highlighted the potential for a game-based approach to resource creation (in this case image tagging). Since then, new games have been developed for different tasks including language resource creation, search verification and media tagging.

The most significant challenges collaborative systems face are how to train users with the necessary expertise and how to encourage participation on a scale required to produce large quantities of high quality data comparable with data produced by "traditional" experts.

In this chapter, we provide insights into GWAP for language resource creation, focusing on the successes and limitations of the approach by investigating both quantitative and qualitative results.

This study will use data from the Phrase Detectives game,[3] developed by the University of Essex (England) to gather annotations on anaphoric co-reference, and the JeuxDeMots game,[4] developed by Laboratoire d'Informatique, de Robotique et de Microelectronique de Montpellier (LIRMM, France) to create a lexico-semantic network.

We first provide a brief overview of collaborative creation and the different approaches that have been used to create language resources. We then provide details of Phrase Detectives and JeuxDeMots, followed by other notable efforts of GWAP for language resource creation. Next we discuss some key issues in using a gaming approach, focusing on task design, player motivation, and data quality. Finally we look at the costs of each approach in terms of development, distribution and ongoing administration. In conclusion, we summarise the benefits and limitations of the games-with-a-purpose approach.

## 1.2 Collaborative Creation and Collective Intelligence

Collaboration is a process where two or more people work together to achieve a shared goal. From the point of view of collaborative creation of language resources, the resources are the goal, and they are created or modified by at least two people, who work incrementally, in parallel or sequentially on the project.

---

[1] http://www.wikipedia.org

[2] http://www.gwap.com/gwap

[3] http://www.phrasedetectives.com

[4] http://www.jeuxdemots.org

In the latter case language resources are developed with people working on the same project but never exactly on the same part of it. Parallel work is necessary to evaluate the validity of the created resource. For example, inter-annotator agreement, using parallel annotations, was used in the Penn Treebank [48]. Incremental work involves adjudication, either by an expert, or by consensus.

Several attempts have been made recently to bring order to the rapidly developing field of collaborative creation on the Internet [46, 62, 80]. Wikipedia showed that allowing users free reign of encyclopaedic knowledge not only empowers mass participation but also that the resulting creation is of a very high quality. This can be seen as a good example of the broad term *collective intelligence* where groups of individuals do things collectively that seem intelligent [46].

Collective intelligence can be shown in many domains including Computer Science, Economics and Biology[5] but here we focus on coordinating collective action in computational systems that overcome the bottleneck in creating and maintaining language resources which would normally have to be done by paid administrators.

The utility of collective intelligence came to the fore when it was proposed to take a job traditionally performed by a designated employee or agent and outsource it to an undefined large group of Internet users through an open call. This approach, called *crowdsourcing* [31], revolutionised the way traditional tasks could be completed and made new tasks possible that were previously inconceivable due to cost or labour limitations.

One use for crowdsourcing can be as a way of getting large amounts of human work hours very cheaply as an alternative to producing a computerised solution that may be expensive or complex. However, it may also be seen as a way of utilising human processing power to solve problems that computers, as yet, cannot solve, termed *human computation* [72]. Human computation has particular appeal for *natural language processing (NLP)* because computer systems still need large resources for training algorithms that aim to understand the meaning of human language.

By combining collective intelligence, crowdsourcing and human computation it is possible to enable a large group of collaborators to work on linguistic tasks normally done by highly skilled (and highly paid) annotators and to aggregate their collective answers to produce a more complex dataset that not only is more robust than an individual answer but allows for linguistic ambiguity. Enabling groups of people to work on the same task over a period of time is likely to lead to a collectively intelligent decision [68].

Three variations of this type of collaboration over the Internet have been successful in recent years and are distinguished by the motivations of the participants.

The first variation is where the motivation for the users to participate already exists. This could be because the user is inherently interested in contributing,

[5] http://scripts.mit.edu/~cci/HCI

for example in the case of Wikipedia or GalaxyZoo,[6] or intrinsically motivated because they need to accomplish a different task, for example the reCAPTCHA[7] authentication system.

Unfortunately, most linguistic tasks are neither interesting (for the majority of people) nor easy to integrate into another system. Therefore, a second variation of crowdsourcing called microworking was proposed, where participants are paid small amounts of money to perform tasks. Although the payments are small, the total cost for a language resource produced in this way will increase proportionately with its size. Therefore, it is being used more in NLP for the fast annotation of small to medium sized corpora and for some types of linguistic evaluation [9].

This approach demonstrates the difficulties in producing the size of resources needed for modern linguistic tools, so a third approach was proposed to make the motivation for the user be entertainment rather than money. The *games-with-a-purpose (GWAP)* approach showed enormous initial potential and has been used for a variety of data collection and annotation tasks where the task has been made fun. In this chapter we focus on games used to create language resources.

## 1.3 Approaches to Creating Language Resources

### *1.3.1 Traditional, Entirely Validated Annotation*

In order to evaluate crowdsourcing approaches to language resource creation it is necessary to also consider more traditional approaches. When we talk about traditional annotation, we think of the methodology used, for example, to create the OntoNotes corpus,[8] containing multilingual annotated news articles, dialogue transcriptions and weblogs, and the SALSA corpus[9] of syntactically annotated German newspaper articles.

In this approach, a formal coding scheme is developed, and often extensive agreement studies are carried out. Every document is annotated twice according to the coding scheme by two professional annotators under the supervision of an expert, typically a linguist, followed by merging and adjudication of the annotations. These projects also generally involve the development of suitable annotation tools or at least the adaptation of existing ones.

[6] http://www.galaxyzoo.org

[7] http://www.google.com/recaptcha

[8] http://www.ldc.upenn.edu/Catalog/catalogEntry.jsp?catalogId=LDC2011T03

[9] http://www.coli.uni-saarland.de/projects/salsa

### 1.3.2 Traditional, Partly Validated Annotation

This type of annotation also involves the development of a formal coding scheme and training of annotators but most items will be typically annotated only once, for example in the ARRAU [57] and GNOME [56] corpora for anaphoric co-reference.

Approximately 10 % of items are double-annotated to identify misunderstandings and improve the annotation guide [8]. In many cases, the annotations will have to be corrected, possibly extensively. Annotation is typically carried out by trained annotators, generally students, under the supervision of an expert annotator.

### 1.3.3 Microwork Crowdsourcing

Amazon Mechanical Turk (AMT)[10] pioneered microwork crowdsourcing: using the Web as a way of reaching very large numbers of workers (sometimes referred to as turkers) who get paid to complete small items of work called *human intelligence tasks (HITs)*. This is typically very little – in the order of 0.01 to 0.20 US$ per HIT.

Some studies have shown that the quality of resources created this way are comparable to that of resources created in the traditional way, provided that multiple judgements are collected in sufficient number and that enough post-processing is done [9, 67]. Other studies have shown that the quality does not equal that provided by experts [6] and for some tasks does not even surpass that of automatic language technology [76]. It is beyond the scope of this chapter to go into great depth about the quality attainable from AMT, rather we simply compare reported results with that reported from other approaches.

A further reported advantage of AMT is that the work is completed very fast. It is not uncommon for a HIT to be completed in minutes, but this is usually for simple tasks. In the case of more complex tasks, or tasks where the worker needs to be more skilled, e.g., translating a sentence in an uncommon language, it can take much longer [55].

AMT is very competitive with traditional resource creation methods from a financial perspective. Whilst AMT remains a very popular microworking platform some serious issues regarding the rights of workers, minimum wage and representation have been raised [25]. Other microworking platforms, such as Samasource,[11] guarantee workers a minimum payment level and basic rights.

Microwork crowdsourcing is becoming a standard way of creating small-scale language resources but even this approach can become prohibitively expensive to create resources of the size that are increasingly required in modern linguistics, i.e., in the order of 100 million annotated words.

[10] http://www.mturk.com

[11] http://samasource.org

### 1.3.4 Games with a Purpose (GWAP)

Generally speaking, a game-based crowdsourcing approach uses entertainment rather than financial payment to motivate participation. The approach is motivated by the observation that every year an estimated nine billion person-hours are spent by people playing games on the Web [72]. If even a fraction of this effort could be redirected towards useful activity that has a purpose, as a side effect of having people play entertaining games, there would be an enormous human resource at our disposal.

GWAP come in many forms; they tend to be graphically rich, with simple interfaces, and give the player an experience of progression through the game by scoring points, being assigned levels and recognising their effort. Systems are required to control the behaviour of players: to encourage them to concentrate on the tasks and to discourage them from malicious behaviour. This is discussed in more detail later.

The GWAP approach showed enormous initial potential, with the first, and perhaps most successful, game called the ESP Game. In the game two randomly chosen players are shown the same image. Their goal is to guess how their partner will describe the image (hence the reference to extrasensory perception or ESP) and type that description under time constraints. If any of the strings typed by one player matches the strings typed by the other player, they both score points. The descriptions of the images provided by players are very useful to train content-based image retrieval tools.

The game was very popular, attracting over 200,000 players who produced over 50 million labels [72]. The quality of the labels has been shown to be as good as that produced through conventional image annotation methods. The game was so successful that a license to use it was bought by Google, who developed it into the Google Image Labeler which was online from 2006 to 2011.

GWAP have been used for many different types of crowdsourced data collection [70] including:

- Image annotation such as the ESP Game, Matchin, FlipIt, Phetch, Peekaboom, Squigl, Magic Bullet and Picture This;
- Video annotation such as OntoTube, PopVideo, Yahoo's VideoTagGame and Waisda;
- Audio annotation such as Herd It, Tag a Tune and WhaleFM;
- Biomedical applications such as Foldit, Phylo and EteRNA;
- Transcription such as Ancient Lives and Old Weather;
- Improving search results such as Microsoft's Page Hunt;
- Social bookmarking such as Collabio.

Links to the GWAP listed above can be found in Appendix A.

GWAP have a different goal to *serious games*, where the purpose is to educate or train the player in a specific area such as learning a new language or secondary school level topics [51]. Serious games can be highly engaging, often in a 3D world, and have a directed learning path for the user as all of the data is known to the system beforehand. Therefore, the user can receive immediate feedback as to their level of performance and understanding at any point during the game.

GWAP aim to entertain players whilst they complete tasks that the system does not know, for the most part, the correct answer, and in many cases there may not even be a "correct" answer. Hence, providing feedback to users on their work presents a major challenge and understanding the motivation of players in this scenario is a key to the success of a GWAP.

## 1.4 Using Games to Create Language Resources

This section looks in detail at the design and reported results from two GWAP for NLP: Phrase Detectives and JeuxDeMots. For completeness, we mention other notable GWAP used for linguistic purposes and a summary, with links where available, is in Appendix B.

### *1.4.1 Phrase Detectives*

*Phrase Detectives (PD)* is a single-player GWAP designed to collect data about English (and subsequently Italian) anaphoric co-reference [12, 61]. The game architecture is structured around a number of tasks that use scoring, progression and a variety of other mechanisms to make the activity enjoyable. The game design is based on a detective theme, relating to the how the player must search through the text for a suitable annotation.

The game uses two styles of text annotation for players to complete a linguistic task. Initially text is presented in Annotation Mode (called *Name the Culprit* in the game – see Fig. 1.1). This is a straightforward annotation mode where the player makes an annotation decision about a highlighted markable (section of text). If different players enter different interpretations for a markable, then each interpretation is presented to more players in Validation Mode (called *Detectives Conference* in the game – see Fig. 1.2). The players in Validation Mode have to agree or disagree with the interpretation.

Players are trained with training texts created from a gold standard (a text that has been annotated by a linguistic annotation expert). Players always receive a training text when they first start the game. Once the player has completed all of the training tasks, they are given a rating (the percentage of correct decisions out of the total number of training tasks). If the rating is above a certain threshold (currently 50 %), the player progresses on to annotating real documents, otherwise they are asked to do a training document again. The rating is recorded with every future annotation that the player makes as the rating is likely to change over time.

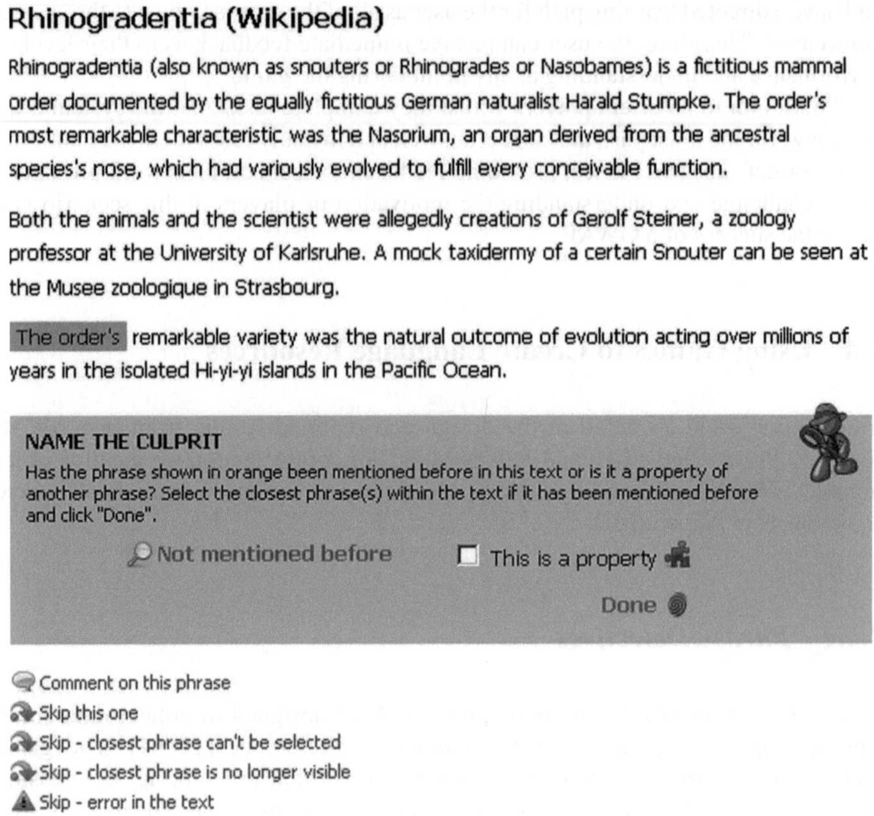


**Fig. 1.1** Detail of a task presented in Annotation Mode in Phrase Detectives on Facebook

The scoring system is designed to reward effort and motivate high quality decisions by awarding points for retrospective collaboration. A mixture of incentives, including personal (achieving game goals and scoring points), social (competing with other players) and financial (small prizes), are employed.

Text used in PD comes from two main domains: Wikipedia articles selected from the 'Featured Articles' page[12] and the page of 'Unusual Articles'[13]; and narrative text from Project Gutenberg[14] including simple short stories (e.g., Aesop's Fables, Grimm's Fairy Tales, Beatrix Potter's tales) and more advanced narratives such as several Sherlock Holmes stories by A. Conan-Doyle, *Alice in Wonderland*, and several short stories by Charles Dickens.

[12]http://en.wikipedia.org/wiki/Wikipedia:Featured_articles

[13]http://en.wikipedia.org/wiki/Wikipedia:Unusual_articles

[14]http://www.gutenberg.org

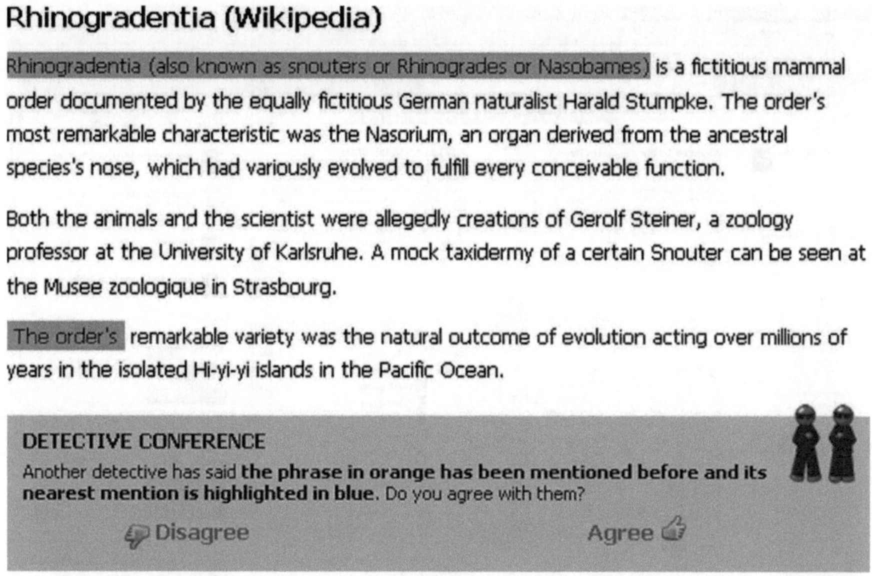


**Fig. 1.2** Detail of a task presented in Validation Mode in Phrase Detectives on Facebook

The goal of the game was not just to annotate large amounts of text, but also to collect a large number of judgements about each linguistic expression. This led to the deployment of a variety of mechanisms for quality control which try to reduce the amount of unusable data beyond those created by malicious users, from validation to tools for analysing the behaviour of players (see Fig. 1.7).

A version of PD was developed for Facebook[15] that maintained the previous game architecture whilst incorporating a number of new features developed specifically for the social network platform (see Fig. 1.3).

The game was developed with PHP SDK[16] (an API for accessing user data, friend lists, wall posting, etc.) and integrates seamlessly within the Facebook site. Both implementations of the game run simultaneously on the same corpus of documents.

This version of the game makes full use of socially motivating factors inherent in the Facebook platform. Any of the player's friends from Facebook, who are also playing the game, form the player's team, which is visible in the left hand menu. Whenever a player's decision agrees with a team member they both score additional points.

Player levels have criteria, including total points scored, player rating and total wall posts made from the game. The player must activate their new level once the criteria are met. In addition to the monthly and all-time leaderboards, the Facebook

[15]http://www.facebook.com

[16]http://developers.facebook.com/docs/reference/php

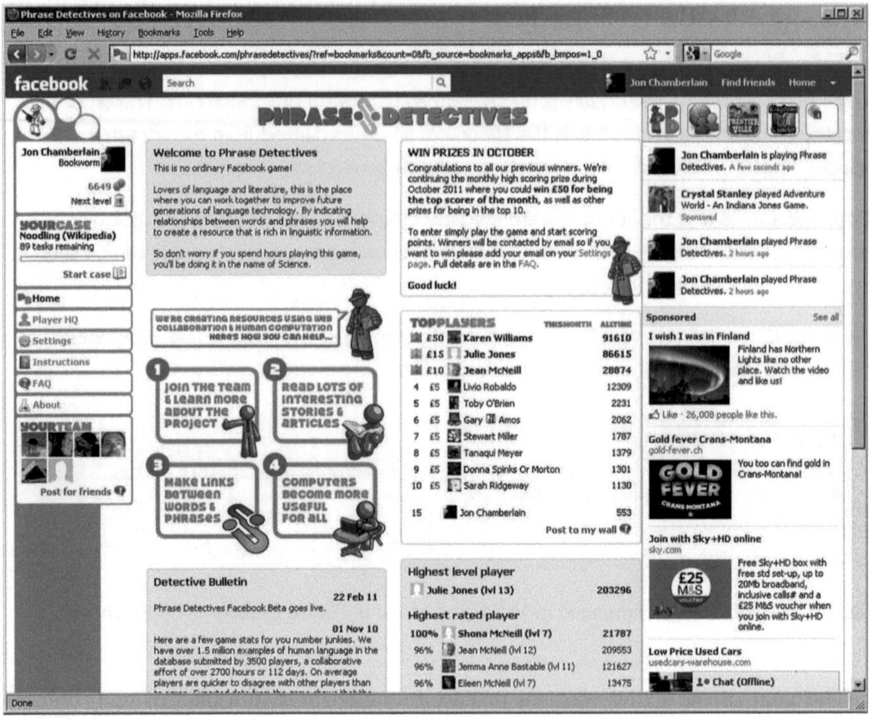


**Fig. 1.3** Screenshot of the Phrase Detectives Facebook homepage

version has leaderboards for the highest level players, highest rated players and the players with the biggest team.

The purpose of redeveloping the game for Facebook was to investigate the utility of social networking sites in achieving high visibility and to explore different ways players can collaborate.

The first game was released in December 2008, with the Facebook version released in February 2011. Both games continue to collect data but results reported here are from December 2008 to February 2012 or are from previously published papers [13–15, 61].

### 1.4.2 JeuxDeMots

*JeuxDeMots (JDM)* is a two player GWAP, launched in September 2007, that aims to build a large lexico-semantic network composed of terms (nodes) and typed relations (links between nodes) [42] – see Fig. 1.4. It contains terms and possible refinements in the same spirit as WordNet [21], although it is organised as decision trees. There are more than 50 different relation types, the occurrences of which are weighted.

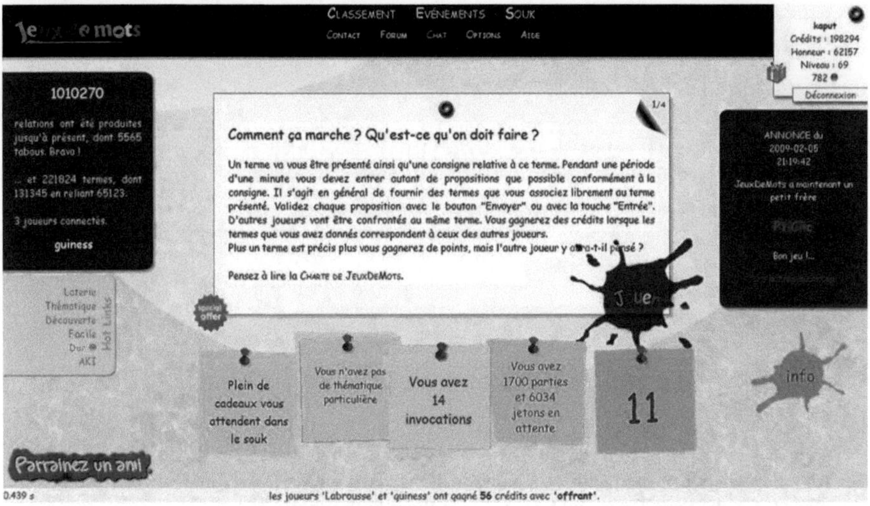


**Fig. 1.4** Screenshot of the JeuxDeMots homepage. From here the player has status information and can launch a game by clicking on the *jouer* (play) button

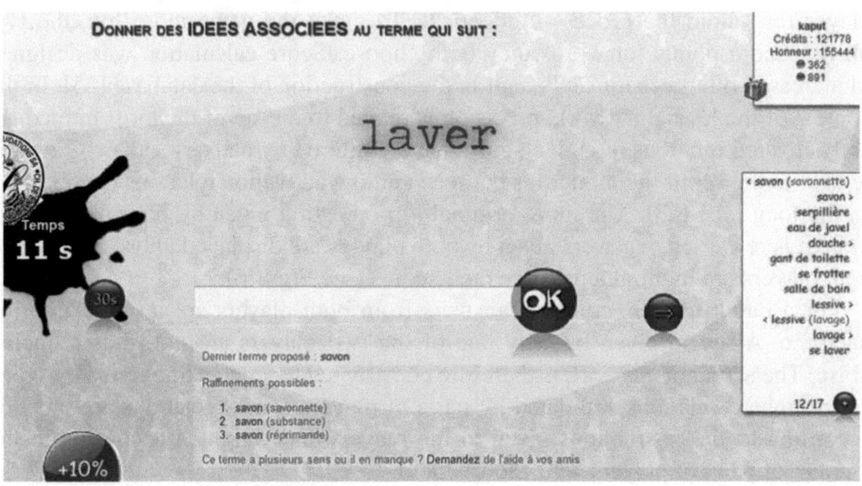


**Fig. 1.5** Screenshot of an ongoing game in JDM with the target word *laver* (to wash). Several propositions have been given by the player and are listed on the right hand side

When a player begins a game, instructions concerning the type of lexical relation (synonyms, antonym, domain, etc.) are displayed, as well as a term *T*, chosen from the database or offered by other players. The player has limited time to enter terms which, to their mind, correspond to term *T* and the lexical relation. The maximum number of terms a player can enter is limited, thus encouraging the player to think carefully about their choices. A screenshot of the game is shown in Fig. 1.5.

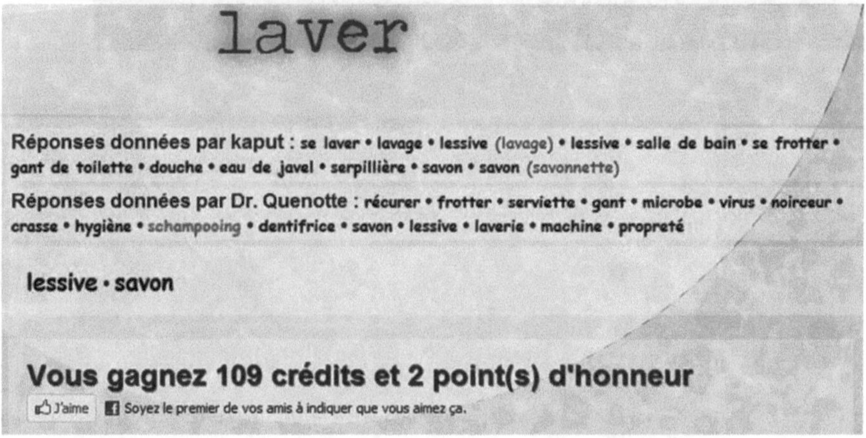


**Fig. 1.6** Screenshot of the result of a game in JDM. Two words *lessive* and *savon* were given by both players for the term *laver* and hence scores them both points

The same term *T*, along with the same instructions, are later given to another player for whom the process is identical. To make the game more fun, the two players score points for words they both choose. Score calculation was designed to increase both precision and recall in the construction of the database [35]. In the context of the lexical network, precision is related to the set of the most immediate and activated relations of a given term that are uttered by native speakers. Recall is related to the set of the numerous but relevant low activation relations (also known as the long tail) [43]. The more original a proposition given by both players, the more it is rewarded. Answers given by both players are displayed, those common to both players are highlighted, as are their scores (see Fig. 1.6).

For a target term *T*, common answers from both players are inserted into the database. Answers given by only one of the two players are not, thus reducing noise. The semantic network is therefore constructed by connecting terms by typed and weighted relations, validated by pairs of players. These relations are labelled according to the instructions given to the players and weighted according to the number of pairs of players who choose them.

Initially, prior to putting the game online, the database was populated with 140,000 terms (nodes) from French dictionaries, however if a pair of players suggest a non-existing term, a new node is added to the database. Since populating the database the players have added 110,000 new terms however these include spelling mistakes, plurals, feminine forms, numbers, dates and foreign words.

In the interest of quality and consistency, it was decided that the validation process would involve anonymous players playing together. A relation is considered valid only if it is given by at least one pair of players. This validation process is similar to the process for indexing images [73] and, more recently, to collect common sense knowledge [45] and for knowledge extraction [65].

The activity of the players in JDM constructs a lexical network which contains over 50 types of ontological relations such as generic relations (hypernyms), specific relations (hyponyms), part and whole relations, matter and substance, domain, synonyms and antonyms (the latter also being strongly lexical). The ongoing process of the network construction leads to the identification of word usages for disambiguating terms of the constructed ontology.

## *1.4.3 Other GWAP for Language Resources*

### 1.4.3.1 Knowledge Acquisition

1001 Paraphrases [16], one of the first GWAP whose aim was to collect corpora, was developed to collect training data for a machine translation system that needs to recognise paraphrase variants. In the game, players have to produce paraphrases of an expression shown at the top of the screen, such as "this can help you". If they guess one of the paraphrases already produced by another player, they get the number of points indicated in the window; otherwise the guess they produced is added to those already collected by the system, the number of points they can win is decreased, and they can try again. Many of the ideas developed in 1001 Paraphrases, and the earlier LEARNER system, are extremely useful, in particular the idea of validation.

Other games for collecting common sense knowledge include FACTory, Verbosity, Categorilla and Free Association.

### 1.4.3.2 Text Annotation

The game most directly comparable with PD is PlayCoref, developed at Charles University in Prague [29]. PlayCoref is a two-player game in which players mark coreferential pairs between words in a text (no phrases are allowed). They mark the coreferential pairs as undirected links. During the session, the number of words the opponent has linked into the coreferential pairs is displayed to the player. The number of sentences with at least one coreferential pair marked by the opponent is displayed to the player as well. A number of empirical evaluations have been carried out showing that players find the game very attractive but the game has not yet been put online to collect data on a large scale.

PhraTris is a GWAP for syntactic annotation developed by Giuseppe Attardi's lab at the University of Pisa using a general-purpose GWAP development platform called GALOAP. PhraTris, based on the traditional game Tetris, has players arrange sentences in a logical way, instead of arranging falling bricks, and won the Insemtives Game Challenge 2010. The game is not online but can be downloaded and installed locally.

PackPlay [28] was another attempt to build semantically-rich annotated corpora. The two game variants *Entity Discovery* and *Name That Entity* use slightly different approaches in multi-player games to elicit annotations from players. Results from a small group of players showed high precision and recall when compared to expert systems in the area of named entity recognition, although this is an area where automated systems also perform well.

#### 1.4.3.3 Sentiment Analysis

Human language technology games integrated into social networking sites such as Sentiment Quiz [63] on Facebook show that social interaction within a game environment does motivate players to participate. The Sentiment Quiz asks players to select a level of sentiment (on a 5 point scale) associated with a word taken from a corpus of documents regarding the 2008 US Presidential election. The answer is compared to another player and points awarded for agreement.

#### 1.4.3.4 Generation

A family of GWAP which have been used to collect data used in computational linguistics are the GIVE games developed in support of the GIVE-2 challenge for generating instructions in virtual environments, initiated in the Natural Language Generation community [40]. GIVE-2 is a treasure-hunt game in a 3D world. When starting the game, the player sees a 3D game window, which displays instructions and allows the players to move around and manipulate objects. In the first room players learn how to interact with the system; then they enter into an evaluation virtual world where they perform the treasure hunt, following instructions generated by one of the systems participating in the challenge. The players can succeed, lose, or cancel the game and this outcome is used to compute the task success metric, one of the metrics used to evaluate the systems participating in the challenge.

GIVE-2 was extremely successful as a way to collect data, collecting over 1,825 game sessions in 3 months, which played a key role in determining the results of the challenge. This is due, in part, to the fact that it is an extremely attractive game to play.

#### 1.4.3.5 Ontology Building

The OntoGame, based around the ESP Game data collection model, aims to build ontological knowledge by asking players questions about sections of text, for example whether it refers to a class of object or an instance of an object. Other Web-based systems include Open Mind Word Expert [52], which aims to create large sense-tagged corpora, and SemKey [47] which makes use of WordNet and Wikipedia to disambiguate lexical forms referring to concepts.

## 1.5 Defining Collaborative Approaches

There have been several recent attempts to define and classify collaborative approaches in collective intelligence and distributed human computation [46, 62]. We focus on 3 dimensions proposed for crowdsourcing projects [77] that are essential considerations when designing GWAP for NLP:

- Task Character
- Player Motivation
- Annotation Quality

### *1.5.1 Task Character*

#### 1.5.1.1 Game Interface

Most GWAP tend to have fairly simple interfaces making it easy for first time users to start playing, with a short timespan (i.e., arcade style) and online delivery. This constrains the game to small tasks in a programmatically simple framework which is suitable for the goal of collecting data. A game deployed on the Web should observe all the normal guidelines regarding browser compatibility, download times, consistency of performance, spatial distance between click points, etc.[17]

Game interfaces should be graphically rich, although not at the expense of usability, and aimed at engaging the target audience (i.e., a game aimed at children may include more cartoon or stylised imagery in brighter colours than a game aimed at adults). The game should also provide a consistent metaphor within the gaming environment. For this PD used a detective metaphor, with buttons stylised with a cartoon detective character and site text written as if the player was a detective solving cases. The game task should be integrated in such a way that task completion, scoring and storyline form a seamless experience.

Three styles of game scenario have been proposed [74]:

1. Output-agreement, where the players must guess the same output from one input;
2. Inversion-problem, where one player describes the input to a second player who must guess what it is;
3. Input-agreement, where two players must guess whether they have the same input as each other based on limited communication.

The Output-agreement game scenario is the most straight forward to implement and collect data from, however, other scenarios can make the game more interesting for the players and increase their enjoyment.

[17]http://www.usability.gov/guidelines

#### 1.5.1.2 Task Design

Whilst the design of the game interface is important, it is the design of the task that determines how successfully the player can contribute data. In PD the player is constrained to a set of predefined options to make annotations, with freetext comments allowed (although this is not the usual mode of play in the game). The pre-processing of text allows the gameplay in PD to be constrained in this way but is subject to errors in processing that also need to be fixed.

JDM requires players to type text into a freetext box which allows for the collection of novel inputs but will also collect more noise from players through spelling mistakes and similar inputs. These can be filtered out using post-processing and validation, however it makes the collection of novel and ambiguous data more difficult.

The task design has an impact on the speed at which players can complete tasks, with clicking being faster than typing. A design decision to use radio buttons or freetext boxes can have a significant impact on performance [1].

The interface of AMT is predefined and presents limitations that constitute an important issue for some tasks, for example to annotating noun compound relations using a large taxonomy [71]. In a word sense disambiguation task considerable redesigns were required to get satisfactory results [30]. These examples show how difficult it is to design NLP tasks for crowdsourcing within a predefined system.

### *1.5.2 Player Motivation*

The motivation of the players is an important issue both in terms of data analysis and of return on investment (and therefore cost).

Incentives that motivate players to participate can be categorised into three groups: personal; social; and financial [14]. These directly relate to other classifications of motivations in previous research: Love; Glory; and Money [46].

Given that GWAP attempts to avoid direct financial incentives (as found in microwork crowdsourcing) the game must motivate the player with entertainment and enjoyment.

There may also be other motivational considerations, such as the desire to contribute to a scientific project or for self enrichment and learning.

All incentives should be applied with caution as rewards have been known to decrease annotation quality [53].

It is important to distinguish between *motivation to participate* (why people start doing something) and *motivation to contribute or volition* (why they continue doing something) [23]. Once both conditions are satisfied we can assume that a player will continue playing until other factors such as fatigue or distraction break the game cycle. This has been called *volunteer attrition*, where a player's contribution diminishes over time [45].

Although incentives can be categorised, in reality they form a complex psychology in participants that is best discussed by focusing on a particular game consideration:

- The concept of enjoyment as a motivator;
- How timing tasks affects player motivation;
- Altruism and citizen science;
- Indirect financial incentives in games;
- Publicity, virality and social networks.

#### 1.5.2.1 Enjoyment as an Incentive

GWAP focuses on one main type of incentive: enjoyment. There is substantial literature on what makes games fun [41] and models of enjoyment in games (called *the game flow*) identify eight criteria for evaluating enjoyment [69] (the model being based on a more generic theory [19]):

1. Concentration – Games should require concentration and the player should be able to concentrate on the game;
2. Challenge – Games should be sufficiently challenging and match the player's skill level;
3. Player skills – Games must support player skill development and mastery;
4. Control – Players should feel a sense of control over their actions in the game;
5. Clear goals – Games should provide the player with clear goals at appropriate times;
6. Feedback – Players must receive appropriate feedback at appropriate times;
7. Immersion – Players should experience deep but effortless involvement in the game;
8. Social interaction – Games should support and create opportunities for social interaction.

The main method used by GWAP to make players enjoy the task is by providing them with a challenge. This is achieved through mechanisms such as requiring a timed response, keeping scores that ensure competition with other players, and having players of roughly similar skill levels play against each other. In JDM, the challenge is both the combination of timed response and word-relation pairs of various difficulties.

For the players of PD, they can choose to read texts that they find interesting and have some control over the game experience. Whilst some texts are straightforward, others can provide a serious challenge of reading comprehension and completion of linguistic tasks. Players can also comment on the gaming conditions (perhaps to identify an error in the game, to skip a task or to generate a new set of tasks) and contact the game administrators with questions.

One of the simplest mechanisms of feedback is scoring. By getting a score the player gains a sense of achievement and some indication as to how well they are doing in the game.

GWAP tend to be short, arcade style games so immersion is achieved by progression through the game: by learning new types of tasks; becoming more proficient at current tasks; and by assigning the player a named level, starting from novice and going up to expert.

Social incentives are also provided by the scoring mechanism. Public leaderboards reward players by improving their standing amongst their peers (in this case their fellow players). Using leaderboards and assigning levels for points has been proven to be an effective motivator, with players often using these as targets [74]. An interesting phenomenon has been reported with these reward mechanisms, namely that players gravitate towards the cut off points (i.e., they keep playing to reach a level or high score before stopping) [75], however analysis of data from PD on Facebook did not support this [15].

#### 1.5.2.2 Time-Based Challenges in Language Tasks

The timing of tasks is usually required in the game format, either as motivational feature or as a method of quality control checking (or both). von Ahn and his colleagues view timing constraints as a key aspect of what makes games exciting [74], and built them into all their games. This is also the case for many other GWAP including JDM.

In PD, however, there are no timing constraints, although the time taken to perform a task is used to assess the quality of annotations. As the task in PD is text based (rather than image based in the ESP Game), it was considered important to give players time to read documents at a relatively normal speed whilst completing tasks.

This was supported by the results of the first usability study of PD. In the game prototype used in that study, players could see how long it had taken to do an annotation. On the contrary to suggestions that timing provides an incentive, the subjects complained that they felt under pressure and that they did not have enough time to check their answers, even though the time had no influence on the scoring. As a result, in all following versions of PD the time it takes players to perform a task is recorded but not shown.

Several players found the timing element of JDM stressful and in one case a player gave up the game for this reason. Most players in this game consider a timed task as normal and exciting and can buy extra time when needed (a game feature).

The time limitation tends to elicit spontaneous answers in a way that is not possible without a time limit where the players can give a more considered response. The design of the task must balance the increase in excitement a timed element can offer with the need to allow players time to give good quality answers.

Related to this are the concepts of "throughput" and "wait time", discussed in more detail later, that are used to assess the efficiency of an interface. By increasing the speed at which the players are working, by using a timed element, you also increase the speed at which you can collect data.

#### 1.5.2.3 Altruism and Participation in a Scientific Community

People who contribute information to Wikipedia are motivated by personal reasons such as the desire to make a particular page accurate, or the pride in one's knowledge in a certain subject matter [79]. This motivation is also behind the success of *citizen science* projects, such as the Zooniverse collection of projects,[18] where the scientific research is conducted mainly by amateur scientists and members of the public.

GWAP may initially attract collaborators (e.g., other computational linguists) by giving them the sense that they are contributing to a resource from which a whole discipline may benefit and these are usually the people that will be informed first about the research. However, in the long term, most of the players of GWAP will never directly benefit from the resources being created. It is therefore essential to provide some more generic way of expressing the benefit to the player.

For example, this was done in PD with a BBC radio interview by giving examples of NLP techniques used for Web searching. Although this is not a direct result of the language resources being created by this particular GWAP, it is the case for efforts of the community as a whole, and this is what the general public can understand and be motivated by.

The purpose of data collection in GWAP has an advantage over microworking in AMT, where the workers are not connected to the requester, in that there is a sense of ownership, participation in science, and generally doing something useful. When players become more interested in the purpose of the GWAP than the game itself it becomes more like a citizen science approach where players are willing to work on harder tasks, provide higher quality data and contribute more.

In JDM, the collected data is freely available and not restricted in use (under the Creative Commons licence). The players do not have to know they are part of a research project although it is written in the rules of the game. Players reported that they were more interested in playing the JDM game than knowing what the data was used for. However, for some players (around 20) the purpose of the GWAP approach became more important than the game. These players played more on incomplete and difficult term-relation couples. The fact that the data constructed is freely available does matter for these types of players.

#### 1.5.2.4 Indirect Financial Incentives

Indirect financial incentives in GWAP are distributed as prizes which are not directly related to the amount of work being done by the player, unlike microworking where a small sum of money is paid for the completion of a particular task.

In PD, financial incentives were offered in the form of a daily or weekly lottery, where each piece of work stood an equal chance of winning, or for high scoring players. These were distributed as Amazon vouchers emailed to the winning player. The ESP Game occasionally offers financial incentives in a similar way. JDM and most GWAP do not offer financial incentives.

[18] https://www.zooniverse.org

Whilst financial incentives seem to go against the fundamental idea behind GWAP (i.e., that enjoyment is the motivation), it actually makes the enjoyment of potentially winning a prize part of the motivation. Prizes for high scoring players will motivate hard working or high quality players but the prize soon becomes unattainable for the majority of other players. By using a lottery style financial prize the hard working players are more likely to win, but the players who only do a little work are still motivated.

Indirect financial incentives can be a cost-effective way to increase participation in GWAP, i.e., the increase of work completed per prize fund is comparable to the cost of other approaches.

#### 1.5.2.5 Attracting Players

In order to attract the number of participants required to make a success of the GWAP approach, it is not enough to develop attractive games; it is also necessary to develop effective forms of advertising. The number of online games competing for attention is huge and without some effort to raise a game's profile, it will never catch the attention of enough players. The importance of this strategy was demonstrated by von Ahn's lab. The ESP Game was constantly advertised in the press and also on TV. Other methods to reach players included blogs and being discussed on gaming forums. Part of the success of PD was down to the advertising of the game on blogs, language lists, conferences, tutorials and workshops as well as traditional media (via press releases). JDM on the other hand relied exclusively on word of mouth.

Not all advertising methods are equally successful and it is important to evaluate which works best for the game task, language or country.

Indirect financial incentives have been shown to be a good way to attract new players, however it is other motivational elements that keep players contributing to a game [66].

#### 1.5.2.6 Virality and Social Networks

Social incentives can be made more effective when the game is embedded within a social networking platform such as Facebook. In such a setting, the players motivated by the desire to contribute to a communal effort may share their efforts with their friends, whereas those motivated by a competitive spirit can compete against each other.

The PD game on Facebook allowed players to make posts to their wall (or news feed). Posting is considered a very important factor in recruiting more players as surveys have shown that the majority of social game players start to play because of a friend recommendation.[19,20]

---

[19]http://www.infosolutionsgroup.com/2010_PopCap_Social_Gaming_Research_Results.pdf

[20]http://www.lightspeedresearch.com/press-releases/it's-game-on-for-facebook-users

Posts were automatically generated in PD and could be created by a player by clicking a link in the game. They could either be *social* in nature, where the content describes what the player is doing or has done, or *competitive*, where the content shows achievements of the player. Results showed that players preferred to make social posts, i.e., about the document they were working on or had just completed (52 %). This compares to competitive posts when they went up a level (13 %), when their rating was updated (10 %) or to post about their position in the leaderboard (12 %). The remaining 13 % of posts were players making a direct request for their friends to join the game. This indicates that social motivations are more important than competitive motivations, at least on this platform.

In JDM, some social network features exist as achievements (scoring, winning some words, etc.) displayed on Facebook however the real impact of such features is uncertain.

## *1.5.3 Annotation Quality*

Whereas the designers of standard online games only need to motivate players to participate, the designers of GWAP also need to motivate the players to contribute good quality work. Obtaining reliable results from non-experts is also a challenge for other crowdsourcing approaches, and in this context strategies for dealing with the issue have been discussed extensively [2, 3, 22, 39].

In the case of microworking, the main strategy for achieving good quality labelling is to aggregate results from many users to approximate a single expert's judgements [67].

However, for the task of word-sense disambiguation, a small number of well-trained annotators produces much better results than a larger group of AMT workers [6] which illustrates that higher quality cannot always be achieved by simply adding more workers.

GWAP for linguistic annotation is not motivated solely by the desire to label large amounts of data. Web collaboration could also be used to gather data about the interpretation of natural language expressions, which all too often is taken to be completely determined by context, often without much evidence [59]. From this perspective it is important to attempt to avoid poor quality individual judgements.

The strategies for quality control in GWAP address four main issues:

- Training and Evaluating Players
- Attention Slips
- Multiple Judgements and Genuine Ambiguity
- Malicious Behaviour

### 1.5.3.1 Training and Evaluating Players

GWAP usually begin with a training stage for players to practice the task and also to show that they have sufficiently understood the instructions to do a real task.

However, the game design must translate the language task into a game task well enough for it still to be enjoyable, challenging and achievable. GWAP need to correlate good performance in the game with producing good quality data, but this is not an easy thing to do.

The level of task difficulty will drive the amount of training that a player will need. Simple tasks like image tagging need very little instruction other than the rules of the game, whereas more complex judgements such as those required by PD may require the players to be either more experienced or to undergo more training. The training phase has been shown to be an important factor in determining quality and improvement in manual annotation [20].

Most GWAP, at least initially, will have a core of collaborators to test and perform tasks and these are most likely to be friends or colleagues of the task designers. It can therefore be assumed that this base of people will have prior knowledge of the task background, or at least easy access to this information. These pre-trained collaborators are not the "crowd" that crowdsourcing needs if it is to operate on a large scale nor are they the "crowd" in the wisdom of the crowd.

Training should assume a layman's knowledge of the task and should engage the participant to increase their knowledge to become a pseudo-expert. The more they participate, the more expert they become. This graduated training is difficult to achieve and makes a rating system (where the user is regularly judged against a gold standard) essential to give appropriately challenging tasks.

As previously discussed, players can be motivated by a myriad of complex reasons. The desire to progress in the game may become more important to the player than to provide good quality work and this may lead to the desire to cheat the system.

#### 1.5.3.2 Attention Slips

Players may occasionally make a mistake and press the wrong button. Attention slips need to be identified and corrected by validation, where players can examine other players' work and evaluate it. Through validation, poor quality interpretations should be voted down and high quality interpretations should be supported (in the cases of genuine ambiguity there may be more than one). Validation thus plays a key role as a strategy for quality control.

Unlike collaboration in Wikipedia, it is not advisable to allow players of GWAP to go back and correct their mistakes, otherwise a player could try all possible variations of an answer and then select the one offering the highest score. In this sense the way players work together is more "collective" than "collaborative".

#### 1.5.3.3 Multiple Judgements and Genuine Ambiguity

Ambiguity is an inherent problem in all areas of NLP [36]. Here, we are not interested in solving this issue, but in using collaborative approaches to capture ambiguity where it is appropriate. Therefore, language resources should not only aim to select the best, or most common, annotation but also to preserve all inherent ambiguity, leaving it to subsequent processes to determine which interpretations are to be considered spurious and which instead reflect genuine ambiguity. This is a key difference between GWAP for NLP and other crowdsourcing work.

Collecting multiple judgements about every linguistic expression is a key aspect of PD. In the present version of PD eight players are asked to express their judgements on a markable. If they do not agree on a single interpretation, four more players are then asked to validate each interpretation.[21]

Validation has proven very effective at identifying poor quality interpretations. The value obtained by combining the player annotations with the validations for each interpretation tends to be zero or negative for all spurious interpretations. This formula can also be used to calculate the best interpretation of each expression, which we will refer to in what follows as the *game interpretation*.

Anaphoric judgements can be difficult, and humans will not always agree with each other. For example, it is not always clear from a text whether a markable is referential or not; and in case it is clearly referential, it is not always clear whether it refers to a new discourse entity or an old one, and which one. In PD we are interested in identifying such problematic cases: if a markable is ambiguous, the annotated corpus should capture this information.

#### 1.5.3.4 Malicious Behaviour

Controlling cheating may be one of the most important factors in GWAP design. If a player is motivated to progress in a game, e.g., by scoring points and attaining levels, they may also become motivated to cheat the system and earn those rewards without completing the tasks as intended.

All crowdsourcing systems attract spammers, which can be a very serious issue [22, 38, 50]. However, in a game context we can expect spamming to be much less of an issue because the work is not conducted on a pay-per-annotation basis.

Nevertheless, several methods are used in PD to identify players who are cheating or who are providing poor annotations. These include checking the player's IP address (to make sure that one player is not using multiple accounts), checking annotations against known answers (the player rating system), preventing players from resubmitting their decisions [17] and keeping a blacklist of players to discard all their data [72].

A method of profiling players was also developed for PD to detect unusual behaviour. The profiling compares a player's decisions, validations, skips, comments and response times against the average for the entire game – see Fig. 1.7. It is

[21] It is possible for an interpretation to have more annotations and validations than required if a player enters an existing interpretation after disagreeing or if several players are working on the same markables simultaneously.

| | System | Good player | Bad player |
|---|---|---|---|
| **ANNOTATIONS** | | | |
| Total Annotations: | 1423078 | 4587 | 11018 |
| Average Annotation Time: | 00:00:07 | 00:00:07 | 00:00:04 |
| Total (Ratio) DN: | 955520 (0.67) | 1495 (0.33) | 10935 (0.99) |
| Total (Ratio) DO: | 378256 (0.27) | 2696 (0.59) | 58 (0.01) |
| Total (Ratio) PR: | 79172 (0.06) | 334 (0.07) | 24 (0) |
| Total (Ratio) NR: | 13395 (0.01) | 64 (0.01) | 2 (0) |
| **VALIDATIONS** | | | |
| Total Validations: | 608982 | 3848 | 5256 |
| Total (Ratio) Agree: | 200174 (0.33) | 1186 (0.31) | 8 (0) |
| Ave Agree Time: | 00:00:09 | 00:00:08 | 00:00:18 |
| Total (Ratio) Disagree: | 408808 (0.67) | 2662 (0.69) | 5248 (1) |
| Ave Disgree Time: | 00:00:08 | 00:00:07 | 00:00:02 |
| **OTHER** | | | |
| Total Skips: | 51616 | 142 | 26 |
| Skip per annotation: | 0.04 | 0.03 | 0 |
| Total Comments: | 26593 | 229 | 0 |
| Comment per annotation: | 0.02 | 0.05 | 0 |

**Fig. 1.7** Player profiling in Phrase Detectives, showing the game totals and averages (*left*), a good player profile (*centre*) and a bad player profile (*right*) taken from real game profiles. The bad player in this case was identified by the speed of annotations and that the only responses were DN in Annotation Mode and Disagree in Validation Mode. The player later confessed to using automated form completion software

very simple to detect players who should be considered outliers using this method (this may also be due to poor task comprehension as well as malicious input) and their data can be ignored to improve the overall quality.

## 1.6 Evaluating the Gaming Approach to Creating Language Resources

Evaluating a gaming approach to collaborative resource creation needs to be done in conjunction with other approaches. In order to make things comparable all costs are converted to US$, the lowest level of linguistic labelling is called an *annotation* and an action that the player is asked to perform (that may result in several annotations at once) is called a *task*. To this end, we compare three main areas:

- Participants – How are participants motivated? How much do participants contribute? Do certain participants contribute more?
- Task – How fast is the data being produced? What is the quality of the contributions when aggregated? What is the upper limit of quality that can be expected?
- Implementation – How much does the data collection cost? Which approach represents the best value for money?

The first two areas of comparison correspond to the elements of collective intelligence [46]: the first covering the "who" and "why"; and the latter covering the "how" and "what". The third area of comparison is a more pragmatic view on the approaches, where the choice of approach may be made based on how much budget there is for the data collection, what level of quality is needed for the data to be of use or how much data is required.

## *1.6.1 Participants*

As previously discussed, participant motivation in collaborative approaches is a very complex issue that has implications on data quality. We consider the case of GWAP without financial incentives, with indirect financial incentives and reported results for other approaches.

### 1.6.1.1 Motivating Participation

We measure the success of advertising and the motivation to join the game by how many players have registered over the period of time the game was online. The first version of PD recruited 2,000 players in 32 months (62 players/month) and PD on Facebook recruited 612 players in 13 months (47 players/month). JDM recruited 2,700 players in 56 months (48 players/month).

This level of recruitment, whilst not in the same league as the ESP Game which enjoyed massive recruitment in its first few months online, could be seen as what you could expect if some effort was made to advertise a GWAP and motivate people to play it.

There are 5,500 registered reviewers of Wikipedia articles,[22] which is the equivalent to a player in a GWAP, however there is an unknown (presumably very large) number of unregistered contributors.

[22]http://en.wikipedia.org/wiki/Wikipedia:Wikipedians

The number of active AMT workers has been estimated as between 15,059 and 42,912 [25]. This explains the difficulty in finding workers with specific skills, such as native speakers for some languages [55], or who can perform large tasks [33].

The total number of participants is useful for evaluating the relative success of recruitment efforts. However, it is not a good predictor of how much work will be done, how fast it will be completed or of what quality it will be. Therefore further analysis of the players themselves is required.

#### 1.6.1.2 Motivating Contributions

Participation (or volition) of collaborators to contribute is another way to assess whether the incentives of an approach are effective. We measure player motivation to contribute by the average lifetime play. In the case of PD it was 35 min (the average sum of all tasks) and in the case of JDM it was 25 min (the average length of a session for approximatively 20 games).

The average weekly contribution for Wikipedia is just over 8 h [54] however this is for contributing users of Wikipedia, not for casual browsers of the website. This indicates that when a user starts contributing to Wikipedia they are highly motivated to contribute. In AMT the contribution rate is a little lower, between 4–6 h [33], and it can also be expected that the user, once registered, will be highly motivated to contribute.

Obviously, there is a huge complexity and spread of user types within the AMT user base, however it is interesting to note that for 20 % of the workers, AMT represents their primary source of income (and for 50 %, their secondary source of income), and they are responsible for completing more that one third of all the HITs [32]. Participating for leisure is important for only 30 % of workers. So the motivations for participating to AMT are very different from that of Wikipedia.

An observation in most crowdsourcing systems is the uneven distribution of contribution per person, often following a Zipfian power law curve. In PD, it was reported that the ten highest scoring players (representing 1.3 % of total players) had 60 % of the total points on the system and had made 73 % of the annotations [14].

In the Facebook version of PD, the ten highest scoring players (representing 1.6 % of total players) had 89 % of the total points and had made 89 % of the annotations – see Fig. 1.8 [15].

Similarly in JDM the top 10 % of the player represents 90 % of the activity and studies of AMT also find that only 20 % of the users are doing 80 % of the work.[23]

These results show that the majority of the workload is being done by a minority of players. However, the influence of players who only contribute a little should not be undervalued as in some systems it can be as high as 30 % of the workload [37] and this is what makes the collective decision making robust.

[23]http://groups.csail.mit.edu/uid/deneme/?p=502

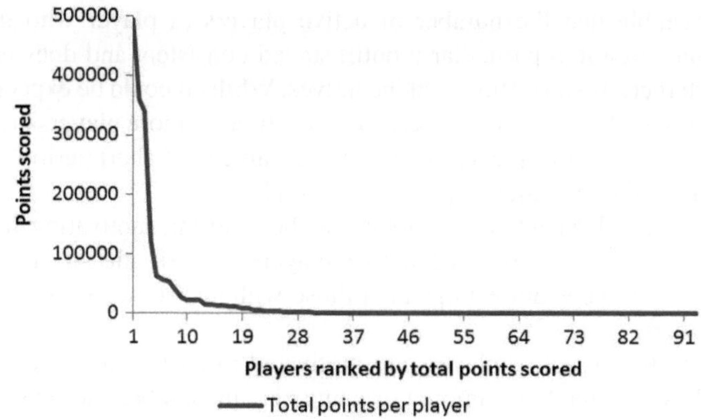


**Fig. 1.8** Chart showing the scores of players (approximately equivalent to workload) in the Phrase Detectives game on Facebook

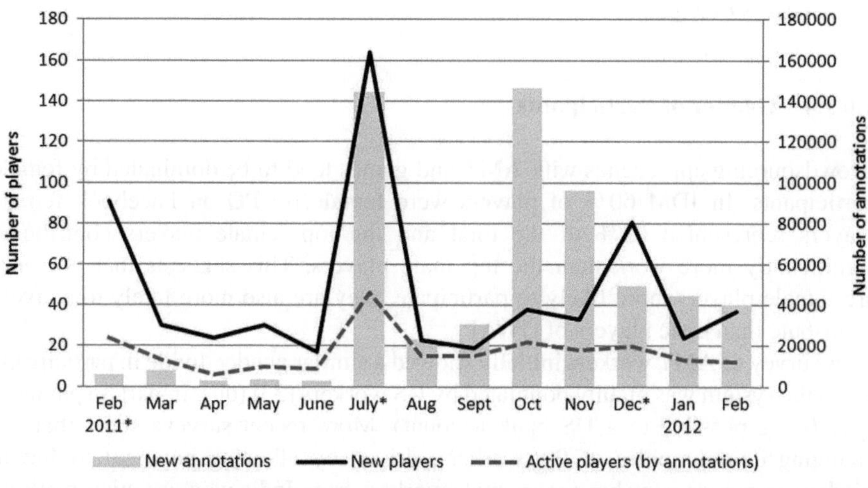


**Fig. 1.9** Chart showing new annotations plotted with new players and active players in Phrase Detectives on Facebook. Prizes were available in the game from July 2011 to February 2012. * indicates a month with active promotion for the game

#### 1.6.1.3 The Effect of Incentives on Participation and Contribution

Further to the figures for motivation and participation, Fig. 1.9 shows the growth of PD on Facebook. Months where there was active promotion of the site (February, July and December 2011) show increases in new players, as one would expect.

Based on the assumption that the first promotion month, when the site went live, was an exception as players of the previous game joined the new version, there is an indication that financial incentives increase recruitment to the game, if sufficiently advertised.

It is noticeable that the number of active players (a player who made more than one annotation in a particular month) stayed consistent and does not seem to increase with recruitment or financial incentives. Whilst it could be expected that the number of active players steadily increases over time as more players are recruited, the results show that most players will play the game for a short period of time and only a small number continue to play every month.

Indirect financial incentives do appear to be a strong motivating factor when considering how much work the active players do. Months with prizes have considerably more new annotations than those without, but with a similar number of active players.

This suggests that active players are motivated to contribute more by financial incentives, however the large amount of game play in October and November 2011 indicates that other motivating factors, such as personal and social incentives are, to some extent, also successful. Whilst financial incentives are important to recruit new players, a combination of all three types of incentives is essential for the long term success of a game.

#### 1.6.1.4 Gender of Participants

Crowdsourcing approaches with AMT and games tend to be dominated by female participants. In JDM 60 % of players were female. In PD on Facebook female players represented 65 % of the total and the top female players contributed significantly more work than the top male players. This suggests that not only are female players more likely to participate, they are also more likely to actively contribute than male players of GWAP.

A survey of AMT workers initially showed a similar gender divide in participants when the system was mainly populated by US workers [33] (due, in part, to payment only being possible to a US bank account). More recent surveys show that the changing demographics of the workers, driven by allowing payment to Indian workers in rupees, now have more male workers from India who use microworking as a primary source of income [64] and the gender split is almost even [33].

The changing demographics of crowdsourcing participants will have an impact on the types of incentives and types of tasks offered. For example a further study of AMT users performing two tasks showed female dominance, but with preference for word puzzle tasks (74 % female) over image sorting tasks (58.8 % female) [50].

Conversely, it has been reported that only 12 % of contributors to Wikipedia are female [27]. This prompted significant research into the gender bias in the authorship of the site [44].

It has been shown that diverse groups are better at solving tasks and have higher collective intelligence (termed *c*) than more homogeneous groups. A balanced gender divide within a group also produces a higher *c* as females demonstrate higher social sensitivity towards group diversity and divergent discussion [78]. However, this may not have such an impact where the collaboration is indirect.

## 1.6.2 Task

### 1.6.2.1 Throughput

A measure of efficiency of the interface and task design is how fast tasks are being completed or annotations being generated. This measure is called *throughput*, the number of labels (or annotations) per hour [74].

The throughput of PD is 450 annotations per human hour, which is almost twice as fast as the throughput of 233 labels per human hour reported for the ESP Game.

There is a crucial difference between the two games: PD only requires clicks on pre-selected markables, whereas in the ESP Game the user is required to type in the labels. However, the throughput for JDM is calculated to be 648, where the players also had to type labels, so throughput may also be an indication of task difficulty and cognitive load on the player.

Designers of GWAP who are considering making their task timed should therefore carefully consider the speed at which the player can process the input source (e.g. text, images) and deliver their response (e.g. a click, typing) in order to maximize throughput and hence the amount of data that is collected without making the game unplayable.

The throughput of AMT has been reported to be close to real time (within 500 ms of a HIT being posted) however this is usually for very simple tasks [7]. More complex tasks can take up to a minute to complete giving a throughput range from 1 to 7,200 labels per hour, while some may never be completed. Whilst these figures are not especially helpful, it highlights the potential speed of this approach if the task can be presented in an efficient way.

Related to throughput is the *wait time* for tasks to be done. Most crowdsourcing systems allow data collection in parallel (i.e., many participants can work at once on the same tasks), although validation requires users to work in series (i.e., where one user works on the output of another user). So whilst the throughput may give us a maximum speed from the system, it is worth bearing in mind that the additional time spent waiting for a user to be available to work on the task may slow the system considerably.

This is where the AMT approach, with a large worker pool, has an advantage and some task requesters even pay workers a retainer to be on demand [5]. With GWAP it is possible to prioritise tasks to maximise completion of corpora, but for open collaboration like Wikipedia it is much more difficult to direct users to areas that need contribution. This can be seen by comparing popular pages that have considerable work, such as for the film Iron Man[24] with 8,000 words, with less popular pages, such as Welsh poetry[25] with only 300 words.

[24]http://en.wikipedia.org/wiki/Iron_Man

[25]http://en.wikipedia.org/wiki/Welsh_poetry

#### 1.6.2.2 Annotation Quality

Annotation quality is usually assessed by comparing the work to a gold standard or to an expert's opinion. However it is worth noting that there is an upper boundary of quality with these resources as gold standards may occasionally contain errors and experts do not always agree.

In PD agreement between experts is very high although not complete: 94 %, for a chance-adjusted $\kappa$ value [11, 18], of $\kappa = 0.87$ which can be considered good for coreference tasks [4, 58]. This value can be seen as an upper boundary on what we might get out of the game.

Agreement between experts and the PD game interpretation is also good. We found 84.5 % agreement between Expert 1 and the game ($\kappa = 0.71$) and 83.9 % agreement between Expert 2 and the game ($\kappa = 0.70$). In other words, in about 84 % of all annotations the interpretation specified by the majority vote of non-experts was identical to the one assigned by an expert.

These values are comparable to those obtained when comparing an expert with the trained annotators (usually students) that are typically used to create *Traditional, Partly Validated Annotation* resources.

For JDM, there is no similar resource that could be used as gold standard and it is difficult to assign an expert role for common sense knowledge acquisition.

AKI,[26] a guessing game, was designed as an indirect evaluation procedure. The goal of the game is to make the system (AKI) guess what the player has in mind from given clues, with the system making a proposal after each clue. The game goes on until the system finds the proper answer or fails to do so. In this task, the AKI system finds the right answer in 75 % of the cases. For the same task, humans get the right answer in 48 % of cases.

The data used as a knowledge base is strictly the lexical network constructed with JDM, without any modification or preprocessing.

#### 1.6.2.3 Task Difficulty

There is a clear difference in quality when we look at the difficulty of the task in GWAP [13]. Looking separately at the agreement on each type of markable annotation in PD (see Table 1.1), we see that the figures for a discourse-new (DN) annotation are very close for all three comparisons, and well over 90 %. Discourse-old (DO) interpretations are more difficult, with only 71.3 % average agreement.

Of the other two types, the 0 % agreement between experts and the game on property (PR) interpretations suggests that they are very hard to identify, or possibly the training for that type is not effective. Non-referring (NR) markables on the other hand, although rare, are correctly identified in every single case with 100 % precision.

---

[26] http://www.jeuxdemots.org/AKI.php

**Table 1.1** Agreement on annotations in Phrase Detectives, broken down by annotation type

| | Expert 1 vs. Expert 2 (%) | Expert 1 vs. Game (%) | Expert 2 vs. Game (%) |
|---|---|---|---|
| Overall agreement | 94.1 | 84.5 | 83.9 |
| Discourse-new (DN) agreement | 93.9 | 96.0 | 93.1 |
| Discourse-old (DO) agreement | 93.3 | 72.7 | 70.0 |
| Non-referring (NR) agreement | 100.0 | 100.0 | 100.0 |
| Property (PR) agreement | 100.0 | 0.0 | 0.0 |

This demonstrates the issue that quality is not only affected by player motivation and interface design but also by the inherent difficulty of the task. As we have seen, users need to be motivated to rise to the challenge of difficult tasks and this is where financial incentives may prove to be too expensive on a large scale.

The quality of the work produced by AMT, with appropriate post-processing, seems sufficient to train and evaluate statistical translation or transcription systems [10, 49]. However, it varies from one task to another according to the parameters of the task. Unsurprisingly, workers seem to have difficulties performing complex tasks, such as the evaluation of summarisation systems [26].

## 1.6.3 Implementation

### 1.6.3.1 Cost

When evaluating the costs of the different approaches to collaboratively creating language resources, it is important to also consider other constraints, namely the speed at which data needs to be produced, the size of the corpus required, and the quality of the final resource. In order to compare the cost effectiveness we make some generalisations, convert all costs to US$ and calculate an approximate figure for the number of annotations per US$. Where we have factored in wages for software development and maintenance we have used the approximate figure of US$ 54,000 per annum for a UK-based post doc research assistant.[27] Additional costs that may be incurred include maintenance of hardware, software hosting, and institutional administrative costs but as these are both difficult to quantify and apply to all approaches they will not be included in the estimates below.

*Traditional, Entirely Validated Annotation* requires in the order of US$ 1 million per 1 million tokens.[28] On average English texts contain around 1 markable every 3 tokens, so we get a cost of 0.33 markables/US$.

[27] http://www.payscale.com/research/UK/Job=Research_Scientist/Salary

[28] This figure was obtained by informally asking several experienced researchers involved in funding applications for annotation projects.

*Traditional, Partly Validated Annotation*, from estimates of projects by the authors in the UK and Italy, are in the order of US$ 400,000 per 1 million tokens, including the cost of expert annotators. This equates to 0.83 markables/US$.

Both of the above figures are generalisations that include the costs for administering the data collection and developing tools for the task if required. The timescale of data collection is usually several years.

Costs with AMT depend on the amount paid per HIT, which is determined by the task difficulty, the availability of workers with sufficient skills to do the task, and on the extent of redundancy. The literature suggests that US$ 0.01 per HIT is the minimum required for non-trivial tasks, and for a more complex linguistic task like anaphoric co-reference or uncommon language translation, the cost is upwards from US$ 0.1 per HIT. Redundancy for simple tasks is usually around five repetitions per task, although in practice we find that ten repetitions is more likely to be required to attain reasonable quality and filter out poor quality decisions. AMT allows requesters of HITs to set a performance threshold for participants based on whether previous work has been acceptable to other requesters. By using this method there would be less need for redundancy, however the cost of the HIT may need to increase to attract the better workers.

In the case of simple tasks where quality is not a priority the cost would be in the region of 20 markables/US$. In the case of more complicated tasks it would be more like a cost of 1 markable/US$. This is a more conservative estimate than what has previously been cited for early studies with AMT at 84 markables/US$ [67] however, we feel this is more realistic given a more developed microwork platform and workforce.

AMT has the advantage that it is fast to develop, deploy and collect data on a small scale. Typically it may take 1 month for a researcher to create an interface for AMT (US$ 4,500), and perhaps 2 months to collect the data (US$ 9,000) for a small resource.

The advantage of GWAP over other approaches is that once the system is set up, annotations do not cost anything to collect data.

PD took approximately 6 months to develop (US$ 27,000) and a further 3 months to develop the Facebook interface (US$ 13,500). Approximately US$ 9,000 was spent over 38 months in prizes and advertising for the game (approximately US$ 235 per month) with a researcher maintaining the system part-time (at 20 %, equivalent to US$ 900 per month, totalling US$ 34,200). 2.5 million annotations have been collected by PD, which is gives a cost of 30 annotations/US$. 84,000 markables were completely annotated (although in reality many more were partially annotated) giving a conservative estimate of 1 markables/US$.

JDM took approximately 4 months to develop (US$ 18,000) and was maintained for 54 months by a researcher part-time (at 10 %, equivalent to US$ 450 per month, totalling US$ 24,300). JDM did not spend any money on prizes or promotion. During this time 1.3 million individual relations were collected (but not validated in this game) giving an estimate of 53.5 unvalidated markables/US$.

From these estimates it is clear that creating language resources using traditional methods is expensive, prohibitively so beyond 1M words, however the quality is high. This approach is best suited for corpora where the quality of the data is paramount.

AMT for simple tasks is quick to set up and collect data and very cheap, however, more complex tasks are more expensive. The quality of such resources needs more investigation and the approach becomes prohibitively expensive when scaling beyond 10M words. Microworking approaches are therefore most suited for small to medium scale resources, or prototyping interfaces, where noisy data can be filtered.

The GWAP approach is expensive compared to AMT to set up, but the data collection is cheap. In a long term project it is conceivable to collect a 10M+ word corpus, with the main problem being the length of time it would take to collect the data. Over a long period of time the data collection would not only need continuous effort for player recruitment, but also the project requirements may change, requiring further development of the platform. With this in mind, this approach is most suited to a long term, persistent data collection effort that aims to collect very large amounts of data.

#### 1.6.3.2 Reducing Costs

One of the simplest ways of reducing costs is to reduce the amount of data needed and to increase the efficiency of the human computation. Pre-annotation of the data and bootstrapping can reduce the task load, increase the annotation speed and quality [24] and allow participants to work on more interesting tasks that are ambiguous or difficult. Bootstrapping has the downside of influencing the quality of usable output data and errors that exist in the input data multiply when used in crowdsourcing.

This was seen in the PD game, where occasional errors in the pre-processing of a document led to some markables having an incorrect character span. The game allowed players to flag markables with errors for correction by administrators (and to skip the markable if appropriate) however this created a bottleneck in itself. Currently there is no mechanism for players to correct the markables as this would have a profound impact on the annotations that have been collected. JDM did not have these problems as there was no preprocessing in the game.

As can be seen from the cost breakdown of PD, more savings can be made by reusing an existing GWAP platform; the development of the Facebook interface cost half that of the original game.

The advantage of GWAP over microworking is that personal and social incentives can be used, as well as financial, to minimise the cost and maximise the persistence of the system. The use of prizes can motivate players to contribute more whilst still offering value for money as part of a controlled budget.

However, we should be aware that the race towards reducing costs might have a worrying side-effect as short term AMT costs could become the standard. Funding agencies will expect low costs in future proposals and it will become hard to justify funding to produce language resources with more traditional, or even GWAP-based methodologies.

#### 1.6.3.3 Data Size and Availability

In JDM, more than 1,340,000 relations between terms have been collected, the sum of the weights being over 150,000,000. More than 150,000 terms have at least one outgoing relation, and more than 120,000 have at least one incoming relation. The current JDM database is available from the game website.

In PD, 407 documents were fully annotated, for a total completed corpus of over 162,000 words, 13 % of the total size of the collection currently uploaded for annotation in the game (1.2M words). The first release of the PD Corpus 0.1 [60] is about the size of the ACE 2.0 corpus[29] of anaphoric information, the standard for evaluation of anaphora resolution systems until 2007/2008 and still widely used.

The size of the completed corpus does not properly reflect, however, the amount of data collected, as the case allocation strategy adopted in the game privileges variety over completion rate. As a result, almost all the 800 documents in the corpus have already been partially annotated. This is reflected, first of all, in the fact that 84,280 of the 392,120 markables in the active documents (21 %) have already been annotated. This is already almost twice the total number of markables in the entire OntoNotes 3.0 corpus,[30] which contains 1 million tokens, but only 45,000 markables.

The number of partial annotations is even greater. PD players produced over 2.5 million anaphoric judgements between annotations and validations; this is far more than the number of judgements expressed to create any existing anaphorically annotated corpus. To put this in perspective, the GNOME corpus, of around 40K words, and regularly used to study anaphora until 2007/2008, contained around 3,000 annotations of anaphoric relations [56] whereas OntoNotes 3.0 only contains around 140,000 annotations.

Most of the reported resources created on AMT are small to medium size ones [25, 32]. Another issue raised by AMT is the legal status of intellectual property rights of the language resources created on it. Some US universities have insisted on institutional review board approval for AMT experiments.[31]

## 1.7 Conclusions

In this chapter we have considered the successes and the limitations of the GWAP approach to collaboratively creating language resources, compared to traditional annotation methods and more recent approaches such as microwork crowdsourcing and Wikipedia-style open collaboration.

---

[29] http://www.ldc.upenn.edu/Catalog/catalogEntry.jsp?catalogId=LDC2003T11

[30] http://www.ldc.upenn.edu/Catalog/CatalogEntry.jsp?catalogId=LDC2009T24

[31] From personal communication with K. Cohen.

### 1.7.1 Game Interface and Task Design

The game interface should be attractive enough to encourage players to start playing and easy enough to use so they keep playing. Before building a GWAP it is essential to have an understanding of game concepts, such as game flow and creating entertaining game scenarios.

The design of the task itself will be determined in part by the complexity of the data being collected. By identifying the difficult or ambiguous tasks, the pre- and post-processing can be improved and the human input can be maximised to produce the highest quality resource possible given the inherent difficulty of the task. Participants may need to be motivated to rise to the challenge of difficult tasks and this is where financial incentives may prove to be too expensive on a large scale.

The task design should be streamlined for efficient collection of data as this is one of the simplest ways of reducing costs: by reducing the amount of data needed. The throughput (annotations per hour) of a GWAP is a good measure of how efficient it is at collecting data, however, it is worth bearing in mind that the additional time spent waiting for a user to be available to work on the task may slow the system.

### 1.7.2 Participants and Motivation

Generally speaking, GWAP will use entertainment as the motivating factor rather than direct financial incentives (as found in microwork crowdsourcing). There may also be other motivational considerations, such as the desire to contribute to a scientific project or for self enrichment and learning.

Most the players of GWAP will not benefit directly from the data being collected, however the player connection to the project and sense of contribution to science are strong motivating factors with the citizen science approach, where players are willing to work on harder tasks, provide higher quality data and contribute more.

Controlling cheating may be one of the most important factors in crowdsourcing design and is especially problematic for microworking.

An advantage of GWAP over microworking is that personal and social incentives can be used, as well as financial, to minimise the cost and maximise the persistence of the system. Indirect financial incentives can be a cost-effective way to increase participation in a game.

It is common for the majority of the workload to be done by a minority of players. Motivating the right kind of players is a complex issue, central to the design of the game interface and the task, and is as important as attracting large numbers of players because, although collective intelligence needs a crowd, that crowd also needs to do some work.

The more a player participates in a GWAP, the more expert they become. A system needs to correlate good performance at the task with good quality data and a ratings system (where the user is regularly judged against a gold standard) is essential to give appropriately challenging tasks.

Crowdsourcing approaches with microworking and games tend to be dominated by female participants, although this is not the case for Wikipedia. If crowdsourcing approaches ever hope to produce high quality data, the gender bias needs to be considered as it has been shown that diverse groups are better at solving tasks and have higher collective intelligence than more homogeneous groups.

### *1.7.3 Annotation Quality and Quantity*

The issue of annotation quality is an area of continuous research. However, results with Phrase Detectives and JeuxDeMots are very promising. The ultimate goal is to show that language resources created using games and other crowdsourcing methods potentially offer higher quality and are more useful by allowing for linguistic ambiguity. By quantifying the complexity of the linguistic tasks, human participants can be challenged to solve computationally difficult problems that would be most useful to machine learning algorithms.

Creating language resources using traditional methods is expensive, prohibitively so beyond 1M words, however the quality is high. Whilst the initial costs of developing a GWAP are high, the game can persistently collect data, making it most suitable for long term, large scale projects. The speed and cost of a microworking approach make it most suitable for collecting small to medium scale resources or prototyping software for larger scale collection, however, some issues of requester responsibility and intellectual property rights remain unresolved.

Approaches that require financial motivation for the participants cannot scale to the size of resources that are now increasingly more essential for progress with human language technology. Only through the contribution of willing participants can very large language resources be created, and only GWAP or Wikipedia-style approach facilitate this type of collaboration.

**Acknowledgements** We would like to thank Jean Heutte (CREF-CNRS) for his help with the concepts of game flow and for the comments of the reviewers of this chapter. The contribution of Karën Fort to this work was realized as part of the Quæro Programme[32], funded by OSEO, French State agency for innovation. The original Phrase Detectives game was funded as part of the EPSRC AnaWiki project, EP/F00575X/1.

---

[32]http://quaero.org/

# Appendix A

Categories of GWAP with links where available.

| | |
|---|---|
| *Image annotation* | |
| ESP Game | http://www.gwap.com/gwap/gamesPreview/espgame |
| Matchin | http://www.gwap.com/gwap/gamesPreview/matchin |
| FlipIt | http://www.gwap.com/gwap/gamesPreview/flipit |
| Phetch | http://www.peekaboom.org/phetch |
| Peekaboom | http://www.peekaboom.org |
| Squigl | http://www.gwap.com/gwap/gamesPreview/squigl |
| Magic Bullet | http://homepages.cs.ncl.ac.uk/jeff.yan/mb.htm |
| Picture This | http://picturethis.club.live.com |
| *Video annotation* | |
| OntoTube | http://ontogame.sti2.at/games |
| PopVideo | http://www.gwap.com/gwap/gamesPreview/popvideo |
| Yahoo's VideoTagGame | http://sandbox.yahoo.com/VideoTagGame |
| Waisda | http://www.waisda.nl |
| *Audio annotation* | |
| Herd It | http://apps.facebook.com/herd-it |
| Tag a Tune | http://www.gwap.com/gwap/gamesPreview/tagatune |
| WhaleFM | http://whale.fm |
| *Biomedical* | |
| Foldit | http://fold.it/portal |
| Phylo | http://phylo.cs.mcgill.ca |
| EteRNA | http://eterna.cmu.edu |
| *Transcription* | |
| Ancient Lives | http://ancientlives.org |
| Old Weather | http://www.oldweather.org |
| *Search results* | |
| Page Hunt | http://pagehunt.msrlivelabs.com/PlayPageHunt.aspx |
| *Social bookmarking* | |
| Collabio | http://research.microsoft.com/en-us/um/redmond/groups/cue/collabio |

# Appendix B

Categories of GWAP used for NLP with links where available.

| | |
|---|---|
| *Knowledge acquisition* | |
| 1001 Paraphrases | |
| LEARNER | |
| FACTory | http://game.cyc.com |
| Verbosity | http://www.gwap.com/gwap/gamesPreview/verbosity |
| Categorilla | http://www.doloreslabs.com/stanfordwordgame/categorilla.html |
| Free Association | http://www.doloreslabs.com/stanfordwordgame/freeAssociation.html |
| *Text annotation* | |
| Phrase Detectives | http://www.phrasedetectives.com |
| Phrase Detectives on Facebook | http://apps.facebook.com/phrasedetectives |
| PlayCoref | |
| PhraTris | http://galoap.codeplex.com |
| PackPlay | |
| *Sentiment analysis* | |
| Sentiment Quiz | http://apps.facebook.com/sentiment-quiz |
| *Generation* | |
| GIVE games | http://www.give-challenge.org |
| *Ontology building* | |
| JeuxDeMots | http://www.jeuxdemots.org |
| AKI | http://www.jeuxdemots.org/AKI.php |
| OntoGame | http://ontogame.sti2.at/games |

# References

1. Aker A, El-haj M, Albakour D, Kruschwitz U (2012) Assessing crowdsourcing quality through objective tasks. In: Proceedings of LREC'12, Istanbul
2. Alonso O, Mizzaro S (2009) Can we get rid of TREC assessors? Using mechanical turk for relevance assessment. In: Proceedings of SIGIR '09: workshop on the future of IR evaluation, Boston
3. Alonso O, Rose DE, Stewart B (2008) Crowdsourcing for relevance evaluation. SIGIR Forum 42(2):9–15
4. Artstein R, Poesio M (2008) Inter-coder agreement for computational linguistics. Comput Linguist 34(4):555–596
5. Bernstein MS, Karger DR, Miller RC, Brandt J (2012) Analytic methods for optimizing realtime crowdsourcing. In: Proceedings of the collective intelligence 2012, Boston
6. Bhardwaj V, Passonneau R, Salleb-Aouissi A, Ide N (2010) Anveshan: a tool for analysis of multiple annotators' labeling behavior. In: Proceedings of the 4th linguistic annotation workshop (LAW IV), Uppsala

7. Bigham JP, Jayant C, Ji H, Little G, Miller A, Miller RC, Miller R, Tatarowicz A, White B, White S, Yeh T (2010) Vizwiz: nearly real-time answers to visual questions. In: Proceedings of the 23nd annual ACM symposium on user interface software and technology, UIST '10, New York
8. Bonneau-Maynard H, Rosset S, Ayache C, Kuhn A, Mostefa D (2005) Semantic annotation of the French media dialog corpus. In: Proceedings of InterSpeech, Lisbon
9. Callison-Burch C (2009) Fast, cheap, and creative: evaluating translation quality using Amazon's Mechanical Turk. In: Proceedings of the 2009 conference on empirical methods in natural language processing, Singapore
10. Callison-Burch C, Dredze M (2010) Creating speech and language data with Amazon's Mechanical Turk. In: CSLDAMT '10: proceedings of the NAACL HLT 2010 workshop on creating speech and language data with Amazon's Mechanical Turk, Los Angeles
11. Carletta J (1996) Assessing agreement on classification tasks: the kappa statistic. Comput Linguist 22:249–254
12. Chamberlain J, Poesio M, Kruschwitz U (2008) Phrase detectives: a web-based collaborative annotation game. In: Proceedings of the international conference on semantic systems (I-Semantics'08), Graz, Austria
13. Chamberlain J, Kruschwitz U, Poesio M (2009) Constructing an anaphorically annotated corpus with non-experts: assessing the quality of collaborative annotations. In: Proceedings of the 2009 workshop on the people's web meets NLP: collaboratively constructed semantic resources, Singapore
14. Chamberlain J, Poesio M, Kruschwitz U (2009) A new life for a dead parrot: incentive structures in the phrase detectives game. In: Proceedings of the WWW 2009 workshop on web incentives (WEBCENTIVES'09)
15. Chamberlain J, Kruschwitz U, Poesio M (2012) Motivations for participation in socially networked collective intelligence systems. In: Proceedings of CI2012. MIT, Cambridge
16. Chklovski T (2005) Collecting paraphrase corpora from volunteer contributors. In: Proceedings of K-CAP '05, Banff
17. Chklovski T, Gil Y (2005) Improving the design of intelligent acquisition interfaces for collecting world knowledge from web contributors. In: Proceedings of K-CAP '05, Banff
18. Cohen J (1960) A coefficient of agreement for nominal scales. Educ Psychol Meas 20(1):37–46
19. Csikszentmihalyi M (1990) Flow : the psychology of optimal experience. Harper and Row, New York
20. Dandapat S, Biswas P, Choudhury M, Bali K (2009) Complex linguistic annotation – No easy way out! a case from Bangla and Hindi POS labeling tasks. In: Proceedings of the 3rd ACL linguistic annotation workshop, Singapore
21. Fellbaum C (1998) WordNet: an electronic lexical database. MIT, Cambridge
22. Feng D, Besana S, Zajac R (2009) Acquiring high quality non-expert knowledge from on-demand workforce. In: Proceedings of the 2009 workshop on the people's web meets NLP: collaboratively constructed semantic resources, Singapore
23. Fenouillet F, Kaplan J, Yennek N (2009) Serious games et motivation. In: George S, Sanchez E (eds) 4ème Conférence francophone sur les Environnements Informatiques pour l'Apprentissage Humain (EIAH'09), vol. Actes de l'Atelier "Jeux Sérieux: conception et usages", p. 41–52. Le Mans
24. Fort K, Sagot B (2010) Influence of pre-annotation on POS-tagged corpus development. In: Proceedings of the 4th ACL linguistic annotation workshop (LAW), Uppsala
25. Fort K, Adda G, Cohen KB (2011) Amazon Mechanical Turk: gold mine or coal mine? Comput Linguist (editorial) 37:413–420
26. Gillick D, Liu Y (2010) Non-expert evaluation of summarization systems is risky. In: Proceedings of the NAACL HLT 2010 workshop on creating speech and language data with Amazon's Mechanical Turk, Los Angeles
27. Glott R, Schmidt P, Ghosh R (2010) Wikipedia survey – overview of results. UNU-MERIT, Maastricht, pp 1–11

28. Green N, Breimyer P, Kumar V, Samatova NF (2010) Packplay: mining semantic data in collaborative games. In: Proceedings of the 4th linguistic annotation workshop, Uppsala
29. Hladká B, Mírovskỳ J, Schlesinger P (2009) Play the language: play coreference. In: Proceedings of the ACL-IJCNLP 2009 conference short papers, Singapore
30. Hong J, Baker CF (2011) How good is the crowd at "real" WSD? In: Proceedings of the 5th linguistic annotation workshop, Portland
31. Howe J (2008) Crowdsourcing: why the power of the crowd is driving the future of business. Crown Publishing Group, New York
32. Ipeirotis P (2010) Analyzing the Amazon Mechanical Turk marketplace. CeDER working papers. http://hdl.handle.net/2451/29801
33. Ipeirotis P (2010) Demographics of Mechanical Turk. CeDER working papers. http://hdl.handle.net/2451/29585
34. Johnson NL, Rasmussen S, Joslyn C, Rocha L, Smith S, Kantor M (1998) Symbiotic intelligence: self-organizing knowledge on distributed networks driven by human interaction. In: Proceedings of the 6th international conference on artificial life. MIT, Cambridge
35. Joubert A, Lafourcade M (2008) Jeuxdemots : Un prototype ludique pour l'émergence de relations entre termes. In: Proceedings of JADT'2008, Ecole normale supérieure Lettres et sciences humaines, Lyon
36. Jurafsky D, Martin JH (2008) Speech and language processing, 2nd edn. Prentice-Hall, Upper Saddle River
37. Kanefsky B, Barlow N, Gulick V (2001) Can distributed volunteers accomplish massive data analysis tasks? In: Lunar and planetary science XXXII, Houston
38. Kazai G (2011) In search of quality in crowdsourcing for search engine evaluation. In: Proceedings of the 33rd European conference on information retrieval (ECIR'11), Dublin
39. Kazai G, Milic-Frayling N, Costello J (2009) Towards methods for the collective gathering and quality control of relevance assessments. In: Proceedings of the 32nd international ACM SIGIR conference on research and development in information retrieval, Boston
40. Koller A, Striegnitz K, Gargett A, Byron D, Cassell J, Dale R, Moore J, Oberlander J (2010) Report on the 2nd NLG challenge on generating instructions in virtual environments (GIVE-2). In: Proceedings of the 6th INLG, Dublin
41. Koster R (2005) A theory of fun for game design. Paraglyph, Scottsdale
42. Lafourcade M (2007) Making people play for lexical acquisition. In: Proceedings SNLP 2007, 7th symposium on natural language processing, Pattaya
43. Lafourcade M, Joubert A (2012) A new dynamic approach for lexical networks evaluation. In: Proceedings of LREC'12: 8th international conference on language resources and evaluation, Istanbul
44. Laniado D, Castillo C, Kaltenbrunner A, Fuster-Morell M (2012) Emotions and dialogue in a peer-production community: the case of Wikipedia. In: Proceedings of the 8th international symposium on Wikis and open collaboration (WikiSym'12), Linz
45. Lieberman H, A SD, Teeters A (2007) Common consensus: a web-based game for collecting commonsense goals. In: Proceedings of IUI, Honolulu
46. Malone T, Laubacher R, Dellarocas C (2009) Harnessing crowds: mapping the genome of collective intelligence. Research paper no. 4732-09, Sloan School of Management, MIT, Cambridge
47. Marchetti A, Tesconi M, Ronzano F, Rosella M, Minutol S (2007) SemKey: a semantic collaborative tagging system. In: Proceedings of WWW 2007 workshop on tagging and metadata for social information organization, Banff
48. Marcus M, Santorini B, Marcinkiewicz MA (1993) Building a large annotated corpus of English : the Penn Treebank. Comput Linguist 19(2):313–330
49. Marge M, Banerjee S, Rudnicky AI (2010) Using the Amazon Mechanical Turk for transcription of spoken language. In: IEEE international conference on acoustics speech and signal processing (ICASSP), Dallas
50. Mason W, Watts DJ (2009) Financial incentives and the "performance of crowds". In: Proceedings of the ACM SIGKDD workshop on human computation, Paris

51. Michael DR, Chen SL (2005) Serious games: games that educate, train, and inform. Muska & Lipman/Premier-Trade
52. Mihalcea R, Chklovski T (2003) Open mind word expert: creating large annotated data collections with web users help. In: Proceedings of the EACL 2003 workshop on linguistically annotated corpora (LINC 2003), Budapest
53. Mrozinski J, Whittaker E, Furui S (2008) Collecting a why-question corpus for development and evaluation of an automatic QA-system. In: Proceedings of ACL-08: HLT, Columbus
54. Nov O (2007) What motivates Wikipedians? Commun ACM 50(11):60–64
55. Novotney S, Callison-Burch C (2010) Cheap, fast and good enough: automatic speech recognition with non-expert transcription. In: Human language technologies: the 2010 annual conference of the North American chapter of the association for computational linguistics, Los Angeles
56. Poesio M (2004) Discourse annotation and semantic annotation in the GNOME corpus. In: Proceedings of the ACL workshop on discourse annotation, Barcelona
57. Poesio M, Artstein R (2008) Anaphoric annotation in the ARRAU corpus. In: LREC'08, Marrakech
58. Poesio M, Vieira R (1998) A corpus-based investigation of definite description use. Comput Linguist 24(2):183–216
59. Poesio M, Sturt P, Arstein R, Filik R (2006) Underspecification and anaphora: theoretical issues and preliminary evidence. Discourse Process 42(2):157–175
60. Poesio M, Chamberlain J, Kruschwitz U, Robaldo L, Ducceschi L (2012) The phrase detective multilingual corpus, release 0.1. In: Proceedings of LREC'12 workshop on collaborative resource development and delivery, Istanbul
61. Poesio M, Chamberlain J, Kruschwitz U, Robaldo L, Ducceschi L (2012 forthcoming) Phrase detectives: utilizing collective intelligence for internet-scale language resource creation. ACM Trans Interact Intell Syst
62. Quinn A, Bederson B (2011) Human computation: a survey and taxonomy of a growing field. In: CHI, Vancouver
63. Rafelsberger W, Scharl A (2009) Games with a purpose for social networking platforms. In: Proceedings of the 20th ACM conference on hypertext and hypermedia, Torino
64. Ross J, Irani L, Silberman MS, Zaldivar A, Tomlinson B (2010) Who are the crowdworkers?: shifting demographics in Mechanical Turk. In: Proceedings of CHI EA '10, Atlanta
65. Siorpaes K, Hepp M (2008) Games with a purpose for the semantic web. IEEE Intell Syst 23(3):50–60
66. Smadja F (2009) Mixing financial, social and fun incentives for social voting. Proceedings of the WWW 2009 workshop on web incentives (WEBCENTIVES'09), Madrid
67. Snow R, O'Connor B, Jurafsky D, Ng AY (2008) Cheap and fast—but is it good?: evaluating non-expert annotations for natural language tasks. In: EMNLP'08: proceedings of the conference on empirical methods in natural language processing, Honolulu
68. Surowiecki J (2005) The wisdom of crowds. Anchor, New York
69. Sweetser P, Wyeth P (2005) Gameflow: a model for evaluating player enjoyment in games. Comput Entertain 3(3):1–24
70. Thaler S, Siorpaes K, Simperl E, Hofer C (2011) A survey on games for knowledge acquisition. Technical Report STI TR 2011-05-01, Semantic Technology Institute
71. Tratz S, Hovy E (2010) A taxonomy, dataset, and classifier for automatic noun compound interpretation. In: Proceedings of the 48th annual meeting of the association for computational linguistics, Uppsala
72. von Ahn L (2006) Games with a purpose. Computer 39(6):92–94
73. von Ahn L, Dabbish L (2004) Labeling images with a computer game. In: Proceedings of the SIGCHI conference on Human factors in computing systems, Vienna
74. von Ahn L, Dabbish L (2008) Designing games with a purpose. Commun ACM 51(8):58–67
75. von Ahn L, Liu R, Blum M (2006) Peekaboom: a game for locating objects in images. In: Proceedings of CHI '06, Montréal

76. Wais P, Lingamneni S, Cook D, Fennell J, Goldenberg B, Lubarov D, Marin D, Simons H (2010) Towards building a high-quality workforce with Mechanical Turk. In: Proceedings of computational social science and the wisdom of crowds (NIPS)
77. Wang A, Hoang CDV, Kan MY (2010) Perspectives on crowdsourcing annotations for natural language processing. Lang Res Eval 1–23
78. Woolley AW, Chabris CF, Pentland A, Hashmi N, Malone TW (2010) Evidence for a collective intelligence factor in the performance of human groups. Science 330:686–688. URL http://dx.doi.org/10.1126/science.1193147
79. Yang H, Lai C (2010) Motivations of Wikipedia content contributors. Comput Human Behav 26(6):1377–1383
80. Yuen M, Chen L, King I (2009) A survey of human computation systems playing having fun. In: International conference on computational science and engineering, Vanoucer (CSE'09), vol 4, pp 723–728. IEEE

# Chapter 2
# Senso Comune: A Collaborative Knowledge Resource for Italian

**Alessandro Oltramari, Guido Vetere, Isabella Chiari, Elisabetta Jezek, Fabio Massimo Zanzotto, Malvina Nissim, and Aldo Gangemi**

**Abstract** *Senso Comune* is an open knowledge base for the Italian language, available through a Web-based collaborative platform, whose construction is in progress. The resource integrates dictionary data coming from both users and legacy resources with an ontological backbone, which provides foundations for a formal characterization of lexical semantic structures (frames). A nucleus of basic Italian lemmas, which have been semantically analyzed and classified, is available for both online access and download. A restricted community of contributors is currently working on increasing the lexical coverage of the resource.

---

A. Oltramari (✉)
Carnegie Mellon University, Pittsburgh, USA
e-mail: aoltrama@andrew.cmu.edu

G. Vetere
IBM Center for Advanced Studies of Rome, Rome, Italy
e-mail: gvetere@it.ibm.com

I. Chiari
Università *La Sapienza* di Roma, Roma, Italy
e-mail: isabella.chiari@uniroma1.it

E. Jezek
Università di Pavia, Pavia, Italy
e-mail: jezek@unipv.it

F.M. Zanzotto
Università di Roma *Tor Vergata*, Province of Rome, Italy
e-mail: zanzotto@info.uniroma2.it

M. Nissim
Università di Bologna, Bologna, Italy
e-mail: malvina.nissim@unibo.it

A. Gangemi
STLab – ISTC-CNR, Roma, Italy
e-mail: aldo.gangemi@cnr.it

I. Gurevych and J. Kim (eds.), *The People's Web Meets NLP*, Theory and Applications of Natural Language Processing, DOI 10.1007/978-3-642-35085-6_2, 

## 2.1 Introduction

*Senso Comune*[1] is the project of building an open knowledge base for the Italian language, designed as a crowd-sourced initiative that stands on the solid ground of an ontological formalization and well-established lexical resources: in this respect, it leverages on Web 2.0 and Semantic Web technologies. The community behind this project is growing and the knowledge base is evolving by integrating collaboratively user-generated content with existing lexical resources. The ontological backbone provides foundations for a formal characterization of lexical meanings and relational semantic structures, such as verbal frames. *Senso Comune* is an "open knowledge project": the lexical resource is available for both online access and download.

In the present contribution we provide an overview of the project, present some initial results, and discuss future directions. We firstly illustrate history and general goals of the project, its positioning with respect to general linguistic issues, and the state-of-the-art of similar resources. We describe the method to merge crowd-sourced development of the lexical resource and existing dictionaries. We provide some insight of the model underlying the knowledge base, from the perspective of its ontological structure. This paper also focuses on the methodological aspects of the knowledge acquisition process, introducing an interactive Q/A system (TMEO) designed to help users assigning ontological categories to linguistic meanings. Finally, we report the results of the experiment on ontology tagging of noun senses in Sect. 2.5, and stress the relevance of the resource to Natural Language Processing 2.6.

### *2.1.1 History and Objectives*

In fall 2006, a group of Italian researchers[2] from different disciplines gathered to provide a vision on the role of semantics in information technologies.[3]

Among other things, the discussion spotted the lack of open, machine-readable lexical resources for the Italian language. This was seen as one of the major hindering factors for the development of intelligent information systems capable of driving business and public services in Italy. Free, high quality lexical resources such as WordNet[4] contribute to the growth of intelligent information systems in English speaking countries. Lexical machine-readable resources for Italian – primarily

[1]www.sensocomune.it

[2]Besides the authors of this paper, the group, lead by Tullio De Mauro, includes Nicola Guarino, Maurizio Lenzerini and Laure Vieu.

[3]IBM Italia Foundation's symposium *La dimensione semantica dell'Information Technology* (The semantic dimension of Information Technology), Rome, November 27, 2006.

[4]http://wordnetweb.princeton.edu/perl/webwn

MultiWordNet,[5] EuroWordNet and the follow-up project SIMPLE[6] – freely available for research purposes, do not seem to play a similar role in the Italian industry of semantic technologies.

From these premises, the group decided to start an open collaborative research initiative, named *Senso Comune* (literally *common sense*, but more specifically intended as "common semantic knowledge"). A non-profit association was then established, which holds regular activities and annual workshops since 2007. Beyond the scope of industrial development, the group recognized that an open lexical resource for Italian is a way for collecting and organizing a body of knowledge which is particularly important in a modern country where, as in the rest of the world, new communication technologies increase the pace of linguistic changes.

From the outset, *Senso Comune* was conceived as a linguistic knowledge base rather than a dictionary. It is actually based on a conceptual apparatus that is not usually present in standard linguistic resources. In particular, each sense is mapped to ontological categories, and is associated with semantic frames.

The starting point to build such a knowledge base has been the acquisition of a high-quality lexical resource, namely De Mauro's 'vocabolario di base' (Basic Vocabulary) , which consists in the 2,071 most frequent Italian words, kindly made available by the author. The Basic Vocabulary of Italian was developed by De Mauro in 1980 [11] and further updated with minor changes up to 2007. It contains three different vocabulary ranges, the first being the so called 'fundamental vocabulary' containing the top 2,000 lemmas with top rank in two frequency lists of Italian written (LIF) and spoken language (LIP) – see [5] and [12].

The legacy resource was digitalized and put into a collaborative platform on the web, ready to be enriched by a vast (but supervised) community of users. An interdisciplinary, cross-organization team hosted at the Center for Advanced Studies of IBM Italia started designing a representational model and developing the related software tools to accommodate and manage the resource. Fitting the textual dictionary source into the model turned out to be very far from trivial; nonetheless, the web platform was made available in 2009, after 1 year of work.

Based on the acquired resource (see Sect. 2.3), the second step of the project consisted in classifying 4,586 senses of basic nouns (the most frequent in Italian textual sources) by means of a small set of predefined ontological categories. That work was carried out by undergraduate students under the supervision of the association researchers (see Sect. 2.5).

The development of *Senso Comune* has followed two main tracks so far. On the one hand, with the aim of providing a large-scale lexical resource, the group focused on how to extend the dictionary to cover thousands of common and less common words. The idea is to blend user contributions with reliable resources in a way that preserves both quality and availability. On the other hand, the group started

[5]See for example: http://multiwordnet.itc.it/english/home

[6]http://www.ub.edu/gilcub/SIMPLE/simple.html

studying how to extend the model to encompass the kind of lexical knowledge that is not usually represented in traditional lexicography. In particular, a study on verbal frames has been undertaken based on the idea of exploiting the usage examples associated with the sense definitions of the most common verbs included in the dictionary as an empirical base [48].

### *2.1.2 General Linguistic Perspective*

The *Senso Comune* research group includes linguists, computer scientists, logicians, and ontologists, who look at natural language from different perspectives and with different orientations. The relationship between expressions, meanings and reality, that is at the core of lexical semantics and conveys deep philosophical issues, is a largely debated issue. Although the research group members do not share all the assumptions, a common view (synthesized in a *Manifesto*) has been put at the basis of the project: the main tenet is that natural languages manifest themselves in actual usage scenarios, while the regularities that those languages show are a consequence of social evolution and consensus. Since languages serve humans in dealing with the world, ontologies (i.e., theories about physical, social or abstract realities) constitute a reference to characterize social evolution and consensus of language with respect to extra-linguistic entities. In other words, although language is far from being a mere "picture of reality", theories about reality are needed to account for lexical semantics, which is where words and entities come into contact.

Lexical semantics and ontology, though being different realms, are thus related, and much of the project's specificity is, in fact, the research of a suitable account of such relationship.

The representation of linguistic knowledge in a context-based approach (i.e., dealing with phenomena such as polysemy and ambiguity) is closely related to representations of other kinds of knowledge in the effort to reduce the gap between the semantic, pragmatic and contextual-encyclopaedic dimensions. The interaction between ontologies, semantics and lexical resources may be established in different ways [33]. In our first experiment we chose to mark linguistic data with concepts of a general formal ontology.

Ontologies represent an important bridge between knowledge representation and computational lexical semantics, and form a continuum with semantic lexicons [20]. The most relevant areas of interest in this context are Semantic Web and Human-Language Technologies: they converge in the task of pinpointing knowledge contents, although focusing on two different dimensions, i.e. ontological and linguistic structures. Computational ontologies and lexicons aim at digging out the basic elements of a given semantic space (domain-dependent or general), characterizing the different relations holding among them.

Nevertheless, they differ with respect to some general aspects: the polymorphic nature of lexical knowledge cannot be straightforwardly related to ontological categories and relations. Polysemy refers to a genuine lexical phenomenon that is generally absent in well-formed ontologies; the formal features of computational lexicons are far from being easily encoded in a logic-based language.[7]

Since the early 1980s, there has been a huge debate in the scientific community on whether the categorical structures of computational lexicons could be acknowledged as ontologies or not (see e.g. [31] for a survey of the issue). The general approach we adopt in *Senso Comune* is to integrate the two dimensions, with no attempt of reducing one to the other.[8] In the following section we quickly survey three of the most important state-of-the-art computational lexicons, i.e. WordNet, FrameNet and VerbNet, providing the general conceptual framework in which *Senso Comune* is rooted.

### 2.1.3 Comparing *Senso Comune* with WordNet, FrameNet, and VerbNet

WordNet was developed in Princeton University under the direction of the famous cognitive psychologist George A. Miller. Christiane Fellbaum, the principal investigator of the project, describes it as "a semantic dictionary that was designed as a network, partly because representing words and concepts as an interrelated system seems to be consistent with evidence for the way speakers organize their mental lexicons" ([13], p.7). WordNet is constituted by synsets (lexical concepts), namely set of synonym terms – e.g. (life form, organism, being, living thing). The idea of representing world knowledge through a semantic network (whose nodes are synsets, and whose arcs are lexical semantic relations[9]) has been characterizing WordNet development since 1985. Over the years, lexicographers have incrementally populated the resource (from the 37,409 synsets in the 1989 to about 120,000 synsets in the most recent releases), and substantial improvements of the entire WordNet architecture, aimed at facilitating hierarchical organization and computational tractability. Accordingly, RDF- and OWL-based implementations have been released (e.g. [1]).

WordNet covers several domains, namely groups of homogeneous terms referring to the same topic (art, geography, aeronautics, sport, politics, biology, medicine, etc.). In recent years there have been fruitful attempts to annotate WordNet with domain/topical information in order to improve the overall accessibility to the dense lexical database. Wordnets have been and are being constructed in dozens of

---

[7]Lexicons often omit any reference to ontological categories that are not lexicalised in a language, although it sometimes happens, as with EuroWordNet's ILIs or FrameNet's non-lexicalised frames.

[8]In this respect, our approach is essentially different from OntoNotes [32], where multi-lingual corpora have been annotated with shallow semantic features based on the Omega ontology. Omega contains"no formal concept definitions and only relatively few interconnections" [33] while *Senso Comune*, conversely, is explicitly grounded on an ontological model (see Sect. 2.2).

[9]Hyponymy, antonymy, troponymy, causality, similarity, etc.

languages. Besides the EuroWordNet project that built wordnets for eight European languages, BalkaNet project,[10] encompassing six languages, and PersiaNet,[11] have been developed. In addition, wordnets are being constructed in Asia and South America.[12] It's also worthwhile to mention the SIMPLE project [19], an evolution of the EuroWordNet project, which implements Pustejovsky's qualia roles [34].

WordNet has been often considered as a lexical ontology or at least as containing ontological information: although synsets can be conceived as lexically grounded counterparts of ontological categories, wordnet-like resources do not rely on any explicit logical infrastructure.

*Senso Comune* has borrowed from WordNet many basic intuitions about lexical ontology. However, *Senso Comune* differs from WordNet in many respects. Firstly, besides focusing on synonymy and hyponymy relations with the aim of bringing out the conceptual structure behind the lexicon, *Senso Comune* also adopts a set of a priori ontological distinctions, to identify the ontological commitments behind each sense. Secondly, *Senso Comune* will also contain a parallel structuring based on frames. A semantic lexicon can be structured from a different perspective, focusing on semantic frames instead of synsets, as in the case of FrameNet [39]. In the AI tradition, frames are data structures for representing a stereotyped situation, like "in a living room", or "going to a child's birthday party". Minsky describes frames as cognitively-grounded constructs carrying several kinds of information: the structure of the frame itself, how to use the frame, what one can expect to happen after the occurrence of that frame, and what to do if these expectations are not confirmed [25]. There is a close kinship between AI or cognitive frames and linguistic-based semantic frames: a comprehensive analysis of their relations is presented in [15].

FrameNet is the most comprehensive repository of semantic frames; it aims at providing a lexical account of this kind of schematic representations of situations. Developed at Berkeley University and based on Fillmore's frame semantics [14], FrameNet aims at documenting"the range of semantic and syntactic combinatorial possibilities (valences) of each word in each of its senses" through corpus-based annotation. For example, the **Discussion** frame, namely an abstraction of situations where discussants talk about something in a given place at a given time, is grounded in several lexical occurrences in the FrameNet corpus, which are lemmatized as "lexemes", which are grouped into "lexical units" – LUs: e.g. the noun *negotiation* or the verb *debate*. A frame also has different semantic roles (or "frame elements" – FEs): e.g. *Interlocutor* or *Topic*. On their turn, semantic roles are grounded, e.g. the nouns *president* and *advisor* ground the *Interlocutor* role in the **Discussion** frame. The same LU may ground distinct frames or semantic roles: the noun *president*, for example, also grounds the **People** frame.

FrameNet contains about 12,000 LUs in about 1,000 frames (grounded in lexemes from about 150,000 annotated sentences). As with WordNet, new projects

[10] http://www.ceid.upatras.gr/Balkanet/

[11] http://persianet.us/

[12] For an updated list of wordnet projects see: http://www.globalwordnet.org/

are under development to yield FrameNet-based computational lexicons for other languages: SALSA project in Germany,[13] Japanese FrameNet,[14] and domain specific resources like the Soccer FrameNet.[15] FrameNet has also been ported to RDF-OWL, and aligned to WordNet for interoperability [26].

*Senso Comune*'s model is being extended to encompass verbal frames (see below (see below and Sect. 2.6), which will make it comparable to existing framenet-like resources. However, existing framenets don't supply a formal characterization of the relations between frames, roles, etc., although FrameNet documentation is more explicit than WordNet's about its possible formal interpretation. In practice, such interpretation has to be reconstructed (cf. [26]). On the contrary, formal interpretation of lexical knowledge is a key feature of Senso Comune.

FrameNet is not the only resource for semantic frames and roles we are reusing for building the frame-oriented structuring of *Senso Comune*. VerbNet [18] is a freely available verb lexicon which encodes syntactic and semantic information for *classes* of verbs, and is linked to WordNet and FrameNet. Verb classes are mainly based on Levin's classification [22], thus implying a strong link between the syntax and the semantics of verbs. Indeed, in VerbNet, the semantics of a verb is associated with its syntactic frames, and information about thematic roles and selectional preferences is also included. Verbs belonging to the same VerbNet class are supposed to share the same subcategorisation frame – information that is not included in FrameNet – and have the same selectional preferences and thematic roles associated with the expected arguments.

While there are a few Italian wordnets available (e.g. MultiWordNet [30] and ItalWordNet [38]),[16] and there have been attempts at automatically inducing an Italian FrameNet [21, 47], there is as yet no VerbNet-like resource for Italian. However, as a starting point, *Senso Comune*'s predicate representation has been based on efforts towards combining theoretical and corpus-derived information for obtaining a verb classification which is meaningful at the syntax-semantics interface: in particular, [35] combines a theoretical approach grounded on Pustejovsky's Generative Lexicon [34] and a corpus-based distributional analysis for representing word meaning.

## 2.2 The Model

The adoption of a full-featured, legacy dictionary as a foundation for the resource construction, has led to modeling *Senso Comune* basing on a clear distinction between lexicographic structures and linguistic facts. Basically, *Senso Comune*'s

[13] http://www.coli.uni-saarland.de/projects/salsa

[14] http://jfn.st.hc.keio.ac.jp/

[15] http://www.kicktionary.de/

[16] There is yet another WordNet for Italian developed by a company, but it is not freely accessible.

notion of LEMMA captures the section of a dictionary where an etymologically consistent bundle of senses (that we call MEANING RECORD) of a given lexeme is described by means of a suitable lexicographic apparatus (e.g. definition, grammatic constraints, usage examples). Thus, although related, it must not be confused with the linguistic notion of *lexeme*. This is a distinguishing feature of *Senso Comune* with respect to other models, such as LMF [6] or Lemon [8], to which, however, *Senso Comune* is strongly connected. The common goal of these models is to provide a structure to accommodate semasiological information, i.e. linguistic resources where lexical units are associated with their acceptations. Separating the description of linguistic senses and relationships (e.g. synonymy, hyponymy, and antinomy) from the formal account of their phenomenal counterparts (e.g. concepts, equivalence, inclusion, disjointness) brings a number of benefits. Primarily, this separation prevents lexicographical artifacts to be directly mapped to logic propositions, thus relieves the dictionary the burden of embodying ontological commitments [48], while preserving the possibility of relating lexicographic records with any suitable ontology.

*Senso Comune*'s model is specified in a set of "networked" ontologies [45] comprising a *top level module*, which contains basic concepts and relations, a *lexical module*, which models general linguistic and lexicographic structures, and a *frame module* providing concepts and axioms for modeling the predicative structure of verbs and nouns. The root of the class hierarchy of *Senso Comune* is ENTITY, which is defined as the class of anything that is identifiable by humans as an object of experience or thought. The first distinction is among CONCRETE ENTITY, i.e. the class of objects located in definite spatial regions, and NON PHYSICAL ENTITY, including objects that don't have proper spatial properties. In the line of [43], CONCRETE ENTITY is further distinguished into CONTINUANT and OCCURRENT, that is, roughly, entities without temporal parts (e.g. artefacts, animals, substances) and entities with temporal parts (e.g. events, actions, states) respectively. The *top level* ontology is inspired by DOLCE (Descriptive Ontology for Linguistic and Cognitive Engineering) [24], which has been developed in order to address core cognitive and linguistic features of common sense knowledge. We kept the basic ontological distinctions: DOLCE's *Endurant* and *Perdurant* match *Senso Comune*'s CONTINUANT and OCCURRENT, respectively. The main difference between *Senso Comune*'s top level and DOLCE is the merging of DOLCE's *Abstract* (e.g. mathematical entities, dimensional regions, ideas) and *Non-physical-endurant* (e.g. social objects) categories into a *Senso Comune* category NON PHYSICAL ENTITY.

Among non physical entities, *Senso Comune*'s top level distinguishes CHARACTERIZATION, which is defined on the basis of the irreflexive, antisymmetric relation CHARACTERIZES, that maps instances of non physical entities to other entities (including collective ones), meaning that the former represent some aspect of the latter in some way and under some respect. SOCIAL OBJECT is the class of non physical entities instituted within (and dependent upon) human societies e.g. by means of linguistic acts [40], while INFORMATION OBJECT is the class of social objects which convey information of any kind.

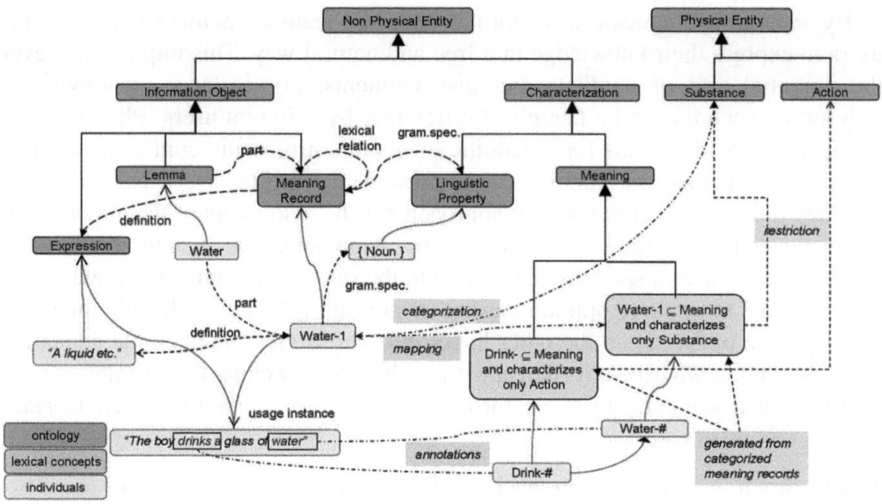


**Fig. 2.1** *Senso Comune* model

The semasiological model of *Senso Comune* (Fig. 2.1) unfolds under the hierarchy of non physical entities. In particular, LEMMA and MEANING RECORD are both information objects, the latter part of the former, whose instances, along with their attributes, form the main body of our lexical resource. On the other hand, MEANING is a social characterization, whose instances occur in the context of linguistic acts. A specific meaning (e.g. *water* in the sense of *liquid substance*) will be a subclass of MEANING, suitably restricted to characterize only liquid substances. The instance of MEANING RECORD where such meaning is described, will be mapped to that class. Mapping between instances of meaning record and meaning classes can be done, in the OWL2 syntax, by annotations, punning, or other structures. In any case, formal semantics of mappings can be specified in different ways, which are out of the scope of this writing. Attributes of meaning record instances (e.g. glosses, grammatic features, usage marks, rhetoric marks, etymology, etc) do not affect the mapped meaning class (if any). Moreover, different meaning records instances (e.g. from different dictionaries) can be mapped to the same meaning class. This way, the model may accommodate meaning records coming from different sources, that might use different sets of attributes (e.g. different usage marks). Also, lexical relations are predicated on meaning records (instead of meanings); hence they are set among information objects and do not have a direct ontological import. Any correspondence (e.g. hyponymy $\mapsto$ inclusion) should be introduced based on suitable heuristics. In sum, both meaning and lexical relation records are purely informative, which could facilitate the process of integrating different (possibly diverging) sources of lexical knowledge.

By separating linguistic from formal semantic features, *Senso Comune* allows users to express their knowledge in a free and natural way. This implies, however, the potential rise of conflicts and disagreements. For instance, synonymy or polysemy of words can be perceived differently by different users. Platforms like Wikipedia provide means for amending errors and arbitrating conflicts, based on self-regulation emerging from large (and presumably well behaved) user communities. We think that a collaborative approach can be also adopted when collecting linguistic and semantic knowledge. At the same time, we recognize that such knowledge requires a specific treatment. On the one hand, linguistic knowledge is less sensitive to emotive opinion clashes or prejudice than encyclopedic one (e.g. about people or facts); on the other hand, in order to take the maximal advantage from user input, we need a formal apparatus that works behind the curtains.

To build a semantic resource through a cooperative process, *Senso Comune* follows two main paths:

- **Top-down** axiomatized top-level ontological categories and relations are introduced and maintained by ontologists in order to constrain the formal interpretation of lexicalised concepts;
- **Bottom-up** language users are asked to enrich the semantic resource with linguistic information through a collaborative approach.

Meanings from De Mauro's core Italian lexicon have been clustered and classified according to concepts belonging to *Senso Comune*'s model, through a supervised process. To enrich the knowledge base, though, language users have been given access to the lexical level only. This access restriction produces an epistemological spread between ontological and linguistic dimensions, but this gap is a necessary requirement if we want to keep control of the ontological layer, while keeping users free from modeling constraints. Filling this gap is the main task of a supervised content revision process. Nevertheless, to make the bottom-up approach plainly effective, users are encouraged to fit their lexical concepts and relations to the basic ontological choices and capture non-trivial aspects of their intended meanings.

For this reason, we designed TMEO, a tutoring methodology to support enrichment of hybrid semantic resources based on *Senso Comune*'s ontological distinctions (see Sect. 2.4). In the rest of this paper we present some aspects related to the population of the *Senso Comune*'s knowledge base, focusing both on the top-down and the bottom-up approach (Sects. 2.3 and 2.4, respectively).

## 2.3 The Acquisition Process

*Senso Comune*'s knowledge base has been populated with approximately 13,000 meaning entries (senses) generated by acceptations of 2,075 lemmas from the De Mauro's core Italian dictionary [10]. Starting from this set of fundamental

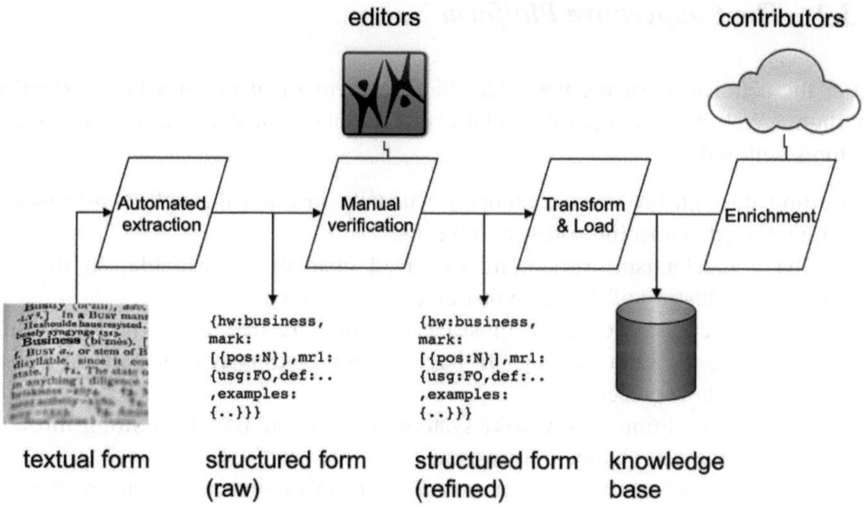


**Fig. 2.2** Acquiring the basic lexicon

senses, the *Senso Comune* knowledge base is developed by the supervised contribution of speakers through a cooperative open platform.

### 2.3.1 *Acquiring the Basic Lexicon*

Starting from plain textual lemmas extracted from De Mauro's dictionary,[17] the acquisition process of *Senso Comune* consisted in producing individuals corresponding to some of the main classes of the *Senso Comune*'s lexical ontology: **LexicalEntry**, **Word**, **MeaningRecord**, and **UsageInstance** classes. This conversion turned out to be less trivial than initially expected, since lexicographers are used to use the same typographic conventions to convey information that is assigned to different portions of the *Senso Comune* target model. For example, senses and usage instances are not always clearly distinguishable, especially in presence of several meaning 'nuances', which is quite common for basic lemmas (Fig. 2.2).

Therefore, after having automatically transformed the dictionary content into an intermediate XML format, a manual revision was needed to amend errors. In many cases, corrections required significant linguistic skills.[18]

[17]Grande dizionario italiano dell'uso (Gradit), Torino, UTET

[18]Two teams of five linguists each, based in Rome and Bologna, under the supervision of Isabella Chiari and Malvina Nissim, were dedicated to this task.

### 2.3.2 The Cooperative Platform

After the acquisition of the basic terminology, *Senso Comune* has been extended through a Web-based cooperative platform. The platform shares a number of key features with wikis:

- Editing through browser: contents are usually inserted through web-browsers with no need of specific software plug-ins.
- Rollback mechanism: versioning of saved changes is available, so that an incremental history of the same resource is maintained.
- Controlled access: even if, in most cases, wikis are free access resources and visitors have the same editing privileges, specific resources (or parts of them) can be somehow preserved.
- Collaborative editing: many wiki systems provide support for editing through discussion forums, change indexes, etc.
- Emphasis on *linking*: resources are usually strongly connected to one another.
- Search functions: rich search functionalities over internal contents.

At the same time, *Senso Comune* shares some critical aspect with wikis:

1. Quality of contents: this aspect focuses on 'bad' or low-level contents.
2. Exposure to "malevolent attacks" that aim at damaging contents or at introducing offensive (or out of scope) information.
3. Neutrality: the difficulty of being completely fair when making statements about questionable matters. Even if linguistic meanings are less sensitive to neutrality than generic wiki contents, moderators are in charge of monitoring contents and behaviors.

With respect to Wiktionary,[19] the Wikimedia project aiming at building open multilingual dictionaries with meanings, etymologies, pronunciations, etc., *Senso Comune* has the following differentiating features:

- Model: Wiktionary encodes each lemma in a wiki page, where different senses are coded as free text without specific identifiers. This choice makes hard to recover the conceptual information associated with lemmas. On the contrary, senses (and their relationships) are first-class citizens in *Senso Comune*.
- Interface: while Wiktionary is based on a generic wiki environment, *Senso Comune* has developed a rich interactive and WYSIWYG Web interface that is tailored to linguistic content (see Fig. 2.3).

Use cases of *Senso Comune*, however, are very close to Wiktionary's ones. After searching a word, and visualizing the information obtained from the platform, users can decide whether to insert a new lemma, a new sense, a new lexical relation, or simply to leave a "feedback" (e.g. their familiarity with available senses and lexical relations). On the contrary, the deep conceptual part of the lexicon (the ontology) is

---

[19] http://www.wiktionary.org/

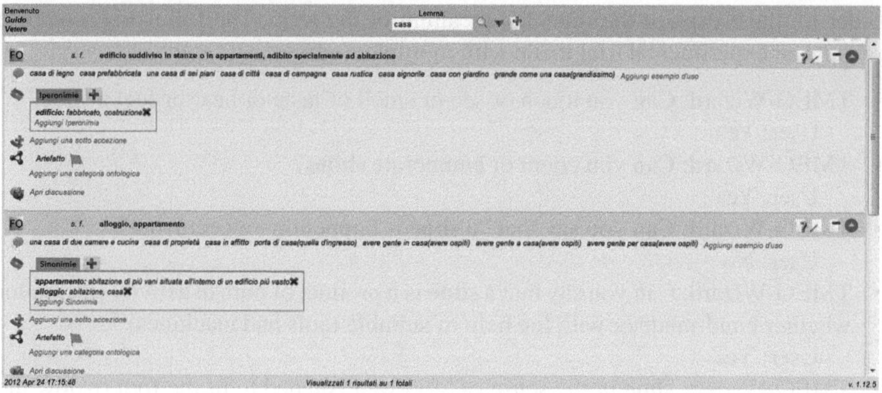


Fig. 2.3 The interface of *Senso Comune*

not accessible to users: when a new sense of a lemma is added, the system creates a corresponding specific concept to be positioned with respect to the ontological layer of the knowledge base. Then, possibly with the help of TMEO (see Sect. 2.4), the user can assign an ontological classification to the new sense. The current prototype of the *Senso Comune* computational lexicon is based on a relational database resulting from the linguistic model (see Sect. 2.2). The database has been also integrated with a DL-Lite reasoner [2], designed and implemented to operate on large ontologies.

## 2.4 The TMEO Methodology

In this section we introduce the general features of TMEO [27], a tutoring methodology to support semi-automatic ontology learning by means of interactive enrichment of ontologies (both from the lexical and the ontological levels).

TMEO is based on the simplified version of DOLCE adopted by *Senso Comune* (see Sect. 2.2). TMEO is inspired by Plato's dialectic (Socratic methodology to drive his disciples to true knowledge, posing questions and arguing on answers [36]): it exploits some suitable ontological properties for posing questions to users in support of domain independent or dependent knowledge modeling. TMEO is an interactive Q/A system based on general distinctions embedded in *Senso Comune*'s ontology.

Consider the case in which a given user is asked to classify the term (shoe), whose WordNet gloss is "footwear shaped to fit the foot (below the ankle) with a flexible upper of leather or plastic and a sole and heel of heavier material".[20] After initializing TMEO wizard, the HCI interface will put the user through a series of intuitive conceptual questions – driven by the underlying ontological model – in

[20]See http://wordnetweb.princeton.edu/perl/webwn

order to make explicit the intended meaning of the term. The following sequence reflects an experimental trial made with multiple users.

- TMEO-Wizard: Can you touch or see or smell or taste or hear or feel **a shoe**?
  User: Yes
- TMEO-Wizard: Can you count or enumerate **shoe**s?
  User: Yes
- TMEO-Wizard: Can you say that "a **shoe** is happening or occurring"?
  User: No
- TMEO-Wizard: Can you say that a **shoe** is a product of human activity (no matter whether hand-made or with the help of suitable tools and machines)?
  User: Yes
- TMEO-Wizard: **shoe** in the sense of "footwear shaped to fit the foot (below the ankle) with a flexible upper of leather or plastic and a sole and heel of heavier material" has been classified as a kind of ARTIFACT.
- TMEO-Wizard: What are **shoe**s commonly used for?
  User: to walk
- TMEO-Wizard: Select the correct sense of **to walk**: [$s_1$ - $s_2$ - $s_3$ - $s_n$].[21]
  User: $s_1$
- TMEO-Wizard: **to walk** in the sense $s_1$ has been classified as the typical FUNCTION of **shoes**.

Here the algorithm drives the user through tracing the following path of knowledge: *shoes as* ARTIFACT *have the common* FUNCTION *of being used in walking events*. As the above-mentioned scenario suggests, TMEO methodology may therefore be adopted not only in the unilateral classification of a given term ('shoe') but also in making related lexical items explicit. This kind of relatedness between terms actually unwraps the inter-categorial relation(s) holding between the corresponding ontological categories. Indeed, from the ontological viewpoint we can say that there is a relation of *Participation* holding between the category ARTIFACT (which is a kind of PHYSICAL OBJECT) and FUNCTION, which is conceptualized in *Senso Comune* as a kind of PROCESS.[22]

TMEO has been implemented as a finite state machine (FSM): in general, the elaboration process of a FSM begins from one of the states (called a 'start state'), goes through transitions depending on input to different states and can end in any of those available (only the subset of so-called 'accept states' mark a successful flow of operation). In the architectural framework of TMEO, the 'start state' is equivalent to the top-most category ENTITY, the 'transitional states' correspond to disjunctions within ontological categories and 'accept states' are played by the most specific categories of the model, i.e. 'leaves' of the relative taxonomical structure.

[21]For the sake of readability, we don't go through the basic senses of the verb 'to walk', also assuming that $s_1$ is adequately selected by the user.

[22]Note that we may wish to distinguish descriptions of functions from actual ones, namely those functions which are performed at a certain time by a given object. In the above example we simplify this distinction only focusing on the latter case.

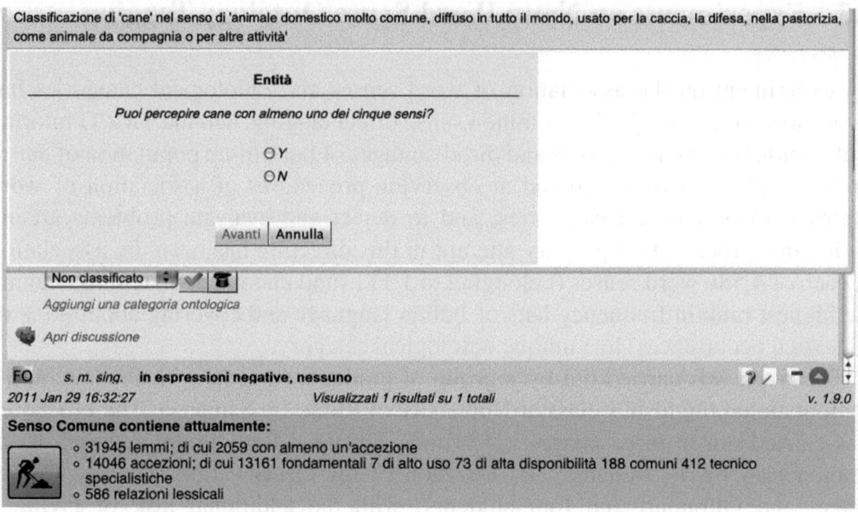


**Fig. 2.4** Senso Comune's interface for TMEO-Wizard. Users can classify word-senses by answering to a logically-interconnected sequence of questions, designed on *top* of *Senso Comune* ontology

In this context, queries represent the conceptual means to transition: this means that, when the user answers to questions like the ones presented in the above-mentioned example (e.g. "can you count or enumerate shoes?"), the FSM shifts from one state to another according to answers driven by boolean logic[23]. If no more questions are posited to the user, then this implies that the operations have reached one of the available final 'accept state', corresponding to the level where ontological categories don't have further specializations (no transitions are left).

TMEO human language interface is very simple and comes in the form of a window where *yes/no* options are presented together with the step-by-step questions: Fig. 2.4 shows an example in Italian for the word 'cane' (=dog), where the Wizard asks whether one can perceive *cane* with the five senses or not. At the end of any single process of enrichment, the system automatically stores the new concept as an OWL class in the knowledge base under the ontological category selected by the user (e.g. in this sense,'shoe' and 'dog' become respectively a subclass of ARTIFACT and of ANIMAL.

Future work on TMEO aims at extending the coverage of the model, adding new 'transitional states' and 'accept states'. We discovered that users, in fact, have a high degree of confidence and precision in classifying the concepts referring to the physical realm, while they face several problems in distinguishing abstract notions like 'number', 'thought', 'beauty', 'duration', etc. (see Sect. 2.5): future releases of TMEO need to be improved both conceptually and heuristically, in this direction.

[23]Uncertainty will be included only in future releases of the TMEO system.

## 2.5 Experiments on Noun Word Sense Ontology Tagging

An experiment on the association of word senses and ontological categories has been carried out using both a common sense direct tagging, and the TMEO tutoring tool in order to test advantages and disadvantages of bottom-up population of *Senso Comune*. The experiment aimed at observing procedures of association of word senses with ontological categories, and to detect and evaluate problems arising during this process. Our primary attempt in this direction has been the association of each of 4,586 word senses (belonging to 1,111 fundamental noun lemmas having the highest rank in frequency lists of Italian language and covering about 80 % of all textual occurrences) to a unique ontological category.

The work was carried out by a group of graduate students of Isabella Chiari's computational linguistics class at University of Rome La Sapienza. The procedure was carried out in three phases: (I) Primary unsupervised common sense classification lead by 12 students; (II) Revision of the classification (lead by Chiari, Vetere and Oltramari and four students) with the additional task of giving a confidence evaluation to the classification using three tags (accepted, controversial, not accepted) and discussion; (III) Final revision of consistency in classification actions.

For the annotation of ontological categories, experienced users directly select a single item from a given list containing all ontological categories. Categories can be also kept "opaque" in order to facilitate those who need guidance in understanding ontological commitments behind specific categorization choices. Thus students who were not confident in direct selection were adviced to rely on TMEO. The *Senso Comune* implementation of TMEO helps the user/editor select the most adequate category of the reference ontology as the super-class of the given lexicalised concept: different answer paths lead to different mappings between the lexicon and the (hidden) ontological layer (Fig. 2.5).

Since ontological categorization is not a simple task and involves complex metalinguistic and cognitive operations a significant control check was introduced by giving experimenters the possibility of associating a confidence label to their choices asserting whether their classification was perceived as fully confident or problematic – especially if the subject was in doubt among different possible categories – or ultimately tentative. We further checked inter-annotator agreement, and observed what categories and association tasks were accepted as common by different annotators, what produced disagreement, and what were perceived as hazardous. Contradictions and disagreements can emerge at the level of language – as stressed in Sect. 2.2 – and even more so in the task of ontological classification. Accordingly, we allowed the users to access to a dedicated 'Forum' room where they could discuss their ontological classification tasks, share their opinions and choices, ask moderators for advise if needed. In general, the Forum became the core tool of support for the experiment and a good instrument to monitor the learning progress of the subjects.

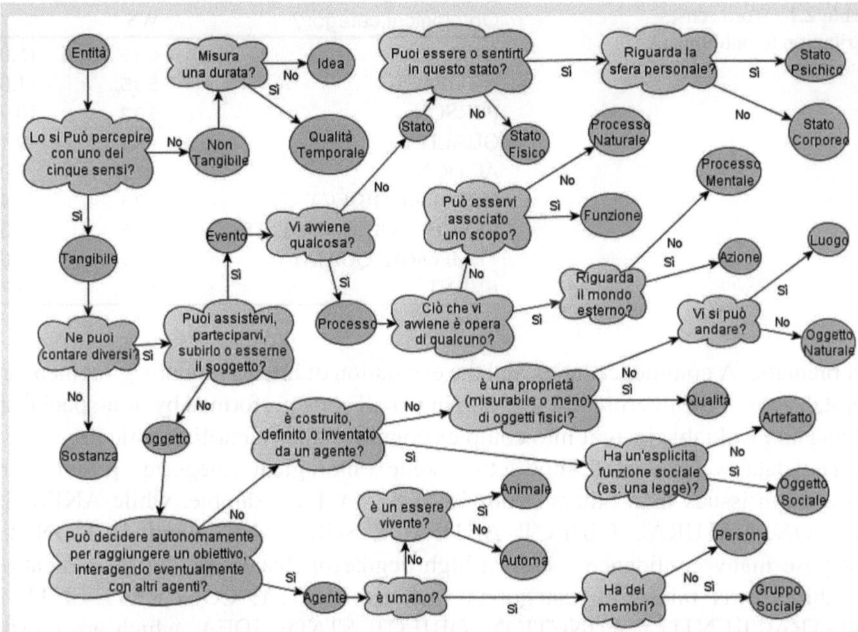


**Fig. 2.5** This conceptual map represents the Q/A mechanism underlying TMEO. *Senso Comune* categories are represented in *yellow circles* with the corresponding Italian labels (literally translated from English, except for 'Tangibile' and 'Non-Tangibile' that maps respectively, to CONCRETE-ENTITY and ABSTRACT-ENTITY). State transitions are driven by "yes-no"' answers (*black arrows*) to questions enclosed in *blue clouds*

After 6 months of work, including supervision, data was analysed to extract information about word sense distribution in ontological categories, data on categorization problems, and information of variety of ontological classes in the fundamental vocabulary nouns examined.

Table 2.1 shows the most populated ontological categories, and the number of word senses attributed to them. The interpretation of this table is very complex and involves the consideration of the hierarchical structure of ontological categories and the observed preference for association of a basic (medium abstraction) level exhibited by the experimenters.[24]

Further issues to be considered carefully are posed by the different degrees of confidence in the association process performed as well as inter-annotator agreement issues: 2,685 (59 %) dictionary word senses were classified with full confidence, while 1,537 (33 %) caused discussions, uncertainty and disagreement among annotators, and 364 (8 %) revealed the ontology to be incomplete or

[24]Please refer to [9] for a more complete version of Table 2.1 and a more detailed analysis of this experiment.

**Table 2.1** Word senses attribution to ontological categories

| Ontological category | WS | % |
|---|---|---|
| IDEA | 689 | 15.02 |
| ARTIFACT | 505 | 11.01 |
| PERSON | 502 | 10.95 |
| QUALITY | 433 | 9.44 |
| ACTION | 413 | 9.01 |
| NATURAL OBJECT | 205 | 4.47 |
| PSYCHOLOGICAL STATE | 185 | 4.03 |
| TEMPORAL QUALITY | 184 | 4.01 |
| EVENT | 172 | 3.75 |

problematic. A confidence index and the evaluation of inter-annotator agreement are capital steps in the interpretation of tagging of all sorts performed by non-specialists giving an invaluable insight into complex cognitive and (meta)linguistic processes.

The data we collected shows that some ontological categories posed more association issues than others (from 68 to 81 %). For example, while ANIMAL, PERSON, NATURAL OBJECT, ARTIFACT, SUBSTANCE, and ACTION did not pose many confidence issues, a high degree of discussion and classification instability was raised by categories such as ENTITY, CONCRETE-ENTITY, ABSTRACT-ENTITY, FUNCTION, OBJECT, STATE, IDEA, which are mostly abstract categories. Further results lead us to observe the complex relationship among word senses as coded in traditional lexical resources as the dictionary used in the experiment and ontological categories: the richness or variety of ontological classes associated with each lemma entry. We have observed that there is a proportional relation between the number of word senses of a lemma and the variety of ontological categories. Most lemmas were associated to two or three different ontological categories while bearing an average of three to five word senses. Lemmas associated to only one ontological category in all the word senses are only 182 (20 % of all fundamental nouns), mostly belonging to PERSON (52), ARTIFACT (27), IDEA (18) and ACTION (14), like in the Italian lemmas *balcone* "balcony", *calza* "socks", *coltello* "knife", *ingegnere* "engineer", etc.

As a result of the experiment, the research group decided to allow multiple classifications of senses in further experiments, in order to evaluate specific patterns in possible associations, and to broaden the list of ontology concepts. Feedback from actual associations, discussions and confidence degree was further used to make some changes in the ontology and discussing some methodological problems that have emerged during the experiment.

## 2.6 Relevance to Natural Language Processing

Resources such as WordNet, FrameNet and VerbNet are in constant development so as to increase their coverage and optimise their internal coherence. These efforts are more than welcome and encouraged within the Natural Language Processing

(NLP) community since such resources constitute a crucial supply of knowledge to be integrated in NLP systems. For instance, automatic word sense disambiguation (WSD) systems, and thus all the higher level NLP tasks that need WSD as a component, heavily rely on WordNet-like resources for creating gold standards and for system development. WordNet has also proved useful, for example in learning information extraction patterns for data mining [44], estimating semantic relatedness of concepts [29], and clustering entities for predicting violations of selectional restrictions [37]. In the latter respect, though, recent work has shown that learning selectional preferences from data using a distributionally-based algorithm can perform better than relying on hand-crafted resources such as WordNet [28].

Another specific NLP task that has hugely benefited from the resources we are discussing, and FrameNet in particular, is semantic role labelling (SRL), i.e. the identification and labelling of predicate arguments in text in an automated way. After the pioneering work of Gildea and Jurafsky [16], who indeed use FrameNet for training their SRL system, several shared tasks have been organised (for an overview see [23]). Interestingly, this task has also been tackled by *combining* WordNet, VerbNet, and FrameNet so as to make up for the shortcomings of each resource since they are complementary in the information they provide [17]. Shi and Mihalcea [42] indeed combine the three resources in order to enhance each of them and show, as a case in point, that they can perform robust semantic parsing this way.

WordNet, VerbNet, and FrameNet have undoubtedly proved a useful source of knowledge for NLP tasks. However, their main drawback is the fact that they are handcrafted, thus requiring a huge amount of manual work and resources, in time and economic terms. But we can look at the other side of the medal: while such resources are crucial for the development of semantically-aware NLP systems, it is also true that NLP tools can be used for *building*, or enhancing, such resources, especially in a semi-automatic, human-assisted setting, thus reducing the amount of human intervention. Inducing FrameNet-like structures has been the successful focus of large-scale projects like SALSA [7], for German, and we have already mentioned the existing efforts for inducing an Italian FrameNet [21, 47]. Work on Italian has also prompted an infrastructure for extending FrameNet induction to other languages [46].

*Senso Comune* lies on both sides of the medal as it will provide a lexical resource along with an annotated corpus associated with it that is used to improve the resource. In line with the rest of the activity, the linguistic annotation on the corpus is done with crowdsourcing methods (cf. Sect. 2.3.2). The target corpus consists of about 8,000 usage examples associated with the fundamental senses of the verb lemmas in the resource. The annotation task involves tagging the usage instances with syntactic and semantic information about the participants in the frame realized by the instances, including argument/adjunct distinction. Specifically, syntactic annotation involves identifying the constituents that hold a relation with the target verb, classifying them as arguments or adjuncts and tagging them with information about the type of phrase and grammatical relation. In semantic annotation, users are asked to attach a semantic role and an ontological category to each participant and to annotate the sense definition associated with the filler. For this aim, we provide them with a hierarchical taxonomy of 27 coarse-grained semantic roles based on [4], together with definitions and examples for each role, as well as decision trees for the roles with rather subtler differences. As in the previous experiment of 'ontologization' of noun senses (Sect. 1.5), the TMEO methodology is used to help them selecting the ontological category in Senso Comune's top-level (Sect. 1.4). For noun sense tagging, the annotator exploits the senses already available in the resource. Drawing on the results of the previous experiment on noun senses, we allow multiple classification, that is, we allow the users to annotate more than one semantic role, ontological category and sense definition for each frame participant. Up to now we annotated about 400 usage examples (about 6 % of the entire corpus) in a pilot experiment we performed to release the beta version of the annotation scheme.

It is interesting to note that in spite of the difficulties related to specialised annotation, such as specific linguistic phenomena, current efforts towards using crowdsourcing methods for gathering linguistic annotation are proving successful (e.g. [3]), although the most technical information is usually added by experts. Also, thanks to regularly increasing amounts of annotated data, NLP tools can be used for inducing some of the annotation, possibly using active learning techniques, successfully employed in minimizing the annotation effort while maximizing accuracy and coverage for several NLP tasks [41]. This bootstrapping setting is already on the other side of the medal, since the resource is being used for developing semantically-aware NLP systems.

By being built collaboratively on the basis of a logically and linguistically motivated paradigm, and by being made freely available to the research community, *Senso Comune* can contribute to the virtuous cycle of using annotating data for developing and/or enhancing NLP systems and and viceversa.

Moreover, by integrating in one resource several levels of representation, it encompasses the kind of information provided by the three different resources WordNet, VerbNet, and FrameNet.

## 2.7 Conclusions and Future Work

This paper presented *Senso Comune*, an open cooperative platform for the Italian language aimed at knowledge acquisition, and we discussed some of the major topics related to linguistic knowledge acquisition.

One of the main features of *Senso Comune* is the semiotic approach used to interface linguistic meanings and ontological concepts. Meanings are not modeled as concepts, but rather as signs. Accordingly, lexical relationships such as synonymy or hyponymy are not mapped into formal relations such as equivalence or inclusion, but are taken as input for the construction of ontological theories.

Thanks to the loose relation between linguistic and ontological data, conflicts and inconsistences in user inputs do not affect the ontology directly; instead, there's room for introducing automatic, semi-automatic, or manual procedures to map linguistic senses to their ontological counterparts.

Current research includes modeling situations by means of frame-like structures, consistently with the formal model that is being developed. Lexical relationships to capture thematic roles will be therefore introduced. Another research direction is toward algorithms for automating the introduction of ontology axioms (e.g. equivalence, inclusion, disjointness, participation) based on linguistic information, by taking both quantitative and qualitative aspects into account.

Hybridisation of manual and crowd-sourced techniques for lexical knowledge acquisition, together with the contribution of NLP methods is also under study. Future efforts will be also devoted to widen the scope of the project, e.g. porting *Senso Comune* into the 'Multilingual Semantic Web' framework,[25] in order to enable cross-linguistic access and queries thorugh Linked Data representations.

Finally, we think that *Senso Comune* as an open source of knowledge of Italian language can make a long way as key enabling factor for business, Web communities, and public services in Italy. The resource will be distributed under Creative Commons license and made available for any kind of use.

## References

1. Assem Mv, Gangemi A, Schreiber G (2006) Conversion of WordNet to a standard RDF/OWL representation. In: Proceedings of the international conference on language resources and evaluation (LREC), Genoa, Italy
2. Baader F, Calvanese D, Mcguinness DL, Nardi D, Patel-Schneider PF (eds) (2003) The description logic handbook: theory, implementation and applications. Cambridge University Press, New York
3. Basile V, Bos J, Evang K, Venhuizen N (2012) Developing a large semantically annotated corpus. In: To appear in proceedings of LREC, Istanbul, Turkey
4. Bonial C, Corvey W, Palmer M, Petukhova V, Bunt H (2011) A hierarchical unification of lirics and verbnet semantic roles. In: 2011 fifth IEEE international conference on semantic computing (ICSC), Palo Alto. IEEE, pp 483–489
5. Bortolini U, Tagliavini C, Zampolli A (1972) Lessico di frequenza della lingua italiana contemporanea. Garzanti, Milano
6. Buitelaar P, Cimiano P, Haase P, Sintek M (2009) Towards linguistically grounded ontologies. In: Aroyo L, Traverso P, Ciravegna F, Cimiano P, Heath T, Hyvönen E, Mizoguchi R, Oren E, Sabou M, Simperl E (eds) The semantic web: research and applications. Lecture notes in computer science, vol 5554, Springer, Berlin/Heidelberg, pp 111–125
7. Burchardt A, Erk K, Frank A, Kowalski A, Pado S, Pinkal M (2006) The SALSA corpus: a German corpus resource for lexical semantics. In: Proceedings of LREC 2006, Genoa
8. Chiarcos C, Nordhoff S, Hellman S (eds) (2012) Integrating WordNet and Wiktionary with lemon. Springer, Frankfurt
9. Chiari I, Alessandro Oltramari GV (2010) Di cosa parliamo quando parliamo fondamentale? In: Atti del Convegno della Società di linguistica italiana (SLI 2010), Viterbo, pp 221–236
10. De Mauro T (1965) Introduzione alla semantica. Laterza, Bari

[25]See for example: http://msw2.deri.ie/

11. De Mauro T (1980) Guida all'uso delle Parole. Editori riuniti, Roma
12. De Mauro T, Mancini F, Vedovelli M, Voghera M (1993) Lessico di frequenza dell'italiano parlato (LIP). Etaslibri, Milano
13. Fellbaum C (1998) WordNet - An Electronic Lexical Database. MIT Press, Cambridge, Massachusetts
14. Fillmore CJ (1968) The case for case. In: Bach E, Harms T (eds) Universals in linguistic theory. Rinehart and Wiston, New York
15. Gangemi A (2010) What's in a schema? A formal metamodel for ECG and FrameNet. Studies in Natural Language Processing. Cambridge University Press, Cambridge
16. Gildea D, Jurafsky D (2002) Automatic labeling of semantic roles. Comput Linguist 28(3):245–288
17. Giuglea AM, Moschitti A (2006) Semantic role labeling via FrameNet, VerbNet and PropBank. In: Proceedings of the COLING-ACL, association for computational linguistics, Stroudsburg, PA, USA, ACL-44, pp 929–936. DOI 10.3115/1220175.1220292
18. Kipper K, Korhonen A, Ryant N, Palmer M (2008) A large-scale classification of English verbs. Lang Resour Eval 42(1):21–40
19. Lenci A, Bel N, Busa F, Calzolari N, Gola E, Monachini M, Ogonowski A, Peters I, Peters W, Ruimy N, Villegas M, Zampolli A (2000) Simple: a general framework for the development of multilingual lexicons. Int J Lexicogr 13(4):249–263
20. Lenci A, Calzolari N, Zampolli A (2002) From text to content: computational lexicons and the semantic web. In: Eighteenth national conference on artificial intelligence; AAAI workshop, "Semantic web meets language resources", Edmonton, Alberta, Canada
21. Lenci A, Johnson M, Lapesa G (2010) Building an Italian framenet through semi-automatic corpus analysis. In: Proceedings of LREC 2010, Valletta, pp 12–19
22. Levin B (2003) English verb classes and alternation: a preliminary investigation. University of Chicago Press, Chicago
23. Màrquez L, Carreras X, Litkowski KC, Stevenson S (2008) Semantic role labeling: an introduction to the special issue. Comput Linguist 34(2):145–159
24. Masolo C, Gangemi A, Guarino N, Oltramari A, Schneider L (2002) WonderWeb deliverable D17: the WonderWeb library of foundational ontologies. Technical report
25. Minsky M (1997) A framework for representing knowledge. In: Winston P (ed) Mind design. MIT Press, Cambridge, pp 111–142
26. Nuzzolese AG, Gangemi A, Presutti V (2011) Gathering lexical linked data and knowledge patterns from FrameNet. In: Proceedings of international conference on knowledge capture (K-CAP2011), Banff
27. Oltramari A, Mehler A, Kühnberger KU, Lobin H, Lüngen H, Storrer A, Witt A (2012) An introduction to hybrid semantics: the role of cognition in semantic resources, vol 370. Springer, Berlin/Heidelberg, pp 97–109
28. Pantel P, Bhagat R, Coppola B, Chklovski T, Hovy EH (2007) ISP: learning inferential selectional preferences. In: Proceedings of HLT-NAACL, Rochester, pp 564–571
29. Patwardhan S, Pedersen T (2006) Using WordNet-based context vectors to estimate the semantic relatedness of concepts. In: Proceedings of the EACL 2006 workshop on making sense of sense: bringing computational linguistics and psycholinguistics together, Trento, Italy, pp 1–8
30. Pianta E, Bentivogli L, Girardi C (2002) MultiWordNet: developing an aligned multilingual database. In: Proceedings of the first international conference on global WordNet. Central Institute of Indian Languages, Mysore
31. Poesio M (2005) Domain modelling and NLP: formal ontologies? Lexica? Or a bit of both? Appl Ontol (1):27–33
32. Pradhan S, Hovy E, Marcys M, Palmer M, Ramshaw L, Weischede R (2002) Ontonotes: a unified relational semantic representation. In: Proceedings of the first IEEE international conference on semantic computing (ICSC-07), Irvine, CA
33. Prevót L, Huang CR, Calzolari N, Gangemi A, Lenci A, Oltramari A (eds) (2010) Ontology and the lexicon. Cambridge University Press, Cambridge/New York

34. Pustejovsky J (1995) The generative lexicon. MIT Press, Cambridge
35. Pustejovsky J, Jezek E (2008) Semantic coercion in language: beyond distributional analysis. Ital J Linguist 20(1):175–208
36. Reale G (1990) A history of ancient philosophy: Plato and Aristotle. SUNY Press, Albany
37. Resnik P (1996) Selectional constraints: an information-theoretic model and its computational realization. Cognition 61:127–159
38. Roventini A, Alonge A, Calzolari N, Magnini B, Bertagna F (2000) ItalWordNet: a large semantic database for italian. In: Proceedings LREC 2000, Athens, pp 783–790
39. Ruppenhofer J, Ellsworth M, Petruck M, Johnson C (2005) FrameNet: Theory and Practice. Available at: http://framenet.icsi.berkeley.edu/
40. Searle J (1995) The construction of social reality. Paperback. Free Press, New York
41. Settles B, Small K, Tomanek K (eds) (2010) Proceedings of the NAACL HLT 2010 workshop on active learning for natural language processing. Association for Computational Linguistics, Los Angeles
42. Shi L, Mihalcea R (2005) Putting pieces together: combining FrameNet, VerbNet and WordNet for robust semantic parsing. In: Proceedings of the 6th international conference on computational linguistics and intelligent text processing, CICLing 2005. Springer, Berlin/Heidelberg/New York, pp 100–111
43. Simons P (ed) (1987) Parts: a study in ontology. Clarendon Press, Oxford
44. Stevenson M, Greenwood M (2006) Learning information extraction patterns using wordnet. In: Third international global WordNet conference (GWNC-2006), Jeju Island, Korea, pp 52–60
45. Suárez-Figueroa MC, Gómez-Pérez A, Motta E, Gangemi A (eds) (2012) Ontology engineering in a networked world. Springer, Berlin
46. Tonelli S (2010) Semi-automatic techniques for extending the framenet lexical database to new languages. Ph.D. thesis, Department of Language Sciences, Università Ca' Foscari Venezia, Venezia
47. Tonelli S, Pighin D, Giuliano C, Pianta E (2009) Semi-automatic development of framenet for italian. In: Proceedings of the FrameNet workshop and masterclass, Milano
48. Vetere G, Oltramari A, Chiari I, Jezek E, Vieu L, Zanzotto FM (2012) Senso comune, an open knowledge base for Italian. TAL – Traitement Automatique des Langues, Special Issue of the on "Free language resources". 52(3):217–243

# Chapter 3
# Building Multilingual Language Resources in Web Localisation: A Crowdsourcing Approach

**Asanka Wasala, Reinhard Schäler, Jim Buckley, Ruvan Weerasinghe, and Chris Exton**

**Abstract** Before User Generated Content (UGC) became widespread, the majority of web content was generated for a specific target audience and in the language of that target audience. When information was to be published in multiple languages, it was done using well-established localisation methods. With the growth in UGC there are a number of issues, which seem incompatible with the traditional model of software localisation. First and foremost, the number of content contributors has increased hugely. As a by-product of this development, we are also witnessing a large expansion in the scale and variety of the content. Consequently, the demand for traditional forms of localisation (based on existing language resources, a professional pool of translators, and localisation experts) has become unsustainable. Additionally, the requirements and nature of the type of translation are shifting as well: The more web-based communities multiply in scale, type and geographical distribution, the more varied and global their requirements are. However, the growth in UGC also presents a number of localisation opportunities. In this chapter, we investigate web-enabled collaborative construction of language resources (translation memories) using micro-crowdsourcing approaches, as a means of addressing the diversity and scale issues that arise in UGC contexts and in software systems generally. As the proposed approaches are based on the expertise of human translators, they also address many of the quality issues related to MT-based solutions. The first example we provide describes a client-server architecture (UpLoD) where individual users translate elements of an application and its documentation as they use them, in return for free access to these applications. Periodically, the elements of

A. Wasala (✉) · R. Schäler · J. Buckley · C. Exton
Localisation Research Centre/Centre for Next Generation Localisation, Department of Computer Science and Information Systems, University of Limerick, Limerick, Ireland
e-mail: Asanka.Wasala@ul.ie; Reinhard.Schaler@ul.ie; Jim.Buckley@ul.ie; Chris.Exton@ul.ie

R. Weerasinghe
University of Colombo School of Computing, 35, Reid Avenue, Colombo, 00700, Sri Lanka
e-mail: arw@ucsc.cmb.ac.lk

I. Gurevych and J. Kim (eds.), *The People's Web Meets NLP*, Theory and Applications of Natural Language Processing, DOI 10.1007/978-3-642-35085-6_3,

the system and documentation translated by the individual translators are gathered centrally and are aggregated into an integral translation of all, or parts of, the system that can then be re-distributed to the system's users. This architecture is shown to feed into the design of a browser extension-based client-server architecture (BE-COLA) that allows for the capturing and aligning of source and target content produced by the 'power of the crowd'. The architectural approach chosen enables collaborative, in-context, and real-time localisation of web content supported by the crowd and generation of high-quality language resources.

## 3.1 Introduction

In the modern world access to the internet and basic computing equipment have become the only technical requirements for sharing and accessing information and knowledge by citizens. While the vast majority of this knowledge is available on the web, it is primarily in English and so there are millions of people worldwide who cannot access this knowledge mainly due to the language barrier. That is, information sharing across languages, while widely considered a fundamental and universal human right [40], has so far been restricted by the lack of adequate language resources and access to adequate language services. The demand from 'rich' language speakers has created a US$31 billion language services industry [11] for commercial languages, but less commercial languages suffer, as illustrated by DePalma's reasons for lack of translation [11]: some of this material, e.g. images, is just not translatable; some does not make sense to translate, e.g. numeric data; and – very significantly – much of the remaining data is "not budgeted for translation". It seems that the translation industry is still living in an age where, "previously, the powerful – companies, institutions, and governments – believed they were in control, and they were" [24]. We observe that the translation and localisation services sector has not caught up with the times where "(wider internet access) allows us to speak to the world, to organise ourselves, to find and spread information, to challenge old ways, to retake control" [24].

Although the English language still dominates the web, the situation is changing. Non-English content is growing rapidly [9, 27, 46]. Additionally, according to Ferranti in 1999 (as cited in [9]) about 30 % of the traffic for an average English website is from foreign visitors, suggesting increased demand for non-English resources. Indeed, China surpassed the USA as the biggest user of the internet in June 2008. Although only 23 % of Chinese people use the web (compared with 73 % in the USA) there are now more internet users in China than there are people living in the USA [23].

Given this increasing demand, localisation can bring enormous benefits to an organisation. Localisation is the "linguistic and cultural adaptation of digital content to the requirements and the locale of a foreign market; it includes the provision of services and technologies for the management of multilingualism across the digital global information flow" [39]. Localisation of a website involves "translating text, content and adjusting graphical and visual elements, content and examples to make them culturally appropriate" [44]. It serves to increase revenue, build credibility and finally help to reach new markets. According to Harvard-based Forrester Research, "English dominated web pages on the internet have resulted in businesses losing sales revenues of $10 million per year" [43,48]. However this observation just serves to reiterate the relative importance commercial organisations place on localising for rich-language locales over poor-language locales.

### *3.1.1 Traditional Localisation*

Traditionally, localisation has been carried out by internal departments within organisations or has been outsourced to specialist localisation companies, creating the US$31 billion language services industry referred to earlier. The generic enterprise localisation processes that emerged, while requiring fine-tuning to particular challenges posed by individual projects, all have core aspects in common: analysis, preparation, translation, engineering, testing, and review [39].

This model supported the globalisation requirements of large digital publishers such as Microsoft, who generates more than 60 % of their revenue from international operations (more than US$5 billion per year) [39]. Yet there is a belief by industry experts that "even today only 90 % of what could be localised, *can* be localised given the still relatively high cost and high dependency on human translators" in traditional main stream localisation settings [39].

It has become evident that current mainstream approaches to localisation are reaching their limits. In 2005, Rory Cowen, the CEO of one of the world's largest localisation service providers and tool developers, Lionbridge, coined the phrase "Localization 2.0" to refer to the "next generation" automated electronic content localisation workflow [10]. However, neither he nor the leading industry analysts, have so far realised what "Localization 2.0" *really* means: Web 2.0 brought content production to the user; Localization 2.0 results from *bringing translation and localisation capabilities to the user*. Control is moving from the enterprise to the citizens. "The illusion of control", as described by the psychologist Ellen Langer, is evaporating. Enterprises are beginning to realise that they will never be able to control a global conversation involving seven billion people, zeta-bytes of data, and hundreds of languages [42].

The traditional process-oriented localisation model begins to break down when considering the nature and scale of material being placed on the web. For example, in cases where a commercial entity publishes the material, the effort expended on localising the material to poor-language locales outweighs the commercial benefit. This difficulty is exacerbated by the amount of locales that may try to access the material. Additionally, in many instances there is no commercial entity behind the published material and thus, no one to pay for this localisation effort. "Social Localisation" describes the emerging concept of users taking charge of their localisation and translation requirements [42]. Instead of waiting for multinational enterprises to take localisation and translation decisions based on short-term financial return-on-investment (ROI) considerations (effectively excluding non-commercially viable content or languages) the users must take charge of localisation and translation. David Brooks, former Director International Product Development, Microsoft, described the consequences of the alternative, contending that "languages not present in the digital world will soon become obsolete" in his contribution to the First International Conference on Language Resources and Evaluation (LREC), Granada, Spain (28–30 May 1998) [6].

### *3.1.2 Web Localisation: Existing Approaches*

Jiménez-Crespo [25] defines web localisation as a "complex communicative, textual, cognitive and technological process by which interactive multimedia web texts are modified in order to be used by a target audience whose language and sociocultural context are different from those of the original production". However, the scope of our research is limited to the translation of text, which is arguably the core component of web content localisation.

The study of web content localisation is a relatively new field within academia [25]. The reported approaches to website localisation are predominantly human [9] and machine-based translation [9, 27, 46], with only very basic collaborative [20] or in-context approaches [5] attempted.

Machine Translation (MT) based website localisation services, for example Google Translator[1] and Bing Translator,[2] allow web content to be translated on the fly. However, the architecture proposed in this chapter is different from these systems, primarily due to the use of Translation Memories (TMs), instead of MT. MT systems already exist for most languages and so do the corresponding language resources. While, according to a report published in July 2012 by Common Sense Advisory, only 21 languages are needed to reach 90 % of the online audience [37], the worldwide internet penetration is just 33 % – with the biggest growth, a staggering 2,988 %, in Africa, the continent with the world's largest linguistic variety [22]. Therefore, the aim of the proposed architecture is to build *new* resources for under-resourced languages, as they are not covered by mainstream localisation services, tools and resources. This approach should help avoid the extinction of whole language populations not just in the electronic world, but also in the physical world, as referred to by Brooks [6].

We also propose for content to be segmented by users, as opposed to segmentation by automated systems, as in current approaches. Finally, the architectures of most existing mainstream tools, resources and services are unpublished and the resources are owned by private enterprises. Here the architecture is made explicit.

---

[1] http://translate.google.com/

[2] http://www.microsofttranslator.com/

In this context it is worth mentioning "Wikibhasha[3]", a browser extension based architecture like ours that facilitates the translation of Wikipedia content between different languages. It shows side-by-side views of the source and the target version of the Wikipedia article that is being translated. It allows pre-translation of content using the Bing MT service. Moreover, using the "Collaborative Translation Framework[4]", alternative translations can be chosen for pre-segmented machine translated text. The Google Translation Toolkit offers a similar interface and functionality to Wikibhasha. It can assist in localising general website content as well as Wikipedia.

Many researchers have reported on the use of MT in web content localisation [16], and the low quality of these translation solutions is known to be a significant drawback [9, 27]. Moreover, the research and development of MT systems for less-resourced languages is still in its infancy [46]. Therefore, MT-based web content localisation solutions are clearly not viable at this stage.

An open source project has been launched similar to the model proposed here, which is based on TMs and browser extensions. It is known as the "World Wide Lexicon[5]". It is an open source system that runs on the Google App Engine and is accessed via a web API or custom-built TransKit libraries. While the code is shared under a BSD 2-Clause license, the system's architecture is unpublished, undocumented and not widely discussed. The same group is also working on a TM-based online website translation system known as "Der Mundo[6]", but again, the architecture is unpublished and lacks discussion.

### 3.1.3 *Elements of an Alternative Approach*

Undoubtedly the constant increase of UGC has led to a higher demand for translation of a greater variety of content at increased speed. In this chapter, we examine a possible solution to cope with the high volume and the high speed of content production based on the concept of micro-crowdsourcing, towards the generation and evolution of community-derived language resources: namely Translation Memories (TMs). These concepts are now discussed in more detail.

#### 3.1.3.1 Crowdsourcing

The phenomenon of crowdsourcing/social translation has come into being in the last few years only. The term 'crowdsourcing' was first coined by Jeff Howe in his blog and now-famous article in Wired magazine [21]:

[3]http://www.wikibhasha.org/ (accessed July 03, 2012)

[4]http://en.wikipedia.org/wiki/WikiBhasha (accessed July 03, 2012)

[5]http://www.worldwidelexicon.org/home (accessed July 03, 2012)

[6]http://www.dermundo.com/ (accessed July 03, 2012)

> Crowdsourcing represents the act of a company or institution taking a function once performed by employees and outsourcing it to an undefined (and generally large) network of people in the form of an open call.

While, the concept initially describes developments in areas such as stock photography, crowdsourcing has also been taken up enthusiastically by the localisation and other communities. It has resulted in almost 8.2 million hits in Google's search engine and has featured as a major topic at recent, seminal localisation-industry events, such as Localization World 2009 and LRC XIII. This is because commercial organisations see it as a mechanism to lower the cost of localisation, enabling them to enter currently inaccessible locales [36], and, perhaps more importantly, to develop enthusiastic user communities around their product lines. In addition, Losse [29] in her keynote at the 2008 LRC conference, stated that organisations like Facebook pursued crowdsourcing because it also produces higher quality translations.

Since her keynote on Facebook's then highly innovative approach to translation, others have reported on the value of involving large communities in translation projects in a variety of contexts. Callison-Burch [7], for example, found that combined non-expert judgments on the quality of translation can be of better quality than existing 'gold-standard' judgments. At RailsConf 2012, Dutro [12] reported on Twitter's highly successful approach to collaborative translation involving 550,000 volunteers, 16,000 active source language phrases, and 1,450,000 translations. Twitter's approach involves the moderation of strings by translators with elevated moderator privileges and a reputation system that not only takes into account the amount of words translated by a particular translator, but also keeps track of their 'karma' and 'phrase maturity'; as an additional mechanism, the Translation Center uses a community voting system to support the identification or the most suitable translation.

What is true of commercial organisations is also true of altruistic organisations. The concept of "social localisation" was first introduced by Reinhard Schäler in the context of his work with The Rosetta Foundation [41]. It describes the services provided by communities of volunteer translators (or people with language skills) not just to their own but also to third party communities, moving the provision of translation and localisation services firmly out of the exclusive realm of profit-driven and tightly controlled organisations [41].

Here we focus on crowdsourcing translation: when the crowd or a motivated part of it, participates in an open call to translate some content, creating highly valuable language resources in the process. In this chapter, we focus on a particular form of crowdsourcing, which we call micro-crowdsourcing. We use the term 'micro' to distinguish it from the more enterprise-focused implementations which are currently in use, where translation activities are carried out in a planned, predictable, and controlled manner, and, generally, by trained, experienced, pre-screened, and paid translators. In our vision of micro-crowdsourcing the translation is carried out by anyone with sufficient language skills (subjectively assessed by the contributors themselves), is ad-hoc, and needs-based; it may only consist of a single line, is carried out in context, and represents a minimal overhead to the translator. This is accomplished by our development of a novel browser extension-based client-server architecture that allows localisation of web content using the power of the crowd, in context. We address many of the issues related to MT-based solutions by proposing this alternative approach based on the generation and refinement of TMs.

Understanding Voluntary Crowdsourcing and User Motivation

For the architectures proposed here to function, there is an assumption that a proportion of the users of the software will be multilingual and in addition will freely provide the translation necessary to localise the software that feeds the TM. Based on existing evidence this assumption is not unfounded: there are a number of existing community-based models, which can be compared to our crowdsourcing approach including Open Source Software (OSS) development and Wikipedia. In relation to Wikipedia, Kuznetsov [26] suggests that the "majority of contributors to collaborative online projects cited motivations that focus on information sharing, learning new skills and communal collaboration" Kuznetsov goes on to list five values which she considers as underlying motivations: Altruism, Reciprocity, Community, Reputation and Autonomy. Compared to Wikipedia, OSS development has been around longer and has been the subject of more studies. As such it gives us additional interesting insights into volunteer communities on the internet. OSS development communities develop software systems that compete with similar offerings from large multinational development organisations such as Microsoft and IBM, yet the organisation of these communities is fundamentally different. For example, the stereotypical OSS developer within these communities receives no financial reward. Thus the development and the organisation of any hierarchies within these communities is not directly associated with financial remuneration but can be more likened to a self organising and emergent process without any formal titles for the participants.

In an attempt to understand the internal structures of OSS communities a number of quantitative studies have attempted to categorise members using a Social Network Analysis (SNA) approach. Valverde et al. [45] suggested that wasp colonies and OSS communities shared similar statistical organisation patterns and that these shared patterns revealed how social hierarchies were formed in OSS communities through self-organising processes. In another SNA based study by Crowston and Howison [8] they considered the communications between project members by studying data extracted from the bug tracking system (a system that enables users and developers to report and discuss software bugs related to the project). They could observe both the regularity of postings and the number of patches (bug repairs) attributed to individual developers. They used this to visualise the social structure of OSS communities for 120 different project teams over a wide range of project types. Both these studies were trying to develop a better understanding of the organisational structures within OSS development communities and how they evolve.

Other studies have focused on the motivational aspects of the OSS development community. Zeitlyn [49] for example suggests that OSS developers' motivation can be explained in terms of existing theories on 'gift economies', where the reciprocal relationship engendered by gift giving forms the moral basis for the society. Thus in the OSS community lasting moral relationships are developed between contributors by the giving and accepting of code. Some researchers suggest that although not immediately apparent, it is economic factors which at least partially motivate OSS developers to contribute to OSS projects [28]. They suggest that the experience and possible fame that the developers amass will eventually result in a larger pay packet in the industrial software development community. This however does not seem to manifest itself in terms of the empirical evidence. For example Hann et al. [19] suggest that the opposite may be true.

Leading members and representatives of the OSS community present a much more chaotic and emotional view of OSS community contribution. For example Raymond [34] compares the Linux OSS community to "a great babbling bazaar of differing agendas and approaches . . . who'd take submissions from anyone". He suggests that most individuals are at first only interested in their own issue and concern; the equivalent of 'scratching an itch'. Linus Torvalds a well known and respected member of the OSS community, feels that "most of the good programmers do programming not because they expect to get paid or get adulation by the public, but because it is fun to program" [17]. In an interview with Ghosh [17] he describes the OSS community as large, diffuse and mainly passive. Torvalds makes it clear however that his notion of community includes the users of the software as he feels they have a central role "in detecting and reporting bugs" and so "are as valuable as developers".

Recent relevant studies around motivation in general include Jarvis [24] and Pink [33]. The latter highlights the fact that there is a significant gap between industrial practice and scientific evidence relating to motivation. He presents the results of studies commissioned by the US Federal Reserve proving that if we really want high performance on the twenty-first century definitional tasks, we need to look at autonomy, mastery and purpose as key motivational factors – and move away from the traditional belief that monetary compensation makes people deliver.

There are a number of interesting research questions that relate to the notion of community within OSS circles and voluntary crowdsourcing in general, but in term of our proposed approach, initiatives like OSS and Wikipedia provide strong precedents of previous success in crowdsourcing contributions that are backed up by scientific evidence in motivational research.

#### 3.1.3.2 Translation Memories (TM)

Since the early 1990s, TMs have changed the way publishers and translators approach translation and localisation [38]. In comparison to Machine Translation (MT) systems, TMs are easier to maintain, produce more predictable translations, require less system resources, are useful for even smaller chunks of content, and are, therefore, very well suited to support the translation of lesser-resourced languages. The first use of TMs in large scale enterprise localisation projects was reported in 1994 and highlighted a number of interesting aspects and effects of the use of TMs in translation projects, e.g. the very different reaction to the technology from novice and professional translators, and the quite vague definition of a "100 % match" [38]. The idea to 'never translate a sentence twice' but instead to retrieve reviewed and quality-checked translations of 'known' segments from databases of bilingual segment pairs, and automatically insert these into the text to be translated has led to very significant savings, especially in the case of highly repetitive or frequently updated source text. It has also led to more consistent translations within and across versions of the same source text [39]. While attempts have been made to set up TM market places, they have failed so far, as in the case of www.tmmarketplace.com. Instead, organisations such as the Translation Automation User Society[7] (TAUS) have been making TMs available under license to their members. These highly valuable language resources have not become available free-of-charge and for open use outside of corporate environments, one of the reasons being the significant investment TMs represent for their owners.

There are a number of very interesting and important issues around the authoring, use and maintenance of TMs that, even after almost 40 years of industrial use, have not been systematically researched. Among these are:

- The shelf-life of TMs: how long can entries into TMs realistically be re-used;
- Risk assessment around the use of TMs: under which circumstances should even 'safe' 100 % matches no longer be used (currency issues, changing in orthographical conventions etc.);
- The use of metadata in TMs: the effect of knowledge around, for example, provenance (who provided the translation) or reputation (how credible is the source of the translation) [31];
- The cost of quality (consistency) or lack of it [30]: what does it mean if up to 5 % of entries in a TM can be inconsistent, as shown in [30].

Parallels could be drawn between content ownership and ownership of the language versions of that content, encased in the TMs. Originally, both content and TM, were fiercely protected by their owners. However, while content ownership on the web has begun shifting from corporations to communities with the advent of UGC, a parallel development has not yet taken place in relation to language resources and, particularly, to TMs. This work addresses this issue.

#### 3.1.3.3 Browser Extensions

Browser extensions enhance the functionality of web browsers. For example, extensions are used to enhance the user interface, improve performance, and integrate with various online services as well as to give a better, personalised browsing experience.

Various browser extensions already exist that are capable of utilising existing Machine Translation (MT) services to translate web content into different lan-

[7] http://www.translationautomation.com/

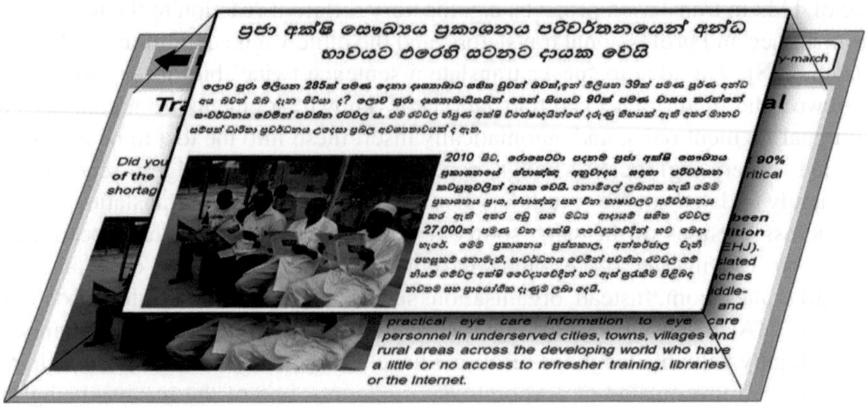


**Fig. 3.1** Conceptual localisation layer

guages. Some extensions even allow selected text to be translated into a different language and searched for on the web. Furthermore, various dictionary extensions are also available. It is worth noting that most of these browser extensions are free and open source.

We exploit the power of browser extensions to design a conceptual localisation layer for the web. Our research is mainly inspired by the works of Exton et al. [13, 14] on real-time localisation of desktop software using the crowd, Wasala and Weerasinghe on a browser-based pop-up dictionary extension [46], and Schäler on information sharing across languages [40] as well as social localisation [41].

The proposed architecture enables in-context real-time localisation of web content by communities sharing not just their content but also their language skills. The architecture is based on Translation Memories (TM) which, as outlined earlier, is particularly suitable for lesser-resourced languages. Therefore, better translation accuracy and greater quality of the translated content can be expected with this approach. The ultimate aim of this work is the collaborative creation and evolution of TMs which will allow for at least partial automatic translation of web content, based on reviewed and quality-checked, human produced translations.

The key feature of the proposed architecture is the separation of the to-be-localised content layer from the source content layer (i.e. the localisation layer is independent of the website). The proposed architecture builds a conceptual localisation layer on the original content separating to-be-localised content from any underlying frameworks (such as Content Management Systems (CMSs)) used to render the original content (see Fig. 3.1).

To the best of the authors' knowledge, this is the first effort of its kind to make explicit the power of browser extensions to build a website independent conceptual localisation layer with the aid of crowdsourced-evolved TMs.

The rest of the chapter is organised as follows: Sect. 3.2 describes the architectural evolution of the proposed system in detail; Sect. 3.3 presents a scenario illustrating how the proposed extensions function; The development of the prototype is discussed in Sect. 3.4; and Sect. 3.5 discusses key outstanding challenges and constraints of the proposed architecture. Finally, this chapter concludes with a summary and discussion of future research directions.

## 3.2 System Architecture

The proposed system architecture is based on earlier work by Exton et al. [13, 14]. They proposed a crowdsourced, client-server architecture for the localisation of applications' User Interfaces (UI) by the crowd. This architecture, while providing a basis for the architecture proposed here, can also be seen as a complimentary approach in that it directs itself at application-interfaces, whereas the approach presented here directs itself at UGC or, with refinement, application content.

### *3.2.1 The UpLoD Architecture*

Consider a software package such as Open Office that has been developed for a purely English speaking audience. Even if this product were designed to facilitate its easy adaptation into other languages it would still require the effort of either a number of altruistic individuals or the coordinated effort of expensive professional localisers to make this product available in another language. The question then is, when there is no/limited immediate economic imperative for the digital content publishers (for example open source software providers or voluntary organisations aiming to bridge the digital divide) how can such applications be localised?

One possibility is to allow the community of users to update the user interface in situ, either directly from the original English version or perhaps working from a less than perfect machine translation. That is, the interface would allow users to localise User Interface (UI) elements like menu-options, error messages and tool-tips as they used the tool. This localisation could be enabled via a simple pop-up micro-localisation editor that would allow them to change UI text in situ simply by ctrl-clicking on any text that is displayed. See Fig. 3.2 for an example where a ctrl-click on the 'Options' menu item brings a localisation editor to the fore. The user is then allowed suggest an alternative lexicon for the menu item, or to utilise other suggestions provided by the community. The localised menu-item would then become part of the application's User Interface.

This editor may have to enforce constraints on the translation, such as restricting string length, and could perhaps include appropriate translation memories, style guides and standards to assist in the translation. As a ctrl-click could be applied to any displayable text area, error messages and help information messages could also be included as translatable material. Indeed, the editor may even go as far as allowing graphical replacement of certain artefacts.

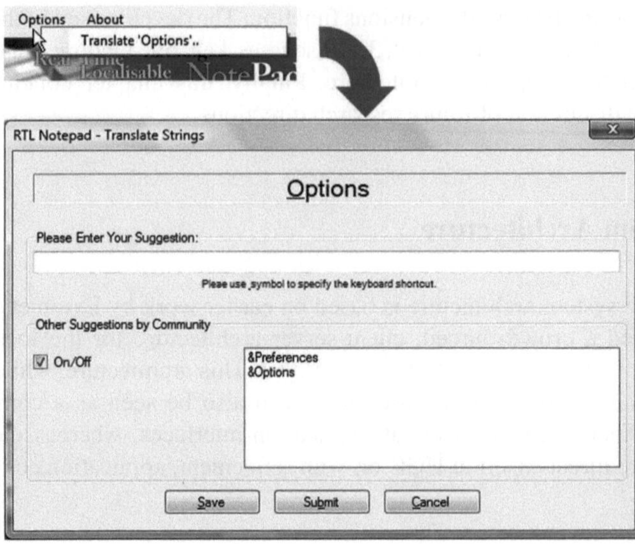


**Fig. 3.2** User-driven localisation with UpLoD [13]

The result would be a set of textual (and possibly graphical) updates for each user. Then suppose that each of these update-sets could be automatically gathered into a central repository that would, in turn, push update events back to the community of users, periodically or on-demand. This would update their product with the latest translations. Imagine that these users could, in turn, quality assure the updates and re-instigate the cycle, in the same way that Web based communities like Wikipedia reach consensus by iterative refinement.

Such an approach would represent a radically novel approach to localisation and thus would require a novel architecture, as demonstrated by a sample application ('Writer') in Fig. 3.3. On the far right of this architecture, we see a 'Central Server' that receives and sends the localisation updates to and from individual deployments of the 'Writer' system.

One deployment of 'Writer', to the left of the figure, is substantially expanded. As can be seen from this expanded 'Writer', the typical three tier architecture of the 'Writer' application (GUI-Business-Database) is augmented by an 'Update-Log-Daemon' (UpLoD) module. This UpLoD module allows the user to update the user interfaces as they use the system and logs the changes in a local audit file. The records in this local audit file contain unique identifiers for the GUI elements that have been changed. These identifiers are associated with the pre-translation and post-translation. Periodically, a Daemon trawls the audit log and, on finding new records, passes updates to the central repository or 'Aggregated Log File' on the 'Central Server' via a SOAP protocol.

These updates can be handled in a number of ways. For example in publicly edited wikis, revision control enables a human editor to reverse a change to its

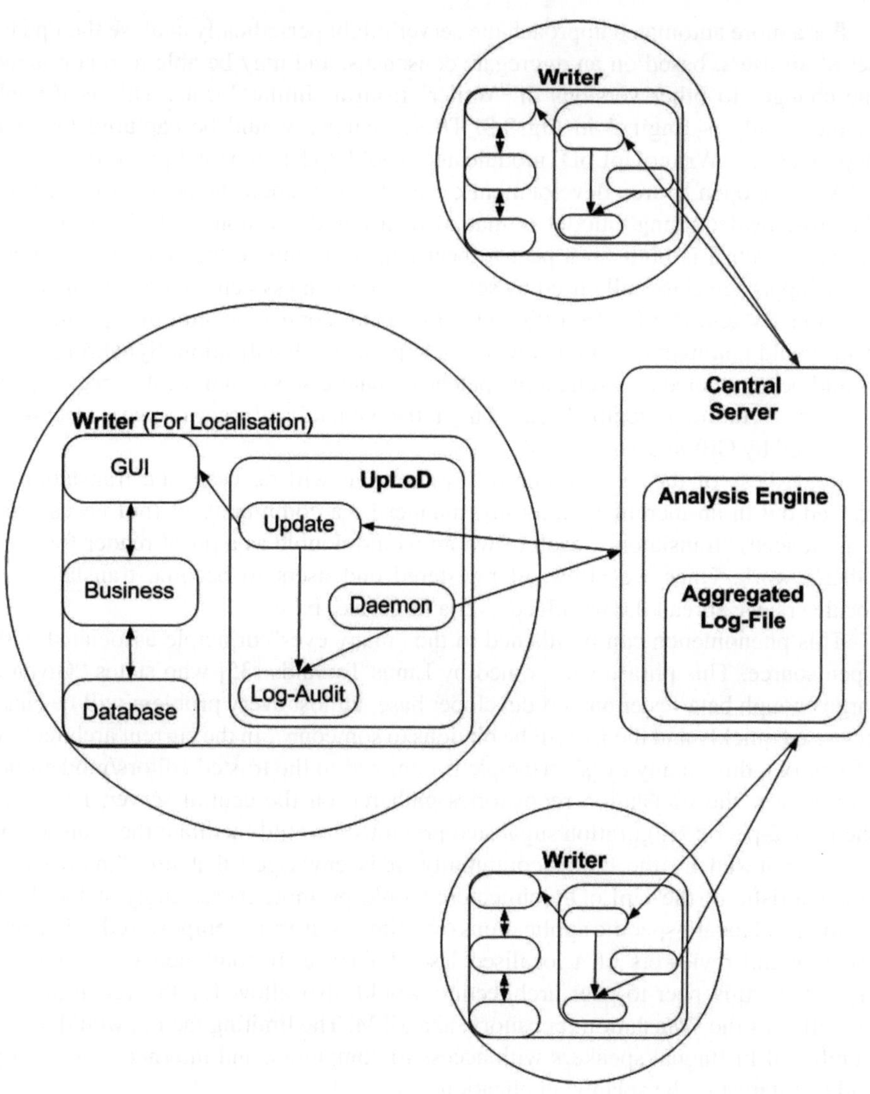


**Fig. 3.3** The UpLoD architecture [13]

previous version. For a "micro-crowdsourcing" system it is possible that a limited number of trusted editors (self moderating) for a specific language group may tidy up the localisation in this fashion. A version control system would then enable editors to build a release package on a periodic basis based on the influx of micro changes from standard users and this could then be released to subscribers. This would serve to drastically decrease the number of changes and updates to the UI and avoid updates with new translations on an ongoing basis.

For a more automated approach the server might periodically analyse the update set of all users, based on an aggregate consensus, and may be able to recommend the changes to other versions of 'Writer' from a similar locale. This is the job of the 'Analysis Engine' in Fig. 3.3. These changes would be captured by each deployment's 'Writer UpLoD' module and would update the GUI proactively.

Another open source development concept which could be adapted to suit the "micro-crowdsourcing" model is that of distributed revision control. Distributed revision control is built on a peer-to-peer approach, unlike the centralised client-server approach classically used by software versioning systems such as Concurrent Versions System (CVS). In a distributed revision control version of UpLoD each peer would maintain a complete working copy of the localisation. Synchronisation would be conducted by exchanging patches (change-sets) from locale-specific peer to peer. A more in-depth discussion of the generalised peer-to-peer process is described by Gift and Shand [18].

Regardless of the version control system that will be used, the translation is carried out in an incremental, ad-hoc manner by a community of (not necessarily experienced) "translators", each of whom would double as a proof reader for each other's work. Once we allow all registered end users to become translators or localisers, we spread the workload over a large user base.

This phenomenon can be likened to the "many eyes" principle associated with open source. This phrase was coined by Linus Torvalds [35] who states "Given a large enough beta-tester and co-developer base, almost every problem will be characterised quickly and the fix will be obvious to someone". In the current architecture of UpLoD, this "many eyes" principle is confined to the trusted editors/moderators who review the translation repositories gathered on the central server. However, the peer-2-peer configuration suggested previously would facilitate the translations being reviewed by the entire community. It is envisaged that this "many-eyes" characteristic of the UpLoD architecture would promote increasingly stable, high quality, and locale-specific applications over time as users are empowered to become creators and reviewers of a localised User Interface. If combined with a central repository, this peer-to-peer architecture would also allow for the retention and evolution of the Translation repository: aka a TM. The limiting factors would be the number of bi-lingual speakers with access to computers, and internet connectivity and an interest in the specific application.

A proof of concept prototype of this architecture was created to validate and refine this approach. The prototype consisted of two components: the central server component and a simple RTL (Real Time Logging) Notepad application which imitates the "Writer" of Fig. 3.3. The UpLoD module was implemented and integrated in the RTL Notepad application in addition to its generic text editing functions. Due to its simplicity and portability, the Portable Object (PO) file format was chosen as the format for the local log-audit file. A more in-depth description can be found in Exton et al. [13].

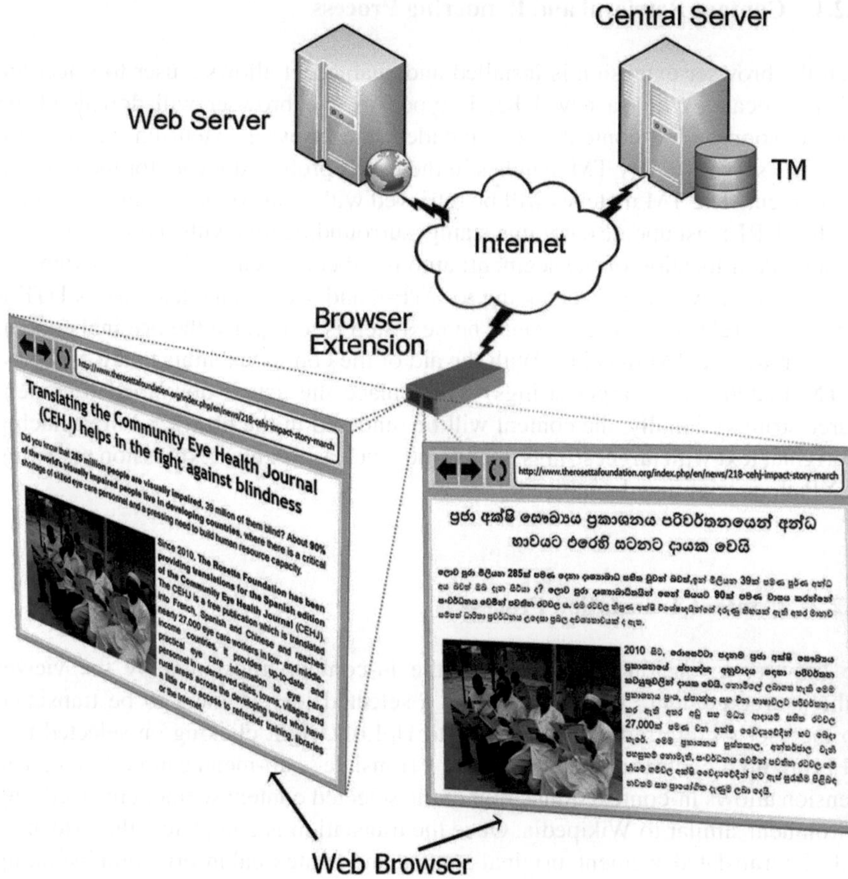


**Fig. 3.4** Proposed system architecture

### 3.2.2 *The Browser-Extension COntent Localisation Architecture*

The UpLoD architecture described in the previous section provides the basis for an architecture that can be applied to generate TMs from web content. This derived architecture: the Browser-Extension COntent Localisation Architecture (BE-COLA) is presented in this section. As the name suggests, the architecture is based on an internet browser extension that wraps the delivery of internet content in a TM localisation envelope (see Fig. 3.4). This envelope allows for the (possibly partial) localisation of internet content while also allowing for the gathering and evolution of the TM resources.

#### 3.2.2.1 Content Retrieval and Rendering Process

When the browser extension is installed and enabled, it allows a user to select the preferred locale. When a new URL is typed in, the browser will download the page. As soon as the content is downloaded, the browser extension will consult the central server for any TM matches in the user's preferred locale for the relevant URL content. The TM matches will be retrieved with contextual information which includes: URL; last update date/time stamp; surrounding text with and without tags; and the XPath location of the segment; among others. Then the browser extension will convert the web page's encoding to UTF-8 and set the character set as UTF-8 in the page's relevant `<meta>` tag. The next step is to replace the original content with the retrieved TM matches. With the aid of the contextual hints that it receives, the TM matches (i.e. target strings) will replace the actual downloaded content (source strings). Finally, the content will be rendered in the browser. For replacing the original text with target strings, techniques such as Regular Expression matching and XPath queries may be utilised.

#### 3.2.2.2 Content Translation Process

The browser extension also facilitates the in-context translation by the viewer of the source content. That is, it allows a selected text segment to be translated into the user's preferred locale. Similar to UpLoD, right clicking on selected text will bring up a contextual menu where a "Translate" sub-menu can be found. The extension allows in-context translation of the selected content segment in an editing environment similar to Wikipedia. Once the translation is completed, the extension sends the translated segment, original content and contextual information including the URL to the central sever. The browser extension keeps track of all translated content. Hovering the mouse pointer over a translation will display the original content as a pop-up (similar to Google's web-based MT system).

Upon receiving translations from a client (browser extension), the central server stores all the information that it retrieves (the locale, the language pair, source string, target string, contextual clues such as XPath location, surrounding text, tags, URL, client identifier, client IP, text positions and text length) in a special TM. Thus TMs get generated for specific locales, over time based on micro-crowdsourcing. Additionally, a mechanism to monitor and uniquely identify browser extensions connected to the central server could be implemented, in order to prevent repeated misuse of the localisation service.

The central server can be scheduled to periodically leverage translations as the TMs grow. Furthermore, later on, MT systems can be trained from the TM data and these trained MT systems can feed back into the system to speed up the translation process and to translate content where TM matches are not entered.

#### 3.2.2.3 Translation Editing and Voting Process

When leveraging the TMs on the central server, a mechanism has to be built to choose the most appropriate translation of a given text segment, as per the UpLoD architecture [13]. To assist in selecting the best translation for a given segment, a voting mechanism is proposed. However, in contrast to the existing UpLoD implementation, this voting mechanism is distributed to the entire community insuring a 'many eyes' philosophy. Additionally, human intervention (mainly the opinions of expert monitors) is also essential to solve potential conflicts.

When a user right clicks on a translated segment it can bring up a context menu where the current translation, along with the top three alternative translations, are displayed. The votes for each translation will also be displayed next to their associated translation. The users are given the opportunity to edit the current translation and/or to vote for any of the alternative translations. Furthermore, clicking on an alternative translation will take the user to a web page where the user can see all the alternative translations that are proposed for the selected segment. In that page users can vote for any of the alternate translations.

Considering the motivation factors related to crowdsourcing, a simple "thumbs up, thumbs down" voting is proposed over complex and confusing rating systems. If the user wishes to edit the existing translation, they can simply go to the in-context edit mode and edit it. Once editing has been performed, the new translation is sent back to the central server. The central server compares the new changes with the existing translations and includes it as an alternative translation. The server needs to keep track of the voters as well as the votes. By keeping track of voters, users can be encouraged to vote for additional translations using ranking systems similar to those implemented in games.

The BE-COLA system resembles the Update-Log-Daemon (UpLoD) based client-server architecture. However, in this architecture, clients (browsers) connect to the central server via a browser extension. The browser extension implements the UpLoD architecture, which acts as a proxy between the browser and the central server.

We also extend the functionality of the central server in this architecture by equipping it with a component to maintain TMs for different language pairs, a desirable feature in Translation Management Systems (TMS).

## 3.3 An Illustrative Scenario

The following section graphically illustrates the BE-COLA architecture in detail. Once enabled, BE-COLA allows the following tasks to be performed by the user:

1. Translating a segment in context;
2. Obtaining translations from the central server and rendering in-context (automatically or manually);
3. Editing existing translations and submitting changes back to the central server;
4. Voting for the most suitable translation;

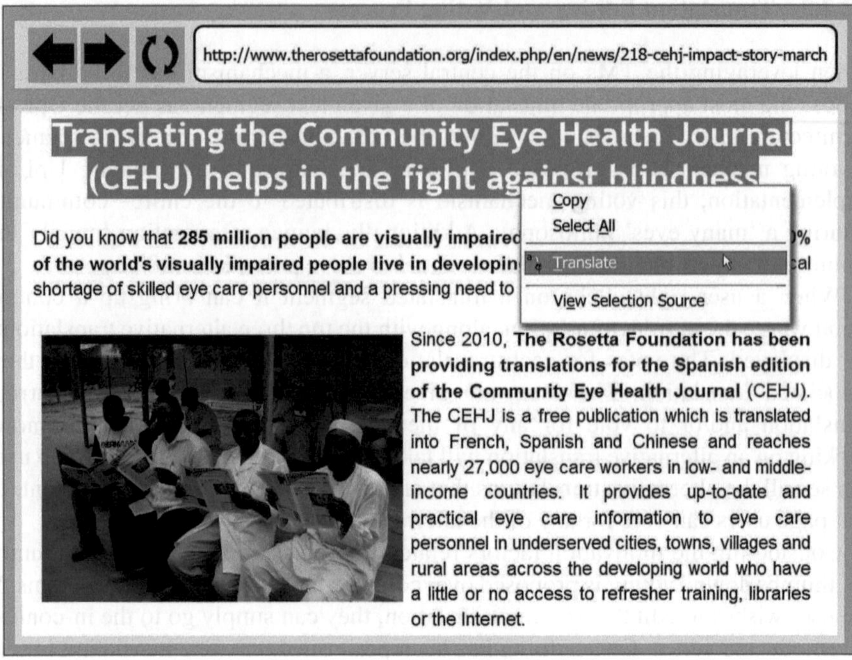


**Fig. 3.5** Step 1: select segment, *right* click and choose translate

## *3.3.1 Translating Text*

If a user right clicks on a selected textual segment BE-COLA will bring up a contextual menu where the translate option can be found (see Fig. 3.5). The next step is to carry out the translation. The text segment will become editable upon clicking the "translate" option (see Fig. 3.6). This mechanism enables in-context translation at a micro-scale.

## *3.3.2 Obtaining Translations from the Web TM and Displaying Them*

When the browser extension is disabled, the browser will simply show the original content in the source language (see Fig. 3.7).

However, when the browser extension is enabled and configured to automatically show the localised page, the browser extension will retrieve locale-specific TM matches from the central server and replace the original source content with them, aided by contextual clues that it retrieves (see Fig. 3.8).

**Fig. 3.6** Step 2: in context translation of a segment (in unicode)

**Fig. 3.7** Extension disabled: original content displayed

**Fig. 3.8** Extension enabled: translated text displayed

### *3.3.3 Editing and Voting for Existing Translations*

Right clicking on a selected translation will bring up a context-menu where the top three alternative translations and corresponding votes are displayed. Furthermore, from this menu, the selected translation can be edited and submitted to the central server (see Fig. 3.9).

## 3.4 Prototype Technologies

To test the above architecture, we developed a prototype with the aid of two open source Firefox Add-ons:

1. Ingiya[8] – a Firefox pop-dictionary add-on similar to the add-on described by Wasala and Weerasinghe [46];
2. FoxReplace[9] – a Firefox add-on that can automatically replace textual content with the aid of a predefined substitution list.

[8]http://subasa.lk/ingiya

[9]https://addons.mozilla.org/en-us/firefox/addon/foxreplace/

**Fig. 3.9** Translation editing and voting

Ingiya is a non-intrusive add-on that shows Sinhala[10] definitions of English terms when the mouse pointer is hovered on top of English words in a website. It is also capable of temporarily replacing English words with Sinhala definitions (i.e. it does so for the web page until it is refreshed. As soon as the page is refreshed, the translations disappear). Currently, the Ingiya add-on only supports individual words. The dictionary entries are stored within a local Ingiya database.

For the purposes of this prototype, the Ingiya add-on was extended to enable capture of a selected lexicon, provide a translation for that lexicon and submit the translation to a central server. The add-on was further modified to support the selection of phrases (text segments) in addition to individual words. The selected text segment, user's translations and the URL of the active tab of the browser is sent via the Ingiya add-on to the central server as an HTTP call. The data is encoded using the Punycode[11] algorithm prior to submission. The Punycode algorithm was chosen in this prototype due to its simplicity, its suitability to encode multilingual text and its text-compression ability.

---

[10]Sinhala is one of the official languages of Sri Lanka and the mother tongue of the majority – 74 % of its population.

[11]http://en.wikipedia.org/wiki/Punycode

In this prototype, the server mainly performs three functions:

1. It accepts data sent via browser add-ons, decodes the data and stores in its local database;
2. Upon a request from a client, it transforms and sends the data in this database into a XML based format understood by the FoxReplace add-on;
3. Upon a request, it can transform and send the data in its aggregated database into an XML Localisation Interchange File Format (XLIFF) file that can be downloaded and used as a TM.

The UpLoD architecture used the Portable Object (PO) file format to store translation data [13]. However, a comparison of the PO format and the XLIFF standard suggests the suitability of using XLIFF for representating localisation data [15]. This is mainly due to specific features of XLIFF such as ability to store metadata (e.g. contextual information, glossaries, segmentation information etc.), binary data, inline markups etc. Work by Morado-Vázquez [31], and Anastasiou and Morado-Vázquez [3] also suggests the suitability of XLIFF over other file formats such as Translation Memory eXchange (TMX) format, again mainly due to XLIFF's ability to store additional metadata, but also because of other interoperability concerns [2]. Moreover, with the bankruptcy of the Localisation Industry Standard Association (LISA), the Translation Memory eXchange (TMX) TMX format is no longer actively developed or maintained. Hence XLIFF was adopted as the standard in this research to represent translation memories.

The FoxReplace add-on is capable of retrieving a regular expression-based substitution list and replacing text in a web page. Different substitutions can be defined for different URLs. The FoxReplace add-on was configured to retrieve translations (i.e. substitution list) from the central server. When combined, these two add-ons can implement most of the BE-COLA architecture described in the previous sections. The exception is the voting mechanism which has not yet been implemented but is part of on-going work by the research group.

## 3.5 Discussion: Outstanding Challenges

### *3.5.1 Empirical Evaluation*

It is an open question whether users of the internet would take it upon themselves to create translations of internet content. For example, Alegria et al. [1] (in Part 1 – Chap. 4) note that one of the chief benefits of the crowdsourced Wikipedia site is the "high number of languages in which it is available", but they also acknowledge that the "rapid growth of the English Wikipedia is leaving most other languages behind". This is well illustrated by a brief review of [47] which shows that there are only four languages with more than a million articles: English, French, German and Dutch. This seems paradoxical when you consider that the top 10 participant languages on Wikipedia are English, simple English, Chinese, Hindi, Arabic, Spanish, Malay, Portuguese, Russian and Indonesian.

Additionally, even if users do take it upon themselves to translate micro-segments of internet content, there is little formal evidence as to the quality of the translations they would provide.

An empirical design has been developed to assess the willingness of users to contribute translations, to assess the quality of those translations and to identify the characteristics of the TMs that evolve. This design is in-vivo in nature, providing us with an ecologically valid context for our work. Likewise the study will be longitudinal in nature allowing us to assess the trends in contributions and quality over time.

A basic requirement for the proposed study is a website that is relevant to an international audience but that is only originally available in English. For this study a "Local Knowledge" website will be created for international students coming to the University of Limerick for the first time. This website will carry 10–15 core pieces of information relevant to international students when they arrive in the region. It will be viewed by students with varying English-speaking ability. The website will carry a message at the top detailing how segments of the text can be translated by the user in accordance with the BE-COLA approach.

Reference translations will be generated in a number of different languages, these languages being representative of the international student, freshman population. These translations will not be made available on the website. At 6 month intervals, over a 2 year period the BE-COLA central server will be analysed to report on the amount of translation that has occurred for each of the international student languages. For each language where a substantial portion of the content has been translated, two measures of the quality of these translations will be undertaken. BLEU [32] will be used to compare the top ranking translations of segments to the reference translation for each language pair. In addition, to cope with the situation where unexpected but valid translations occur, the top ranked translations will be independently rated by a team of fluent, bilingual post-editors. These editors will rank each segment's translation as 'exact semantic match', 'close semantic match', 'plausible semantic match', 'implausible semantic match', and 'no semantic match'. This set of measures will give a quantitative feedback on users' willingness to contribute and the quality of their contributions.

#### 3.5.1.1 Empirical Characterisation of the TMs

In terms of the TMs generated by the approach, the quality measures proposed for individual translations, if analysed cumulatively, will give feedback on the quality of the TMs. However many more analyses will need to be carried out on the TMs to assess the efficacy of this approach. For example, by recording the sequence of translations submitted, we can characterise how TM entries evolve over time. Such analyses can give us an indication as to whether translation consensus is reached over time or whether there is thrashing between users on the optimum translation.

At a more basic level, analysis of the TMs will also serve to identify general crowdsource-micro-translation characteristics. In a novel approach like this one, many questions remain un-answered, such as: what sort of segmentation will crowdsource-micro-translators adopt? Will they select a phrase, a sentence, a paragraph or the entire page? If the former alternatives, and they chose not to translate all the material on a page, will they focus on the most relevant material, the initial material or material defined by some other, as yet undetermined, characteristic.

We see this characterisation of crowdsourced-micro-translations as an important area of future work that will define the feasibility of this approach as a paradigm.

### 3.5.2 *Translator Focus*

As discussed in Sect. 3.5.1.1, there is a strong possibility that volunteer user-translators would focus their efforts on only a small proportion of the user interface. This proposition is based on Pareto's Principle [4] which, to paraphrase for this context, suggests that most users of a large system (both applications and web-content) will only use a small proportion of it. If translators choose to translate as they use, or choose to do the translations that others will see, rather than translating holistically, it is likely that translation coverage will be patchy and will result in a 'pidgin' system made up of translated 'frequently-used' elements and un-translated 'infrequently-used' elements. More relevant for this research, the TMs generated will not be as widely encompassing. Specifically, our intuition is that they will have more translation alternatives for specific text, but much lesser coverage over the entire text. For example, it is unlikely that pages such as legal disclaimers, terms of agreements, licensing information, privacy policies and the like will be localised by the crowd, due to infrequent use of such pages contained in websites. Yet, because this material is generic, coverage of this material would be an important contribution for the TMs generated, and would broaden the domain-relevance of this work.

Thus, while the TMs provided by the BE-COLA architecture may prove sufficient for general content users, they run the risk of frustrating users who have less-normalised, but yet generic, requirements. However, frustration can, in turn, become a motivating factor if the user is empowered to subsequently translate the associated lexicons or strings. And ultimately, if they do end up translating less-frequently visited content, their contribution to the TMs will become available to the community.

### 3.5.3 *Vote Thrashing*

Another potential challenge to the TMs generated is the voting mechanism adopted when evaluating alternative translations. It may prove insufficient and ineffective; Specifically, there is the possibility of 'thrashing', where two individual translators, or groups of translators, have very strong and conflicting ideas about the translation required for specific elements. In such instances, an 'Analysis Engine' on the central server, or trusted monitors would need to intervene, analysing the central logs, deriving the appropriate translation, and locking future changes.

Indeed, we see a specific instance of this 'thrashing' problem as being core to adoption of this crowdsourcing approach to generating TMs. Imagine as a user, you localise some content and then send your changes to the server. Imagine then, retrieving the server-side localisation and finding that very few of your changes had survived. This is a micro-form of thrashing that would probably be prevalent. It would be particularly prevalent in instances when the BE-COLA architecture was successful: i.e. when there was a large number of users and your translation contribution was a relatively small proportion of the whole (i.e. had low impact in the voting). Such negative feedback might discourage the user from making further changes to the interface, resulting in a fall off in localisation activity over time and ultimately, affect contributions to the TMs. Indeed, it might discourage them from using the web-content itself, as the interface they strove to create has been destroyed by server-side customisations. Hence, we see a strong role for 'change alerts' and the option to opt out of server-side customisations when local changes have been made.

### *3.5.4 Non Textual and Textual Translations*

Of course, localisation is not as simple as portrayed in these prototypes. For example in application-interface localisation (as per UpLoD) holding boxes have to be resized, and images may have to be replaced. These same issues spill over to web content as well. One such significant issue, in the context of website content, is non-textual content such as images, audio clips, videos and various embedded objects (e.g. Java, Flash, PDF or Silver-light content) [9, 44]: content known as secondary genres [25]. Textual content represented in graphics such as banners is also very common in websites. These issues are so common that Yunker in 2000 (as cited in [9]) states that "Language is often the least challenging aspect of customising, or localising, a website for a foreign audience. The hard part is all the technical challenges". These technical challenges include the secondary genres, dealing with issues like locale specific parameters such as date/currency formats, and bandwidth capabilities [9]. The current architecture however does not deal with localisation of these technical issues.

Similarly various ethical, cultural and regional issues have to be taken into account when localising a website. This includes the use of appropriate colour schemes, graphics, navigational structures, symbols, layouts etc. [9, 44]. Inappropriate, culturally insensitive or offensive content has to be avoided. Therefore, a reviewing mechanism such as observed in the Wikipedia community, has to be built in to this model.

Even with the textual content, font and rendering problems may surface in the localised version. For example, assume an English phrase displayed in Times New Roman font size 12, which has to be localised into Sinhala. The browser extension may find a TM match. When rendering the Sinhala text (i.e. the TM match), it has to display it in the correct font and in a suitable font size. Font size 12 might not be the optimum font size in Sinhala to display this particular localised text in the context. Furthermore, issues might arise with various font styles (such as italic, bold or headings) as well. Other issues that need to be considered include text direction (e.g. top to bottom, right to left), text justification, text sort order, hyphenation, bullet items and the layout of GUI controls [9].

### *3.5.5 User Quality*

The voting mechanism currently implemented in the prototypes takes no account of user quality, an attribute that could easily be calculated from the available data (a simple measure would be the percentage of each user's suggestions that equate to the localisations with higher votes). This additional information could be used to resolve ties, where equal numbers of votes were obtained for two or more different translations, to resolve thrashing or, more generally, to identify if suggested localisations are appropriate for distribution.

### *3.5.6 Update Capabilities*

Future work might include the development of a suitable light-weight localisation model that includes an appropriate container that could facilitate a new and ongoing micro versioning capability. To accompany this a micro versioning workflow model would have to be developed that could facilitate and address many of the features described throughout this paper: for example the capability to facilitate a 24 h micro update capability that could cover up to 100 + languages on a 24 h basis.

### *3.5.7 BE-COLA Specific Issues*

While most of the issues and challenges emphasised in the preceding sections are relevant to both architectures, additional, unique technical challenges exist for the BE-COLA architecture.

While software localisation mainly deals with the translation of individual terms, web content localisation needs to deal with translation of text segments. A segment can be a word, a sentence, a paragraph or even an entire document. Automatic sentence boundary detection and word boundary detection are still challenging research problems in the area of Natural Language Processing (NLP), which may apply in the scenarios described here. Additionally, in the BE-COLA scenario, the user drives the selection of the text and thus defines the segment of interest. This means that the TMs will have to match arbitrary selected segments of text. However, it is anticipated that, as time goes on, selections will aggregate around locale-specific norms, as discussed in Sect. 3.5.1.1, and this may alleviate the "matching-arbitrary-segments" problem. Indeed, elucidation of these norms may help to identify the most informative contextual segmentation and may influence segmentation research in turn.

Deployment issues, such as the use of server farms, may need to be explored for scalability. Data security is another key factor that has to be considered. Security of TMs as well as transmission channels (i.e. between server and browser extensions) may have to be implemented.

Another important factor is the design of a methodology for coping with constant updates of websites. Especially with the evolving UGC websites and technologies, dynamic websites are becoming more and more common and popular. Dynamically generated content such as content drawn from a database and content displayed with the aid of scripts (e.g. Javascript) cause difficulties in the localisation process using the proposed approach. Therefore coping with constant changes to source content is a challenging aspect that needs to be focused on in future research. We would expect that large TMs would provide a basis for localisation changes but the technology used to apply those changes may have to evolve for these contexts.

The TMs can be further fine-tuned and leveraged more accurately if they can be categorised into different domains. The domain of the source web content can be taken as another contextual parameter. Therefore, the automatic detection of the domain of the web content might prove helpful in optimising the TM leverage process.

One of the advantages of the above methodology is that, once the entire web page is completely translated, the translated page can be cached in the central server for improved performance. This will enable the browser extension to directly render the localised layer without further processing. Furthermore, the browser extension can keep track of the user's most frequently visited websites and cache the localised versions locally to save bandwidth as well as to improve the efficiency, by avoiding frequent calls to the central server. On the other hand, the localisation layer is conceptual and it is only accessible via the browser extension. Therefore, users are not able to interact with the website using their native language (e.g. perform a search in their native language), nor would these pages be indexed by search engines (i.e. the localised version). Thus, when a new page is opened or the users let their cursor hover over a link, the browser extension has to let them know if there is a crowdsourced localised version available. This idea is consistent with Daniel Brandon's [9] idea to warn users ahead of time, if all pages are not translated in a website.

In addition to the various technical issues discussed above, legal issues could potentially be encountered which need to be thoroughly examined, identified and addressed prior to the deployment of the proposed solution. The first question that needs to be answered is, whether people have a right to localise websites without the consent of the website owners (e.g. can the crowd localise Microsoft's website without the company's consent). Moreover, the TMs (for each language pair) will keep on growing once the crowd starts using this framework. Legal implications, regarding who owns the TMs, have to be thoroughly considered. The accuracy of the translations is one of the crucial aspects that need to be considered, especially due to the fact that websites often reflect the public image of organisations. It is essential to investigate the necessary steps to prevent possible misuse. For example, a group of people should not be able to provide fallacious terminology in the localised version by deliberately making incorrect translations.

Misuse of the service can be alleviated to a certain extent by developing a log-on mechanism where users have to be authenticated by the central server to access the localisation service. Individuals who misuse the service can then be blocked or even legally prosecuted. Furthermore, individuals who contribute translations as well as individuals who vote for translations can be tracked and rewarded. Thus, these individuals can be further motivated with the use of public announcements and ranking (or medal offering) systems as in games.

In cases where lower user quality is suspected, their submissions should be reviewed by human experts (preferably a pool of linguists) prior to committing to the server's database. This additional step would ensure the quality of the translations used in terms of criteria such as relevancy, accuracy, suitability and consistency. However larger scale deployments, where a bigger community is involved, may well counteract this potential issue by weight of voting.

## 3.6 Conclusions and Future Work

Crowdsourcing and social localisation are approaches that address the shortcomings of current mainstream localisation, allowing the localisation decision to be shifted from large enterprises to the users, making content available in more languages to more citizens.

In the past, localisation decisions were taken based mainly on short-term, financial return-on-investment (ROI) considerations. Localisation was largely an issue of budgeting and took place if the expected sales exceeded the budgeted effort [11, 40]. It was a matter for large multinational content publishers who owned and controlled the content development process, including its localisation. Language resources, including tools, technologies and data, were developed and deployed within these controlled scenarios. However, as we have seen, this control no longer exists when we consider a website content that is more and more generated by the users.

The rise in UGC has not just brought a fundamental change in how we view the content creation and publishing process: it also offers huge opportunities for a radically new approach to localisation and the creation of open language resources. Collaborative approaches allow us not just to *create* and *publish* content based on community efforts, they also allow us to implement new strategies and approaches to *localisation* and the creation of *language resources*. User-translators can be given powers similar to those given already to content creators who can share their views and information freely on the web – now, they can do this not just in their own language, but potentially across all languages (as long as they engage the corresponding language community). Overall, this will lead to a more heterogeneous body of digital content, and break the link between the availability of specific content in a specific language, and its dollar-value in that language.

In this chapter, we have discussed the development of a browser extension-based, website-independent, client-server architecture (BE-COLA) that facilitates the collaborative creation and evolution of TMs used for the localisation of web content. This architecture is based on an earlier architecture called UpLoD that facilitated localisation of an applications' user interface. As this BE-COLA approach uses a TMs approach, constructed with the aid of the crowd and reviewed by experts where necessary, rather than an MT system, better quality translations can be expected. The development of the prototype as a single Firefox add-on has proven the viability of the proposed approach. Future research will focus mainly on addressing the issues related to characterising the derived TMs, as discussed in Sect. 3.5 above.

The current architecture will be especially useful in the case of less-resourced languages where MT systems are not (yet) viable. The proposed system focuses on the building of language resources, such as translation memories as parallel corpora, which could be used for the development of MT systems in the future.

**Acknowledgements** This research is supported by the Science Foundation Ireland (SFI) (Grant 07/CE/I1142) as part of the Centre for Next Generation Localisation (CNGL) at the University of Limerick. The prototype was implemented based on Ingiya and FoxReplace add-ons. The authors would like to thank the authors of and the contributors to the above add-ons. The authors would also like to thank Aram Morera-Mesa for his helpful comments and suggestions.

## References

1. Alegria I, Cabezon U, de Betoño UF, Labaka G, Aingeru M, Sarasola K, Zubiaga A (2012) Reciprocal enrichment between basque Wikipedia and machine translation. In: Gurevych I, Kim J (eds) The People's web meets NLP: collaboratively constructed language resources. Theory and applications of natural language processing. Springer, Berlin/Heidelberg
2. Anastasiou D (2011) The impact of localisation on semantic web standards. Eur J ePractice (12):42–52
3. Anastasiou D, Morado-Vázquez L (2010) Localisation standards and metadata. In: Sánchez-Alonso S, Athanasiadis IN (eds) Metadata and semantic research. Communications in computer and information science, vol 108. Springer, Berlin/Heidelberg, pp 255–274. doi:10.1007/978-3-642-16552-8_24
4. Bookstein A (1990) Informetric distributions, part I: unified overview. J Am Soc Inf Sci 41(5):368–375. doi:10.1002/(sici)1097-4571(199007)41:5<368::aid-asi8>3.0.co;2-c
5. Boxma H (2012) RIGI localization solutions. https://sites.google.com/a/rigi-ls.com/www/home. Cited 1 Apr 2012

6. Brooks D (1998) Language resources and international product strategy. Paper presented at the first international conference on language resources and evaluation (LREC), Granada, Spain, 28–30 May 1998
7. Callison-Burch C (2009) Fast, cheap, and creative: evaluating translation quality using Amazon's mechanical turk. In: Proceedings of the 2009 conference on empirical methods in natural language processing: volume 1, Singapore
8. Crowston K, Howison J (2005) The social structure of free and open source software development. First Monday 10(2–7). http://firstmonday.org/htbin/cgiwrap/bin/ojs/index.php/fm/article/view/1207/1127. Cited 20 July 2012
9. Daniel Brandon J (2001) Localization of web content. J Comput Small Coll 17 (2):345–358
10. DePalma DA (2007) Lionbridge announces 2006 results. http://upgrade.globalwatchtower.com/2007/06/lionbridge-2006-results/. Cited 20 Jul 2012
11. DePalma DA (2012) Most content remains untranslated. http://www.tcworld.info/tcworld/translation-and-localization/article/most-content-remains-untranslated/. Cited 02 Apr 2012
12. Dutro C (2012) i18n on rails: a Twitter approach. RailsConf 2012, Austin (Texas), 23–25 Apr 2012. https://github.com/newhavenrb/conferences/wiki/i18n-on-Rails:-A-Twitter-Approach. Cited 23 Aug 2012
13. Exton C, Wasala A, Buckley J, Schäler R (2009) Micro crowdsourcing: a new model for software localisation. Localis Focus 8(1):81–89
14. Exton C, Spillane B, Buckley J (2010) A micro-crowdsourcing implementation: the Babel software project. Localis Focus 9(1):46–62
15. Frimannsson A (2005) Adopting standards based XML file formats in open source localisation. Queensland University of Technology, Queensland
16. Gaspari F (2007) The role of online MT in webpage translation. University of Manchester, Manchester
17. Ghosh RA (1998) Interviews with linus torvalds: what motivates software developers. First Monday 3(3–2). http://firstmonday.org/htbin/cgiwrap/bin/ojs/index.php/fm/article/viewArticle/583/504. Cited 20 July 2012
18. Gift N, Shand A (2009) Introduction to distributed version control systems. https://www.ibm.com/developerworks/aix/library/au-dist_ver_control. Cited 07 Apr 2009
19. Hann IH, Roberts J, Slaughter S, Fielding R (2002) Why do developers contribute to open source projects? First evidence of economic incentives. In: Meeting challenges and surviving success: 2nd workshop on open source software engineering, international conference on software engineering, Orlando
20. Horvat M (2012) Live website localization. Paper presented at the W3C workshop: the multilingual web – the way ahead, Luxembourg, 15–16 Mar 2012
21. Howe J (2006) The rise of crowdsourcing. Wired 14(6):1–4
22. Internet-World-Stats (2012) World internet usage and population statistics (as per December 31, 2011). Miniwatts marketing group. http://www.internetworldstats.com/stats.htm. Cited 21 Jul 2012
23. Jacobs A (2009) Internet usage rises in China. The New York Times. http://www.nytimes.com/2009/01/15/world/asia/15beijing.html?_r=1. Cited 21 Jul 2012
24. Jarvis J (2009) What Would Google Do? HarperCollins, New York
25. Jiménez-Crespo MA (2011) To adapt or not to adapt in web localization: a contrastive genrebased study of original and localised legal sections in corporate websites. J Spec Transl 15:2–27
26. Kuznetsov S (2006) Motivations of contributors to Wikipedia. SIGCAS Comput Soc 36(2):1. doi:10.1145/1215942.1215943
27. Large A, Moukdad H (2000) Multilingual access to web resources: an overview. Program Electron Libr Inf Syst 34(1):43–58. doi:10.1108/EUM0000000006938
28. Lerner J, Tirole J (2002) Some simple economics of open source. J Ind Econ 50(2):197–234. doi:10.1111/1467-6451.00174
29. Losse K (2008) Keynote. Paper presented at the LRC XIII: localisation4All, Dublin, Ireland, 2–3 Oct 2008

30. Moorkens J (2011) Translation memories guarantee consistency: truth or fiction? Paper presented at the ASLIB translating and the computer 33, London, UK, 17–18 Nov 2011
31. Morado-Vázquez L, Rey JTd (2011) The relevance of metadata during the localisation process – an experiment. Paper presented at the internacional T3L conference: tradumatica, translation technologies and localization, Universitat Autonoma de Barcelona, Spain, 21–22 June 2011
32. Papineni K, Roukos S, Ward T, Zhu WJ (2002) BLEU: a method for automatic evaluation of machine translation. In: Proceedings of the 40th ACL. Association for Computational Linguistics, Philadelphia, pp 311–318
33. Pink D (2009) The surprising science of motivation. TED conferences, LLC. http://www.ted.com/talks/dan_pink_on_motivation.html. Cited 20 July 2012
34. Raymond ES (1999) The cathedral and the bazaar. Knowl Technol Policy 12(3):23–49. doi:10.1007/s12130-999-1026-0
35. Raymond ES (2001) The cathedral and the bazaar: musings on Linux and open source by an accidental revolutionary. O'Reilly, Beijing/Cambridge
36. Rickard J (2009) Translation in the community. Paper presented at the LRC XIV: localisation in the cloud, Limerick, Ireland, 24–25 Sept 2009
37. Sargent BB (2012) ROI lifts the long tail of languages in 2012. Common Sense Advisory, Inc. http://www.commonsenseadvisory.com/AbstractView.aspx?ArticleID=2899. Cited 20 July 2012
38. Schäler R (1994) A practical evaluation of an integrated translation tool during a large scale localisation project. In: The 4th conference on applied natural language processing (ANLP-94), Stuttgart, Germany
39. Schäler R (2010) Localization and translation. In: Handbook of translation studies, vol 1. John Benjamins Publishing Company, Amsterdam/Philadelphia, pp 209–214
40. Schäler R (2012) Information sharing across languages. In: Computer-mediated communication across cultures: international interactions in online environments. IGI Global, Hershey, pp 215–234. doi:10.4018/978-1-60960-833-0.ch015
41. Schäler R (2012) Introducing social localisation. Paper presented at the workshop, localization world, Silicon Valley
42. Schäler R (2012) The illusion of control and next generation localisation. http://www.therosettafoundation.org/index.php/en/archive/251-st-peter-the-illusion-of-control-and-next-generation-localisation-a-message-to-the-localisation-industry. Cited 20 July 2012
43. Shannon P (2000) Including language in your global strategy for B2B E-commerce. http://www.worldtradewt100.com/articles/print/83222. Cited 24 Apr 2012
44. Stengers H, Troyer OD, Baetens M, Boers F, Mushtaha AN (2004) Localization of web sites: is there still a need for it? Paper presented at the international workshop on web engineering (held in conjunction with the ACM HyperText 2004 conference), Santa Cruz, USA
45. Valverde S, Theraulaz G, Gautrais J, Fourcassie V, Sole RV (2006) Self-organization patterns in wasp and open source communities. Intell Syst IEEE 21(2):36–40. doi:10.1109/mis.2006.34
46. Wasala A, Weerasinghe R (2008) EnSiTip: a tool to unlock the English web. Paper presented at the 11th international conference on humans and computers, Nagaoka University of Technology, Japan, 20–23 Nov 2008
47. Wikipedia (2012) Wikipedia statistics. http://stats.wikimedia.org/EN/Sitemap.htm. Cited 28 Aug 2012
48. WorldLingo (2000) Increase global sales with WorldLingo. http://www.worldlingo.com/en/company/pr/pr20000223_02.html. Cited 24 Apr 2012
49. Zeitlyn D (2003) Gift economies in the development of open source software: anthropological reflections. Res Policy 32(7):1287–1291

# Chapter 4
# Reciprocal Enrichment Between Basque Wikipedia and Machine Translation

**Iñaki Alegria, Unai Cabezon, Unai Fernandez de Betoño, Gorka Labaka, Aingeru Mayor, Kepa Sarasola, and Arkaitz Zubiaga**

**Abstract** In this chapter, we define a collaboration framework that enables Wikipedia editors to generate new articles while they help development of Machine Translation (MT) systems by providing post-edition logs. This collaboration framework was tested with editors of Basque Wikipedia. Their post-editing of Computer Science articles has been used to improve the output of a Spanish to Basque MT system called Matxin. For the collaboration between editors and researchers, we selected a set of 100 articles from the Spanish Wikipedia. These articles would then be used as the source texts to be translated into Basque using the MT engine. A group of volunteers from Basque Wikipedia reviewed and corrected the raw MT translations. This collaboration ultimately produced two main benefits: (i) the change logs that would potentially help improve the MT engine by using an automated statistical post-editing system, and (ii) the growth of Basque Wikipedia. The results show that this process can improve the accuracy of an Rule Based MT (RBMT) system in nearly 10 % benefiting from the post-edition of 50,000 words in the Computer Science domain. We believe that our conclusions can be extended to

I. Alegria (✉)
Ixa Group, University of the Basque Country UPV/EHU, San Sebastián, Spain
e-mail: i.alegria@ehu.es

U. Cabezon · G. Labaka · A. Mayor · K. Sarasola (✉)
Ixa Group, University of the Basque Country, San Sebastián, Spain
e-mail: ucabezon001@ikasle.ehu.es; gorka.labaka@ehu.es; aingeru@ehu.es; kepa.sarasola@ehu.es

U. Fernandez de Betoño
Basque Wikipedia and University of the Basque Country, San Sebastián, Spain
e-mail: unai.fernandezdebetono@ehu.es

A. Zubiaga
Basque Wikipedia and Queens College, CUNY, CS Department, Blender Lab, New York, NY, USA
e-mail: arkaitz.zubiaga@gmail.com

I. Gurevych and J. Kim (eds.), *The People's Web Meets NLP*, Theory and Applications of Natural Language Processing, DOI 10.1007/978-3-642-35085-6_4,

MT engines involving other less-resourced languages lacking large parallel corpora or frequently updated lexical knowledge, as well as to other domains.

## 4.1 Introduction

One of the key features on the success of Wikipedia, the popular and open online encyclopedia, is that it is available in more than 200 languages. This enables the availability of a large set of articles in different languages. The effort of Wikipedia editors to keep contents updated, however, increases as the language has a smaller community of editors. Because of this, less-resourced languages with smaller number of editors cannot keep pace with the rapid growth of top languages such as English Wikipedia. To reduce the impact of this, editors of small Wikipedias can take advantage of contents produced in top languages, so they can generate large amounts of information by translating those. To relax such process of translating large amounts of information, machine translation provides a partially automated solution to potentially facilitate article generation [13]. This presents the issue that current machine translation systems generate inaccurate translations that require substantial post-editing by human editors. We argue that creatively combining machine translation and human editing can benefit both article generation on Wikipedia, and the development of accurate machine translation systems.

In this chapter, we introduce our methodology to enable collaboration between Wikipedia editors and researchers, as well as the system we have developed accordingly. This system allows to generate new articles by editing machine translation outputs, while editors help improve a machine translation system. Specifically, editors of the Basque Wikipedia have used this system to collaborate with the University of the Basque Country (UPV/EHU) producing articles in Basque language while helping improve an existing Spanish-Basque machine translation (MT) system called Matxin [1, 9]. We believe that amateur translators can benefit from MT rather than professional translators.

To perform such a collaboration between editors and researchers, a set of 100 articles were selected from Spanish Wikipedia to be translated into Basque using the MT engine. A group of volunteers from Basque Wikipedia reviewed and corrected these raw translations. In the correction process, they could either post-edit a text to fix errors, or retranslate it when the machine-provided translation was inaccurate. We logged their changes, and stored the final article generated. This process ultimately produced two main benefits: (i) the change logs potentially help improve the MT engine by using an automated statistical post-editor [11], and (ii) the generated articles help expand the Basque Wikipedia. The results show that this process can improve the accuracy of an Rule Based MT (RBMT) system in nearly 10 % benefiting from the post-edition of 50,000 words in the Computer Science domain.

The remainder of the chapter is organized as follows: Sect. 4.2 provides an overview of the most representative features of Basque language, as well as a summary of previous research on statistical post-edition (SPE), and collaborative

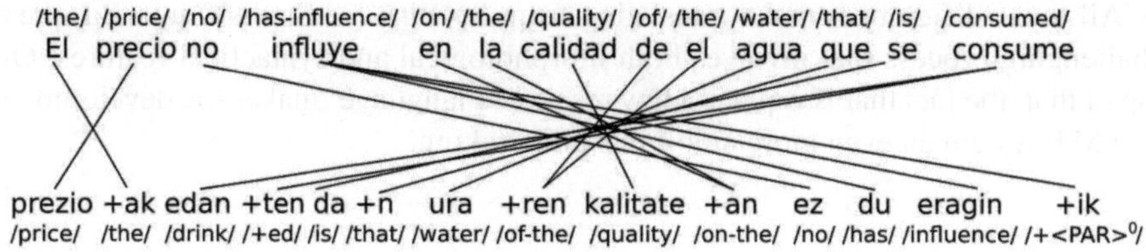


**Fig. 4.1** Comparison of word alignment for a sentence in Spanish and Basque (their transcription to English is "The price does not affect the quality of the drinking water")

work and MT; Sect. 4.3 describes the methodology used to build the post-editing system; Sect. 4.4 outlines and discusses the results and resources obtained through the collaborative work; finally, Sect. 4.5 concludes the chapter and sketches our future research plans.

## 4.2 Background

In this section we briefly describe features of Basque language, and summarize previous research on collaborative work for machine translation and automated statistical post-edition.

### *4.2.1 Basque Language*

Basque language presents particular characteristics, making it different from most European languages. This also makes translating into Basque a challenging task compared to other languages that share some sort of similarities. As an agglutinative language, many morpho-syntactic information that most European languages express in multiple words are expressed in a single word using suffixes in Basque. For instance, while Spanish and English use prepositions and articles, in Basque, suffixes are added to the last word of the noun-phrase; similarly, conjunctions are attached at the end of the verbal phrase.

Additionally, syntactic differences can also be found when looking into word orderings. These include: (i) modifiers of both verbs and noun-phrases are ordered differently in Basque and in Spanish; (ii) prepositional phrases attached to noun-phrases precede the noun phrase instead of following it; (iii) having very flexible ordering of sentence constituents, a neutral ordering suggests placing the verb at the end of the sentence and after the subject, object and any additional verb modifiers.

Figure 4.1 shows an example that compares word alignment for the Spanish sentence "El precio no influye en la calidad del agua que se consume" (The price does not affect the quality of the drinking water) and its Basque translation "Prezioak edaten den uraren kalitatean ez du eraginik".

All those differences make translating from Spanish (or English) into Basque a challenging process that involves both morphological and syntactical features. On top of that, the fact that Basque is a low resourced language[1] makes the development of a MT system an even more ambitious undertaking.

## 4.2.2 Related Work on Collaboration Initiatives and Machine Translation

Most MT engines make use of translations produced by humans. Specifically, translation repositories (usually referred to as translation memories,TMs) or parallel corpora are harnessed to learn translation models [13]. The use of public TMs has helped in the development and improvement of MT engines, and many companies have shared their memories to this end (e.g. TAUS[2]).

The chapter *Building Multilingual Language Resources in Web Localisation: A Crowdsourcing Approach* of this book describes a client-server architecture to share and use translation memories, which can be used to build (or improve) MT systems.

An alternative solution for improving MT engines is taking advantage of post-edition, i.e., the process of correcting MT outputs. The outcome of a post-editing process can be used in several ways:

- As a quality baseline to evaluate MT engines.
- As a resource that provides new TMs to help improve an MT engine.
- As a set of *automatic output/post-edited output* pairs that enables to learn an automatic post-editor (see Sect. 4.2.3). We use post-editing to this end on our system.

Popular MT engines include a post-edition interface to fix translations. For instance, Google Translate[3] allows its users to post-edit translations by replacing or reordering words. These corrections, which are only internally available to Google, provide valuable knowledge to enhance the system for future translations.

Asia Online is leading a project which aims to translate contents from English Wikipedia into Thai. In 2011, the company translated 3.5 million Wikipedia articles using MT and they are planning to improve them collaboratively.[4] Further details on the selected methodology are not available yet.

---

[1]There are around 700,000 speakers, around 25 % of the total population of the Basque Country.

[2]www.translationautomation.com

[3]http://translate.google.com

[4]http://www.commonsenseadvisory.com/Default.aspx?Contenttype=ArticleDetAD&tabID=63&Aid=1180&moduleId=390

Other companies such as Lingotek,[5] sell *Collaborative Translation Platforms* that include post-edition capabilities.[6]

For our collaborative work, we use OmegaT,[7] an open source Computer Aided Translation (CAT) tool.

### *4.2.3 Related Work on Training a Post-editing System*

Statistical post-editing (SPE), as described by Simard et al. [11], is the process of training a Statistical Machine Translation (SMT) system to translate from rule-based MT (RBMT) outputs into manually post-edited counterparts. They use SYSTRAN as the RBMT system, and PORTAGE as SMT system. They report a reduction in post-editing effort of up to a third when compared to the output of the RBMT. Isabelle et al. [7] conclude that an RBMT+SPE system effectively improves the output of a vanilla RBMT system as an alternative to manual adaptations. Experiments show that a SPE system using a corpus with 100,000 words of post-edited translations can outperform a lexicon-enriched baseline RBMT system while reducing the cost.

Dugast et al. [5, 6] show that a combination of SYSTRAN and an SMT system trained for SPE significantly improves the lexical choice of the final output, even if little improvement is observed in word ordering and grammar. Their comparative analysis suggests ways to further improve these results by adding "linguistic control" mechanisms. Lagarda et al. [8] show that an SPE system built with the Europarl corpus complements and improves their RBMT system in terms of suitability in a real translation scenario (average improvement of 59.5 %). Improvements were less significant (6.5 %) for a more complex corpus.

Potet et al. [10] experiment with a small corpus of 175 post-edited sentence pairs (around 5,000 words). These data were used at three different stages of the translation process: (a) extending the training corpus, (b) automatically post-editing the RBMT outputs, and (c) adjusting the weights of the log-linear model. Their experiments show that the use of this small corpus is helpful for correcting and improving the system to retranslate the same data, but it is challenging to propagate these corrections to new data.

Previous experiments for Basque [4] differ from similar work in that a morphological component is used in both RBMT and SMT translations, and in that the size of available corpora is small. The post-edition corpus was artificially created from a bilingual corpora, creating new RBMT translations for the source sentences and taking the corresponding target sentences as *the post-edited sentences* (see Fig. 4.2). They reported improvements when using an RBMT+SPE approach on a restricted domain but a smaller improvement when using more general corpora. In order to

[5]http://lingotek.com

[6]http://www.translingual-europe.eu/slides/WillemStoeller.pdf

[7]http://www.omegat.org

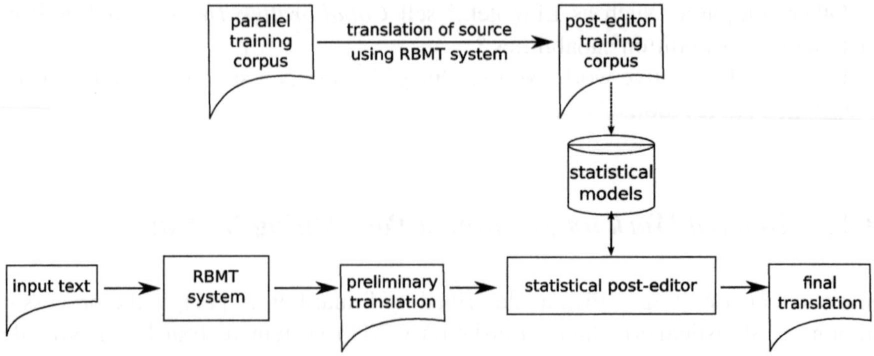


**Fig. 4.2** Architecture of a typical statistical post-editor

improve the MT system, the training material for the post-editing layer of our system consists of a text corpus in two parallel versions: raw machine translation outputs and manually post-edited versions of these translations. Since few resources are available [11], we built the training material from collaboratively constructed language resources.

## 4.3 Methodology

This section describe the collaborative post-editing framework. Figure 4.3 shows the overall architecture of our translation system (RBMT+SPE) that first uses the rule-based system and then the statistical post-edition. Firstly, we describe the aim of the overall system. Secondly, we describe OmegaT, the general human post-editing environment used in this work, as well as the extensions implemented to adapt the tool to the translation of Wikipedia entries. Thirdly, we tweak the translation system to customize it for the Computer Science domain. Finally, the most relevant aspects of the design of the collaboration initiative are described.

### *4.3.1 The Aim*

The main objective of this work is to build and test an MT system based on the RBMT+SPE approach using manually post-edited corpora from Basque Wikipedia editors. We chose articles in the Computer Science domain, both because it is suitable as a domain that does not highly depend on cultural factors and because it allows to focus improvements on a domain-specific scenario as a first step.

We expected editors to extend Basque Wikipedia by post-editing Basque RBMT translations of Spanish Wikipedia articles from the Computer Science domain. At the same time, Basque Wikipedia editors would be providing post-edition logs to

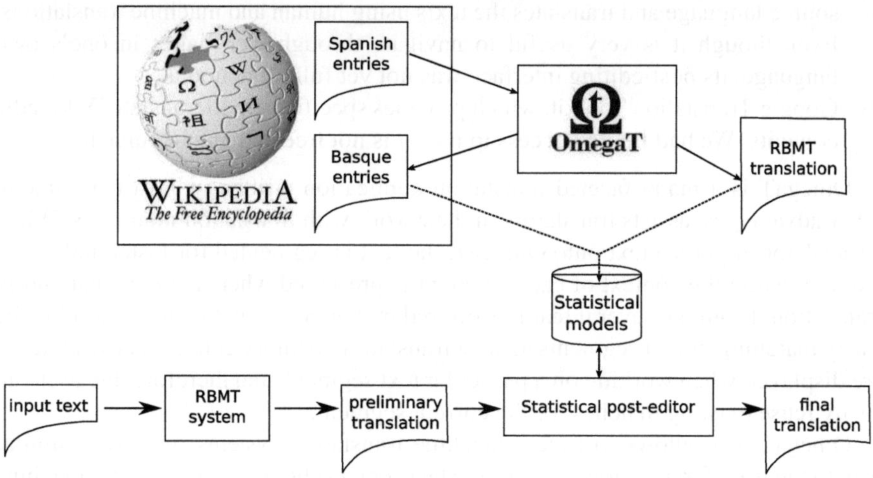


**Fig. 4.3** Architecture of our post-edition environment

feed an MT engine. Since Basque and Spanish belong to different language families, we hypothesized that amateur translators (unlike perhaps professional ones) would find the MT output of substantial help.

With the aim of facilitating the post-edition task for editors, we adapted the well-known open-source tool OmegaT. We stored the post-edited translations they provided as a resource to train a SPE system and evaluate the RBMT+SPE engine.

## *4.3.2 Modifications to OmegaT*

We considered several alternatives to OmegaT when selecting the translation platform to be used during the project, with priority toward open source solutions. We explored a number of tools such as Lokalize, Pootle, Virtaal and OmegaT. Lokalize and Pootle are localization tools that are overly complex for the translation of general texts. Virtaal[8] was initially developed as a specialized tool for translating software but has since moved towards a more graphic-based translation tool. OmegaT is a popular tool among translators and we found interesting features in it that made it suitable for translating general texts. Therefore, OmegaT was selected as the translation platform to be used in our project. Other alternatives that were discarded include:

(a) World Wide Lexicon (WWL) Translator, a Firefox add-on that makes browsing foreign-language sites easy and automatic. When browsing a URL, it detects the

[8] http://translate.sourceforge.net/wiki/virtaal

source language and translates the texts using human and machine translations. Even though it is very useful to navigate through web pages in one's own language, its post-editing interface was not yet fully functional.

(b) Google Translation Toolkit, which provides specific help to translate Wikipedia contents. We had limited access to it as it is not free and open-source tool.

OmegaT is a many-faceted translation application written in Java that, among other advantages, assists translators in their work with translation memories. When a translator imports a text onto OmegaT, the text is segmented for faster and easier reading, while the context of each segment is preserved when a context can aid in translation. From the several features offered by the program, the most useful is the fuzzy matching of text segments against translation memory entries. These matches are displayed when working on a particular text segment, and therefore, the user can easily reuse existing matches for the current segment.

OmegaT also allows to access machine translation systems, in a very similar way to the use of translation memories. The user can choose among several machine translation services (e.g., Google Translate, Apertium and Belazar). The translations produced by the selected systems are shown to the user as alternatives to choose from.

Other features of OmegaT include creating glossaries and dictionaries, importing dictionaries and translation memories, good compatibility with a variety of third-party software, and support for several file types and encodings as well as for different languages. OmegaT is open source and freely available (from www.omegat.org), and is supported by extensive documentation and an active community of users and developers.

To make it easier to use for editors, we adapted the interface of OmegaT with a number of additional features:

- *Integration of our Spanish to Basque MT engine.* OmegaT includes a class that connects several machine translation services, making it relatively easy to customize by adding more services (see Fig. 4.4). We used this class to integrate Matxin [9] within OmegaT. In order to reduce the integration effort, we made Matxin's code simpler, lighter and more readable so that it could be implemented as a web service to be accessed by single API calls using SOAP. Therefore, OmegaT could easily make use of a Spanish to Basque machine translation system.
- *Import/export of Wikipedia articles to/from OmegaT.* We implemented a new feature to upload the translated article to the Basque Wikipedia to OmegaT's existing capability of importing MediaWiki documents from their URL encoded as UTF8. To enable this new feature, we also implemented a new login module and some more details. When uploading an article to Wikipedia, the editor is also required to provide a copy of the translation memory created with the article. We use these translation memories in the process of improving the machine translation service, Matxin. The new upload is language-independent, and can be

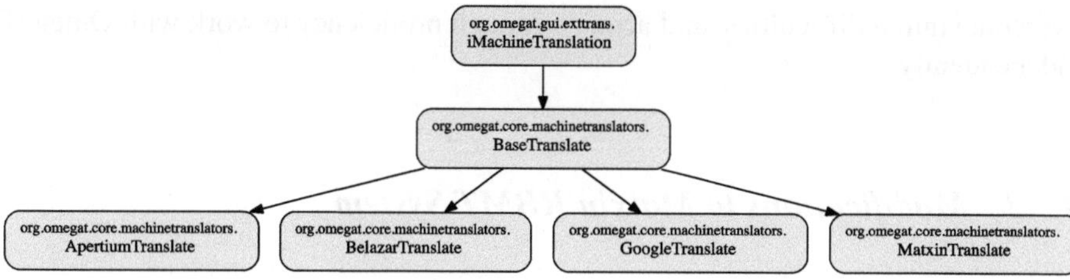


**Fig. 4.4** OmegaT extended with a module to enable the use of the Matxin MT system

used for languages other than Basque. However, this feature has not been tested yet on languages that rely on different character sets such as CJK or Arabic.

- *Integration of the Basque spell-checker to facilitate post-editing.* Thanks to OmegaT's flexible support for third-party applications, we also integrated a Basque spell-checker to assist users during translation.
- *Other improvements related to the translation of metadata in Wikipedia.* As an example of translation of Wikipedia metadata, let us take the translation of the internal Wikipedia link [[*gravedad* | *gravedad*]] in the Spanish Wikipedia (equivalent to the link [[*gravity* | *gravity*]] in the English Wikipedia). Our system translates it as [[GRABITAZIO | LARRITASUNA]], so it translates the same word in a different way when it represents the entry Wikipedia and when it is the text shown in such a link. On the one hand, the link to the entry *gravedad* in the Spanish Wikipedia is translated as GRABITAZIO (gravitation) making use of the mechanics of MediaWiki documents which include information on the languages in which a particular entry is available, and their corresponding entries. And on the other hand, the text word *gravedad* is translated as LARRITASUNA (seriousness) using the RBMT system. Therefore, this method provides a translation adapted to Wikipedia. Offering this option allows the post-editor to correct the RBMT translation with the usually more suitable "Wikipedia translation".

The fact that OmegaT needs to be locally installed and configured is inconvenient when the application is going to be used by a large community of users. Our project would have benefited from having access to an on-line contributive platform like Google Translation Toolkit or platforms based on the concept of Interactive Multilingual Access Gateway [3]. To address this shortcoming in existing tools, we are planning to adapt or develop a suitable platform to be used in future projects. Another issue with OmegaT is the somewhat steep learning curve. A new user may feel overwhelmed with the large number of features of the application, and even after gaining a basic familiarity may find it challenging to locate the most appropriate functionalities for the task at hand. Fortunately, there are several tutorials available that help with this. We have also written some user guides to satisfy the needs of our collaborators; this documentation serves both for understanding the features we have added and for getting the most of the features we deem particularly appropriate for this specific project. In our experience, with some guidance, users quickly overcome initial difficulties, and acquire enough proficiency to work with OmegaT independently.

### *4.3.3 Modifications to Matxin RBMT System*

The Matxin RBMT system was adapted to the Computer Science domain. The bilingual dictionary was customized in two ways:

**Adaptation of lexical resources from dictionary-systems.** Using several Spanish/Basque on-line dictionaries, we performed a systematic search for word meanings in the Computer Science domain [2]. We included 1,623 new entries in the lexicon of the original RBMT system. The new terms were mostly multi-words, such as *base de datos* (database) and *lenguaje de programación* (programming language). Some new single words were also obtained; for example, *iterativo* (iterative), *ejecutable* (executable) or *ensamblador* (assembly). In addition, the lexical selection was changed for 184 words: e.g. *rutina*-ERRUTINA (routine) before *rutina*-OHITURA (habit).

**Adaptation of the lexicon from a parallel corpus.** We collected a parallel corpus in the Computer Science domain from the localized versions of free software from Mozilla, including Firefox and Thunderbird (138,000 segments, 600,000 words in Spanish and 440,000 in Basque). We collected the English/Basque and the English/Spanish localization versions and then generated a new parallel corpus for the Spanish/Basque language pair, now publicly available. These texts may not be suitable for SMT but they are useful for extracting lexical relations. Based on Giza++ alignments, we extracted the list of possible translations as well as the probability of each particular translation for each entry in the corpus. In favour of precision, we limited the use of these lists to the lexical selection. The order was modified in 444 dictionary entries. For example, for the Spanish term *dirección*, the translated word HELBIDE (address) was selected instead of NORABIDE (direction).

### *4.3.4 Design of the Collaborative Work*

The collaboration between Basque Wikipedia editors and the University of the Basque Country started in 2010. In November 2010 we launched a Wikiproject[9] to collect and disseminate information about the project. Besides the links for downloading the adapted version of OmegaT, the Wikiproject included a list of target articles to be translated from the Spanish Wikipedia (which had no equivalents available at the time in Basque). These articles were classified by length into three

[9] http://eu.wikipedia.org/wiki/Wikiproiektu:OpenMT2_eta_Euskal_Wikipedia

subsets: short (less than 600 words), intermediate (between 600 and 1,000 words) and long (over 1,000 words). The short articles were intended to help editors learn the overall process of downloading an article from Wikipedia, translating and post-editing it, to finally upload the result back to Wikipedia.

Our initial plan was that each editor who "practiced" with a short article would also translate one of the 60 long articles. However, translating a long article represented a substantial amount of work (we estimate that the editors spent more than 8 h translating some long articles). The translation of the 60 long articles was thus taking too long, and therefore we created a tool to help us search for short untranslated Wikipedia entries. This tool is a perl script named *wikigaiak4koa.pl* (http://www.unibertsitatea.net/blogak/testuak-lantzen/2011/11/22/wikigaiak4koa) that, given a Wikipedia category and four languages, returns the list of articles contained in the category with their corresponding equivalents in those four languages and their length (1 Kb ~ 1,000 characters).

For instance, the following command:

```
$ perl wikigaiak4koa.pl "ca" "eu" "en" "es" "Informática"
```

searches for entries in the *Informática* (computer science) category on the Catalan Wikipedia ("ca"), looks for corresponding articles in Basque ("eu"), English ("en") and Spanish ("es"), and finally produces a text file like the following:

```
...

@     31.30 Kb
    eu     A_bildu                               25.25 Kb
    en     At_sign                              113.23 Kb
    es     Arroba_(símbolo)                      45.20 Kb
Acord_de_Nivell_de_Servei      22.21 Kb
    en     Service-level_agreement               18.96 Kb
    es     Acuerdo_de_nivel_de_servicio          23.39 Kb
Actic     23.25 Kb
Govern_electrónic      23.69 Kb
    en     E-Government                          18.82 Kb
    es     Gobierno_electrónico                  23.18 Kb

...
```

This example examines the Catalan entries for @, *Acord_de_Nivell_de_Servei*, *Actic* or *Govern_electrónic*. We can observe that there are equivalent entries in Basque (*A_bildu*, 25.25 Kb), English (*At_sign*, 113.23 Kb) and Spanish (*Arroba_(símbolo)*, 45.20 Kb) and that there is no Basque equivalent for the other three articles in Catalan. The script also shows that these entries are not very long, except the entry for *At_sign* in English, which size is 113.23 Kb.

Using this perl script we identified 140 entries that: (1) were included in the Catalan and Spanish Wikipedias, (2) were not in the Basque Wikipedia, and (3) the size in the Spanish Wikipedia was smaller than 30 Kb (~30,000 characters). These 140 intermediate size entries were included in the Wikiproject. The script can be used to examine the contents of any Wikipedia category for any language.

The size of the Catalan Wikipedia (378,408 articles) is midway between the Spanish (902,113 articles) and the Basque (135,273 articles). Therefore, we consider that a Wikipedia article that is present in the Catalan Wikipedia but not in the Basque Wikipedia should be included in the latter before other non-existing articles that are not in the Catalan version.

## 4.4 Results and Discussion

During the first months of 2012 the post-edited texts were processed in order to train a SPE engine and the RBMT+SPE pipeline system was evaluated.

Drawing on previous experience [4] and taking into account the morphology of Basque, we implemented a new automated statistical post-editing system. In this new experiment, the SPE corpus is a real post-edition corpus built from the raw RBMT translation outputs and their corresponding post-editions.

### *4.4.1 Evaluation*

Tables 4.1 and 4.2 show the scoring for different metrics for MT evaluation [12]. The MBLEU, BLEU, NIST and METEOR metrics measure the intersection between the output of the MT system and the human translation; TER, WER and PER express the number of changes necessary to get from the output of the MT system to the human translation. For example, TER (Translation Edit Rate) measures the amount of post-editing that a human would have to perform to change a system output so it exactly matches a reference translation. Possible edits include insertions, deletions, and substitutions of single words as well as shifts of word sequences. All edits have equal cost.

For the former metrics, a higher value represents a higher correlation with human judgments, whereas for the latter metrics lower values are optimal.

The original RBMT system and the RBMT system adapted to the Computer Science domain were tested with the whole set of sentences in selected Spanish Wikipedia articles, and their corresponding sentences after manual correction of RBMT outputs (see Table 4.1). The improvement is marked for all the metrics when the RBMT system is adapted to the domain. The highest relative improvement is for the BLEU and MBLEU metrics (15 %), and the lowest for WER (3.5 %).

The final aim of our experiments was to improve the output of the customized RBMT system using statistical post-editing (see Fig. 4.3). However, the corpus used to train the SPE system contained only 50,000 words. This is not an optimal size for statistical training and therefore we had to explore different ways to use the corpus successfully. We performed a five-fold cross-validation to evaluate different versions of the RBMT+SPE pipeline optimized by training with different subsets of the post-edition corpus. The sentence pairs were reordered using their TER scores, so that the most similar sentence pairs were promoted to the beginning of the corpus.

**Table 4.1** Evaluation of the RBMT systems

| SYSTEM | MBLEU | BLEU | NIST | METEOR | TER | WER | PER |
|---|---|---|---|---|---|---|---|
| Original RBMT | 18.89 | 19.50 | 6.17 | 43.94 | 65.11 | 68.69 | 52.08 |
| Adapted RBMT | **21.84** | **22.38** | **6.58** | **47.20** | **62.40** | **66.31** | **49.24** |

**Table 4.2** Evaluation of the RBMT+SPE systems

| SYSTEM | | | | | | | | | | |
|---|---|---|---|---|---|---|---|---|---|---|
| Test | Optim. | Train. | | MBLEU | BLEU | NIST | METEOR | TER | WER | PER |
| 1/5 | 1/5 | (3/5) | 50 % | 21.57 | 22.24 | 6.41 | 46.48 | 63.25 | 67.04 | 50.36 |
| 1/5 | 1/5 | (3/5) | 75 % | 22.54 | 23.26 | 6.52 | 47.55 | 62.28 | 66.20 | 49.59 |
| 1/5 | 1/5 | (3/5) | 100 % | 23.66 | 24.61 | 6.62 | 48.06 | 61.44 | 65.48 | 48.98 |
| 1/5 | 0 | (4/5) | 50 % | 22.14 | 22.82 | 6.50 | 47.26 | 62.49 | 66.56 | 49.77 |
| 1/5 | 0 | (4/5) | 75 % | 23.37 | 24.10 | 6.60 | 48.35 | 61.67 | 65.76 | 49.05 |
| 1/5 | 0 | (4/5) | 100 % | 24.24 | 25.10 | 6.69 | 48.94 | 60.97 | 65.08 | 48.58 |

Three different RBMT+SPE systems were trained using subsets of the corpus with the top 50, 75 and 100 % of this list ordered by TER.

The evaluation on these three systems was repeated for the other three RBMT+SPE systems where the SPE systems were optimized via MERT using a fifth of the corpus (see Table 4.2).

All the RBMT+SPE systems significantly obtained a better quality than the original RBMT system, and all but the system optimized and trained using only 50 % of the corpus achieved a better quality than the customized RBMT system.

The use of a smaller subset of the post-editing corpus with only the most similar sentence pairs produces no improvement in performance; in contrast, a greater number of sentence pairs always leads to improved results, even when the sentence pairs contain greater divergence. This is probably the result of the limited size of our training corpus, and indicates that a larger post-edition corpus might lead to better results.

The best system does not use any subset of the corpus for MERT optimization and uses 100 % of the sentences for training. It gets an improvement of 1.82 points for BLEU and 3.4 for MBLEU with respect to the customized RBMT. If compared to the original RBMT system, there is an improvement of 5.6 BLEU points or 5.35 MBLEU points. The other metrics confirm these improvements.

The use of a subset of the corpus for MERT optimization is not a good investment. When using only 50 % of the sentences the results are slightly worse, while using 75 % of the sentences only brings a small improvement. Finally, using all the post-edited sentences does produce an improvement, although note that the improvement is higher for the non-optimized system.

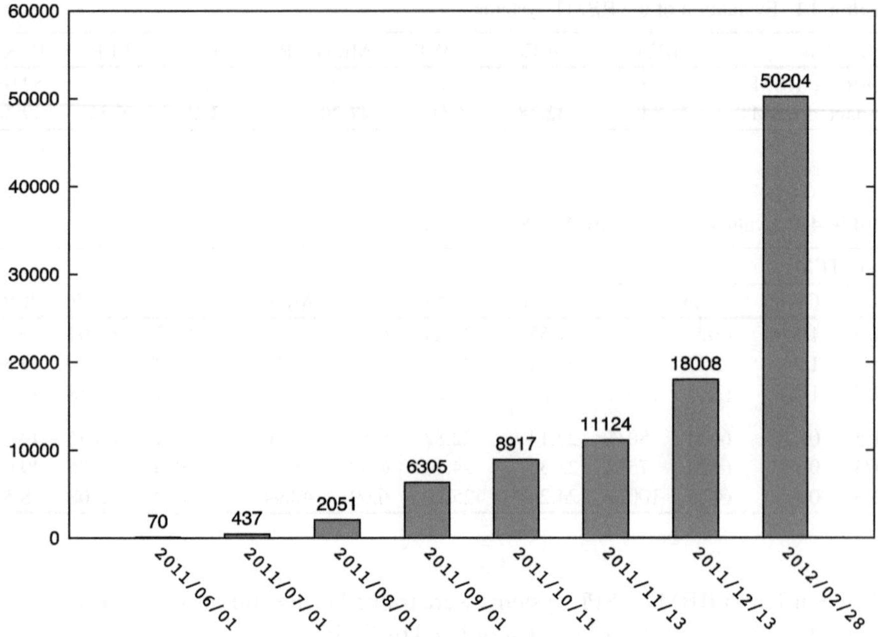


**Fig. 4.5** Evolution of the number of words translated by editors

### *4.4.2 Resources Obtained from the Collaborative Work*

When the public collaboration campaign had been running for 9 months, from July 2011 to February 2012, 100 new entries and 50,204 words had been added to the Basque Wikipedia. Figure 4.5 shows the evolution of the number of words translated by editors in that period.

The current state of our work is described on the web site of the Wikiproject.[10] One hundred new entries were added to Basque Wikipedia (the complete list[11] is available looking for articles in Basque Wikipedia defined with the "OpenMT-2" template) and the corpus created by manual post-editing of the RBMT outputs of these new 100 entries is publicly available.[12]

This data and the interviews with the Wikipedia editors collaborating in the project allow us to draw the following conclusions:

- The use of a MT system, even when its quality is not high, does help editors.
- Short Wikipedia articles are more appropriate to incorporate new collaborators that are sometimes not very motivated to participate in work excessively long.

[10] http://eu.wikipedia.org/wiki/Wikiproiektu:OpenMT2_eta_Euskal_Wikipedia

[11] http://eu.wikipedia.org/w/index.php?title=Berezi:ZerkLotzenDuHona/Txantiloi:OpenMT-2

[12] http://ixa2.si.ehu.es/glabaka/OmegaT/OpenMT-OmegaT-CS-TM.zip

- More than 30 different users have collaborated in the project so far and almost 20 of them have finished a long article.
- The metadata included in the Wikipedia articles is a challenge for the MT engine and for the users.
- Creating and coordinating a community to produce this type of material in a less resourced language is not an easy task, it can be a substantial task.

### *4.4.3 Discussion*

An analysis of the post-edited texts helps to better understand the quantitative evaluation, as well as identify the cases where the machine translation works well and can be improved. To perform such an analysis, we first sorted all the translation hypotheses created by the RBMT system (HYP) and their corresponding post-edition outputs (PDT) depending on their TER score. Next, we manually analyzed sentence pairs to validate the usefulness of this corpus. We observed that many of the post-editions with a TER score between 0 and 50 suggested reasonable lexical translation alternatives to the output of the RBMT system. Even though some of those suggestions were for single words, most of them were for multi-word terms. In many cases the post-edited terms appeared with their most frequent inflection suffixes, and that produced several errors. We identified three main problems that can be improved using a statistical post-editing system:

**Lexical gaps.** When a word is not an entry in the RBMT system's bilingual lexicon this word is not translated by the RBMT system. For example, *video* is not in the lexicon, but its equivalent in Basque (BIDEO) was proposed by editors:

HYP: 3GP **VIDEOA** GORDETZEN DU MPEG – LAU EDO 263 H. ...
PDT: 3GP **BIDEOA** GORDETZEN DU MPEG – LAU EDO 263 H. ...

**Lexical selection.** A better lexical selection can be achieved as a result of training the SPE system with simple contexts. For example, HEDAPEN and LUZAPEN are correct Basque translations for *extensión* (in Spanish). But only LUZAPEN is used for the specific meaning "file extension".

HYP: **HEDAPENA** TXTA HERRITARRA EGIN DA AZKEN GARAIETAN ...
PDT: TXT **LUZAPENA** ASKO ZABALDU EGIN DA AZKEN GARAIETAN ...

**Ambiguous syntactic structures.** Syntactic ambiguities are not always correctly resolved by the parser in the RBMT system. Those corrections that were often registered by the post-editors could be used by the statistical post-editing system to recover correct translations of short chunks. For instance, the translation of *lenguaje de restricciones* (constraint language) could be translated into Basque as MURRIZTEEN HIZKUNTZA (the language of the constraints) or MURRIZTE HIZKUNTZA (constraint language). Of these, the latter is the correct translation, but the RBMT system provides only the former.

HYP: OCL (object constraint language – OBJEKTUEN **MURRIZTEEN** HIZKUNTZA)
PDT: OCL (object constraint language – OBJEKTUEN **MURRIZTE** HIZKUNTZA)

Depending on the amount of post-editing data, some of these features will be learned by the SPE without the need of modifying the quite complex structure of the RBMT engine. For instance, the first two examples above were properly corrected by the SPE system, while the third one remained unchanged.

## 4.5 Conclusions and Future Work

Creating and coordinating a community to produce materials for a less resourced language can be a substantial task. We have defined a collaboration framework that enables Wikipedia editors to generate new articles while they help development of machine translation systems by providing post-edition logs. This collaboration framework has been experimented with editors of Basque Wikipedia. Their post-editing on Computer Science articles were used to train a Spanish to Basque MT system called Matxin. The benefits were twofold: improvement of the outputs of the MT system, and extension the Basque Wikipedia with new articles.

Various auxiliary tools developed as part of this research can also be considered as valuable resources for other collaborative projects: (i) a perl script that, given a Wikipedia category and four languages, returns the list of articles contained in the category with equivalents in those four languages and their length. The script is therefore useful to search short untranslated Wikipedia entries; (ii) the method used to translate Wikipedia links making use of the mechanics of MediaWiki documents which include information on the languages in which a particular entry is available, and their corresponding entries. This allows the post-editor to correct the RBMT translation with a more suitable "Wikipedia translation"; (iii) a new feature added to OmegaT to import/export Wikipedia articles to/from OmegaT. This new upload feature, although used for Basque, was developed as a language-independent tool.

The complete set of publicly available resources created in this project includes the following products:

- The 100 new entries added to Basque Wikipedia.[13]
- The new Spanish/Basque version of the parallel corpus[14] created from the localized versions of free software from Mozilla.
- The corpus[15] created by manual post-editing of the RBMT outputs of the new 100 entries.

[13] http://eu.wikipedia.org/w/index.php?title=Berezi:ZerkLotzenDuHona/Txantiloi:OpenMT-2
[14] http://ixa2.si.ehu.es/glabaka/lokalizazioa.tmx
[15] http://ixa2.si.ehu.es/glabaka/OmegaT/OpenMT-OmegaT-CS-TM.zip

- The perl script *wikigaiak4koa.pl*[16] that returns the list of articles contained in a Wikipedia category (for four languages).
- The improved version of OmegaT,[17] and its user guide.[18]
- The new version of the Matxin RBMT system[19] customized for the domain of Computer Science available as a SOAP service.

We logged the post-editions performed by Wikipedia editors by translating 100 articles from the Spanish Wikipedia into Basque using our MT engine. At the beginning of this work, we set forth the hypothesis that MT could be helpful to amateur translators even if not so much to professionals. After a qualitative evaluation, we can confirm our hypothesis, as even when the quality of the MT output was not high, it was enough to prove useful in helping the editors perform their work. We also observed that Wikipedia metadata makes more complicated both the MT and the post-editing processes, even if the use of Wikipedia's interlanguage links effectively help translation.

Integrating the outcome of collaborative work previously performed in software localization produced a significant enhancement in the adaptation of the RBMT system to the domain of Computer Science. In turn, incorporating the post-editing work of our Wikipedia collaborators into an RBMT system (50,000 words) produced an additional important improvement, despite the fact that the size of this corpus is smaller than those referenced in the major contributions to SPE (for example, Simard et al. [11] used a corpus of 100,000 words). Thus, there may be room for further improvement by the simple expedient of using a larger post-edition corpus. As short Wikipedia articles are more appropriate to incorporate new collaborators, search tools to look for candidate articles in Wikipedia become extremely useful.

The quantitative results show that the contributions can improve the accuracy of a combination of RBMT-SPE pipeline at around 10 %, after the post-edition of 50,000 words in the Computer Science domain. We believe that these conclusions can be extended to MT engines involving other less-resourced languages lacking big parallel corpora or frequently updated lexical knowledge.

In addition, the post-editing logs can function in an intermediate fashion as a valuable resource for diagnosing and correcting errors in MT systems, particularly lexical errors that depend on a local context.

Further improvements could be achieved using several tuning techniques. In the near future we plan to study the use of a combination of real post-edition and parallel texts as a learning corpus for SPE.

---

[16] http://www.unibertsitatea.net/blogak/testuak-lantzen/2011/11/22/wikigaiak4koa

[17] http://ixa2.si.ehu.es/glabaka/OmegaT/OpenMT-OmegaT.zip

[18] http://siuc01.si.ehu.es/~jipsagak/OpenMT_Wiki/Eskuliburua_Euwikipedia+Omegat+Matxin.pdf

[19] http://ixa2.si.ehu.es/matxin_zerb/translate.cgi

**Acknowledgements** This research was supported in part by the Spanish Ministry of Education and Science (OpenMT2, TIN2009-14675-C03-01) and by the Basque Government (Berbatek project, IE09–262). We are indebted to all the collaborators in the project and especially to the editors of the Basque Wikipedia. Elhuyar and Julen Ruiz helped us to collect resources for the customization of the RBMT engine to the domain of Computer Science.

## References

1. Alegria I, Diaz de Ilarraza A, Labaka G, Lersundi M, Mayor A, Sarasola K (2007) Transfer-based MT from Spanish into Basque: reusability, standardization and open source. In: CICLing 2007. Lecture notes in computer science, vol 4394. Springer, Berlin/New York, pp 374–384
2. Alegria I, Diaz de Ilarraza A, Labaka G, Lersundi M, Mayor A, Sarasola K (2011) Matxin-Informatika: Versión del traductor Matxin adaptada al dominio de la informática. In: Proceedings of the XXVII Congreso SEPLN, Huelva, Spain, pp 321–322
3. Boitet C, Huynh CP, Nguyen HT, Bellynck V (2010) The iMAG concept: multilingual access gateway to an elected web sites with incremental quality increase through collaborative post-edition of MT pretranslations. In: Proceedings of Traitement Automatique du Langage Naturel, TALN, Montréal
4. Diaz de Ilarraza A, Labaka G, Sarasola K (2008) Statistical post-editing: a valuable method in domain adaptation of RBMT systems. In: Proceedings of MATMT2008 workshop: mixing approaches to machine translation, Euskal Herriko Unibersitatea, Donostia, pp 35–40
5. Dugast L, Senellart J, Koehn P (2007) Statistical post-editing on SYSTRAN's rule-based translation system. In: Proceedings of the second workshop on statistical machine translation, Prague, pp 220–223
6. Dugast L, Senellart J, Koehn P (2009) Statistical post editing and dictionary extraction: Systran/Edinburgh submissions for ACL-WMT2009. In: Proceedings of the fourth workshop on statistical machine translation, Athens, pp 110–114
7. Isabelle P, Goutte C, Simard M (2007) Domain adaptation of MT systems through automatic post-editing. In: Proceedings of the MT Summit XI, Copenhagen, pp 255–261
8. Lagarda AL, Alabau V, Casacuberta F, Silva R, Díaz-de-Liaño E (2009) Statistical post-editing of a rule-based machine translation system. In: Proceedings of NAACL HLT 2009. Human language technologies: the 2009 annual conference of the North American chapter of the ACL, Short Papers, Boulder, pp 217–220
9. Mayor A, Diaz de Ilarraza A, Labaka G, Lersundi M, Sarasola K (2011) Matxin, an open-source rule-based machine translation system for Basque. Mach Transl J 25(1):53–82
10. Potet M, Esperança-Rodier E, Blanchon H, Besacier L (2011) Preliminary experiments on using users' post-editions to enhance a SMT system. In: Forcada ML, Depraetere H, Vandeghinste V (eds) Proceedings of the 15th conference of the European association for machine translation, Leuven, Belgium, pp 161–168
11. Simard M, Ueffing N, Isabelle P, Kuhn R (2007) Rule-based translation with statistical phrase-based post-editing. In: Proceedings of the second workshop on statistical machine translation, Prague, pp 203–206
12. Snover M, Dorr B, Schwartz R, Micciulla L, Makhoul J (2007) A study of translation edit rate with targeted human annotation. In: Proceedings of the 7th Biennial conference of the association for machine translation in the Americas (AMTA), Cambridge, Massachusetts, USA, pp 223–231
13. Way A (2010) Machine translation. In: Clark A, Fox C, Lappin S (eds) The handbook of computational linguistics and natural language processing. Wiley-Blackwell, Oxford, pp 531–573

# Part II
# Mining Knowledge from and Using Collaboratively Constructed Language Resources

# Chapter 5
# A Survey of NLP Methods and Resources for Analyzing the Collaborative Writing Process in Wikipedia

**Oliver Ferschke, Johannes Daxenberger, and Iryna Gurevych**

**Abstract** With the rise of the Web 2.0, participatory and collaborative content production have largely replaced the traditional ways of information sharing and have created the novel genre of collaboratively constructed language resources. A vast untapped potential lies in the dynamic aspects of these resources, which cannot be unleashed with traditional methods designed for static corpora. In this chapter, we focus on Wikipedia as the most prominent instance of collaboratively constructed language resources. In particular, we discuss the significance of Wikipedi's revision history for applications in Natural Language Processing (NLP) and the unique prospects of the user discussions, a new resource that has just begun to be mined. While the body of research on processing Wikipedia's revision history is dominated by works that use the revision data as the basis for practical applications such as spelling correction or vandalism detection, most of the work focused on user discussions uses NLP for analyzing and understanding the data itself.

## 5.1 Introduction

Over the past decade, the paradigm of information sharing in the web has shifted towards participatory and collaborative content production. In the early days of the Internet, web content has primarily been created by individuals and then shared with

O. Ferschke (✉) · J. Daxenberger
Ubiquitous Knowledge Processing Lab, Technische Universität Darmstadt, Darmstadt, Germany
e-mail: ferschke@ukp.informatik.tu-darmstadt.de; daxenberger@ukp.informatik.tu-darmstadt.de

I. Gurevych
Ubiquitous Knowledge Processing Lab, Technische Universität Darmstadt, German Institute for Educational Research and Educational Information, Darmstadt, Germany
e-mail: gurevych@ukp.informatik.tu-darmstadt.de

I. Gurevych and J. Kim (eds.), *The People's Web Meets NLP*, Theory and Applications of Natural Language Processing, DOI 10.1007/978-3-642-35085-6_5,

the public. Today, online texts are increasingly created collaboratively by multiple authors and are iteratively revised by the community.

When researchers first conducted surveys with professional writers in the 1980s, they found not only that the majority of them write collaboratively, but also that the collaborative writing process differs considerably from the way individual writing is done [25]. In collaborative writing, the writers have to externalize processes that are otherwise not made explicit, like the planning and the organization of the text. The authors have to communicate *how* the text should be written and *what* exactly it should contain.

Today, many tools are available that support collaborative writing for different audiences and applications, like *EtherPad*,[1] *Google Docs*,[2] *Zoho Writer*[3] or *BookType*.[4] A tool that has particularly taken hold is the *wiki*, a web-based, asynchronous co-authoring tool, which combines the characteristics of traditional web media, like email, forums, and chats [7]. Wiki pages are structured with lightweight *markup* that is translated into *HTML* by the wiki system. The markup is restricted to a small set of keywords, which lowers the entry threshold for new users and reduces the barrier to participation. Furthermore, many wiki systems offer visual editors that automatically produce the desired page layout without having to know the markup language. A unique characteristic of wikis is the automatic documentation of the revision history which keeps track of every change that is made to a wiki page. With this information, it is possible to reconstruct the writing process from the beginning to the end. Additionally, many wikis offer their users a communication platform, the *Talk pages*, where they can discuss the ongoing writing process with other users.

The most prominent example of a successful, large-scale wiki is *Wikipedia*, a collaboratively created online encyclopedia, which has grown considerably since its launch in 2001, and which contains over 22 million articles in 285 languages and dialects, as of April 2012. In this chapter, we review recent work from the area of Natural Language Processing (NLP) and related fields that aim at processing Wikipedia. In contrast to Medelyan et al. [19], who provide a comprehensive survey of methods to mine lexical semantic knowledge from a static snapshot of Wikipedia articles, we concentrate on the dynamic aspects of this resource. In particular, we discuss the significance of Wikipedia's revision history for applications in NLP and the unique prospects of the user discussions, a new resource that has just begun to be mined. Figure 5.1 gives an overview of the topics covered in this chapter. While the body of research on processing Wikipedia's revision history is dominated by works that use the revision data as the basis for practical applications such as spelling correction or vandalism detection, most of the work focused on user discussions uses NLP for analyzing and understanding the data itself. Furthermore, there are increasing efforts to build tools and resources for enabling research on Wikipedia, which are discussed in the final section of this chapter.

---

[1] http://etherpad.org/

[2] https://docs.google.com

[3] https://writer.zoho.com

[4] http://www.sourcefabric.org/en/booktype/

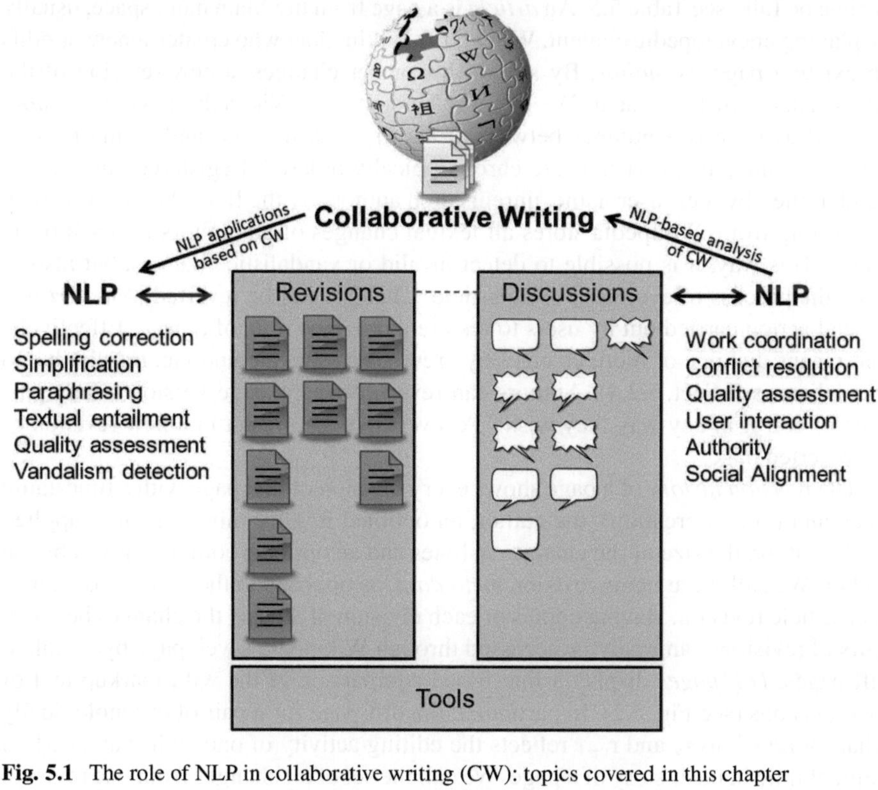


**Fig. 5.1** The role of NLP in collaborative writing (CW): topics covered in this chapter

## 5.2 Revisions in Wikipedia

Wikipedia's revision history lets us track the collaborative writing process in every single page in the encyclopedia. This section will explain the concept of revisions in Wikipedia and their uses for research in computational linguistics. After a short introduction to the concept of revisions in Wikipedia, we describe different NLP tasks that can benefit from the enormous data resulting from storing each single version of an article. Furthermore, we analyze applications of information coming from revisions with respect to article quality and trustworthiness. This is of general interest to computational linguistics, as the concepts and methods used can be applied to other collaboratively constructed discourse that uses revisions, in particular, wiki-based platforms.

### *5.2.1 The Concept of Revisions in Wikipedia*

Throughout this chapter, we will use the term *page* to refer to a document in Wikipedia from any namespace, including articles, stubs, redirects, disambiguation pages, etc. The Wikipedia *namespace* system classifies pages into categories like

Article or Talk, see Table 5.5. An *article* is a page from the Main namespace, usually displaying encyclopedic content. We call the Wikipedian who creates a new or edits an existing page its *author*. By storing his or her changes, a new revision of the edited page will be created. We call any version of a Wikipedia page a *revision*, denoted as $r_v$. $v$ is a number between 0 and $n$, $r_0$ is the first and $r_n$ the present version of the page, revisions are chronologically ordered. Registered authors can be identified by their user name, unregistered authors by the IP of the machine they are editing from. Wikipedia stores all textual changes of all authors for each of its pages. This way, it is possible to detect invalid or vandalistic changes, but also to trace the process of evolution of an article. Changes can be reverted. A *revert* is a special action carried out by users to restore a previous state of a page. Effectively, that means that one or more changes by previous editors are undone, mostly due to Vandalism (see Sect. 5.2.4). Authors can revert the latest page version to any past state or edit it in any way they wish.[5] A revert will also result in a new revision of the reverted page.

The *revision history* of a page shows every revision of that page with a timestamp (date and time of creation), the author, an optional flag for minor changes applied by the author, the size of the changes in bytes and an optional comment given by the author. We call these items revision *meta data*, as opposed to the textual content of each article revision. Having copies of each revision of a page, the changes between pairs of revisions can easily be accessed through Wikipedia's web page by so called diff pages. *Diff pages* display a line-based comparison of the wiki markup text of two revisions (see Fig. 5.2). In particular, the diff page for a pair of chronologically adjacent revisions $r_v$ and $r_{v-1}$ reflects the editing activity of one author at a certain point of time in the history of a page. We call the set of all changes from one revision to another a *diff*. A single diff in an article's revision history can be reverted if subsequent changes do not conflict with it, i.e. modify text affected by the reverted diff. This special kind of revert is usually referred to as *undo*.[6] As changes can affect one or several parts of a page, a diff can consist of various *edits*. An *edit* is a coherent local change, usually perceived by a human reader as one single editing action. In Fig. 5.2, two consecutive revisions $r_v$ and $r_{v-1}$ are displayed in a diff page, consisting of two edits inserting internal links. With respect to the meta data, the revisions in Fig. 5.2 have different authors. Both $r_v$ and $r_{v-1}$ are accompanied by comments. The timestamps indicate that the two versions have a time difference of approximately 9 days.

The remainder of this section will discuss applications of the information that is encoded in revisions and how it is used as resource in NLP research. A list of tools to access revision data in Wikipedia can be found in Sect. 5.4.1.

---

[5]However, pages can be protected from editing by privileged users, as stated in the Wikipedia Protection Policy, see http://en.wikipedia.org/wiki/WP:Protection_policy.

[6]http://en.wikipedia.org/wiki/WP:UNDO

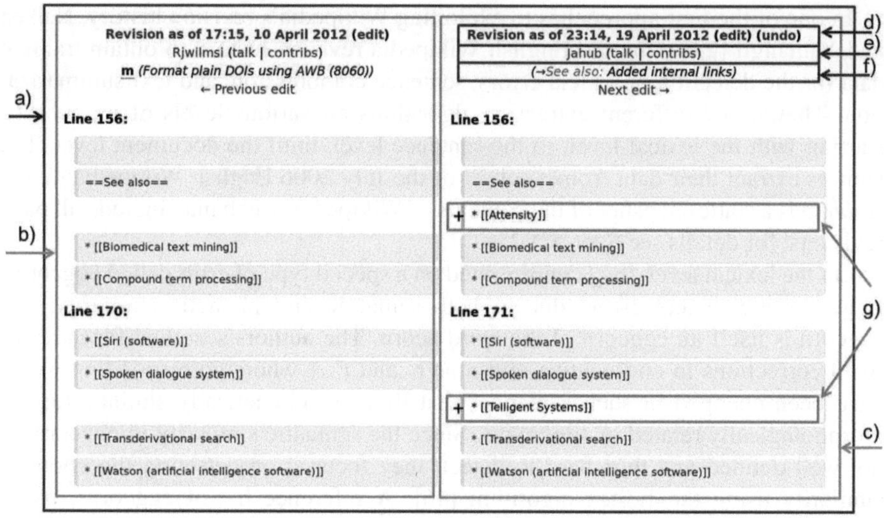


**Fig. 5.2** A diff page: (*a*) entire diff, (*b*) older revision $r_{v-1}$, only the changed part and its context are displayed, (*c*) newer revision $r_v$ (*d*) timestamp with edit and revert button, (*e*) author, *f*) comment, (*g*) edits. (*d*)–(*f*) are meta data of $r_v$

**Table 5.1** Previous approaches using the Wikipedia revision history as source of training data

| Reference | Type of data | Size of labeled data | Language | Publicly avail. |
|---|---|---|---|---|
| [23] | Eggcorns (malapropisms) | 108[a] | English | No |
| [39] | Sentence compression pairs | 380,000 | English | No |
| [43] | Real-word spelling error pairs | 686 | English, German | Yes |
| [18] | Spelling errors and paraphrases | 146,593[b] | French | Yes |
| [40] | Lexical simplifications pairs | 4049[c] | English | Yes |
| [38] | Lexical simplifications pairs | 14,831 | English | No |
| [41] | Textual entailment pairs | 1,614 | English | Yes |

[a] This number is based on a statement in [23] saying that they "successfully found 31 % of the reference eggcorns", the latter summing up to 348

[b] Refers to the spelling error corpus, v2.0

[c] Sum of pairs from the edit model and the meta data model

## 5.2.2 NLP Applications

This section explains how computational linguistics can benefit from analyzing the revisions in collaboratively created discourse. We will present different approaches that are based on data from the Wikipedia revision history. These can be divided into three groups: *error detection*, *simplification* and *paraphrasing*. All of them benefit from the abundance of human-produced, near-parallel data in the Wikipedia revision history, as they employ it to extract task-specific training corpora on demand. See Table 5.1 for an overview.

In one of the first approaches to exploiting Wikipedia's revision history, Nelken and Yamangil [23] mine the English Wikipedia revision history to obtain training data for the detection of lexical errors, sentence compression, and text summarization. They apply different extraction algorithms on various levels of granularity, starting with the lexical level, to the sentence level, until the document level. The authors extract their data from a subset of the July 2006 English Wikipedia dump. A *dump* is a static snapshot of the contents of Wikipedia which may include all page revisions; for details see Sect. 5.4.2.

On the lexical level, they concentrated on a special type of error called eggcorns. Eggcorns are lexical errors due to both semantic and phonetic similarity, e.g. eggcorn is itself an eggcorn of the word acorn. The authors searched for cases of word corrections in consecutive revisions $r_v$ and $r_{v-1}$ where corresponding words have been changed in such a manner that they are phonetically similar, but not morphologically related or synonyms. Since the semantic similarity in eggcorns is not well defined and thus hard to detect, they focused on detecting the phonetic similarity using the Editex algorithm [46]. A reference list of eggcorns, based on the so-called Eggcorn Database[7] which contains misspelled and correct word forms, serves to limit the article search space to those documents containing correct forms of one of the reference examples. For these articles, the authors harvested the revision history for pairs of revisions where $r_v$ contains the correct form of an eggcorn. In the next step, they calculated edit distances between $r_v$ and $r_{v-1}$, first, to identify similar sentences, and second, to find similar words within sentence pairs. Finally, the phonetic similarity of word pairs is measured. As for the resulting data, the authors report low precision, i.e. many false positives like typos or profanity. They justify that with their main goal to optimize the recall.

In another approach, Yamangil and Nelken [39] focus on the problem of data sparsity for applying a noisy channel model to sentence compression. First, they measure the edit distance between all sentences from pairs of adjacent revisions to find related sentences. In the resulting pairs of sentences, the authors look for edits adding or dropping words. That way, they detect sentence pairs being compressions of one another, assuming that all such edits retain the core meaning of the sentence. They verify the validity of the syntax of the extracted examples with a statistical parser, resulting in 380,000 parsed sentence pairs. This data is used within a syntax-based noisy channel compression model, which is based on an approach to sentence compression by Knight and Marcu [14]. In this model, a short sentence $s$ is ranked by a source language model $p(s)$ and expands to a long sentence $l$ with a certain probability $p(l|s)$. In [14], $p(l|s)$ corresponds to the probability of the syntax tree of $l$ to be transformed into the syntax tree of $s$. For a long sentence $l$, the model seeks the short sentence $s$ that is most likely to have generated $l$, that is to maximize $p(s) \cdot p(l|s)$. Yamangil and Nelken's model benefits from the mass of training data as it offers enough examples to add lexical information to the syntactic model. The model thereby learns probabilities not only based on the syntax trees

[7] http://eggcorns.lascribe.net/

of the example sentence pairs, but based on their words. Their result shows an improvement in compression rate and grammaticality over the approach of Knight and Marcu. However, they experience a slight decrease in the importance of the resulting sentences. *Importance* measures the quality of information preservation in the compressed version of a sentence with regard to the original. The authors explain this drop with the training data originally coming from both compressions and expansions (i.e. decompressions) by authors in Wikipedia, where the latter seems to frequently add important information that should not be skipped.

Nelken and Yamangil [23] also present a method for summarization of whole articles or texts. It is based on the assumption that a sentence with high persistence throughout the edit history of an article is of significant importance for it, i.e. it can be used for summarization purposes. They define a weak sentence identity, which allows for small changes in persistent sentences and is defined by a threshold of the edit distance. The authors tested the usability of their approach on two Wikipedia articles and found that the first sentence of a section as well as structural markup (such as link collections) have a higher persistence. As Nelken and Yamangil state, their methods cannot replace a full summarization application, but would be useful as part of a larger system.

In conclusion, Nelken and Yamangil present a series of promising applications for data extracted from the Wikipedia revision history. Their proposals for error detection, sentence compression and text summarization lay a foundation for further approaches working with this kind of data. Advanced systems can benefit from Nelken and Yamangil's insights. A first step would be to normalize data extraction from revision history, e.g. to classify any edits between a pair of revisions into categories like vandalism, spelling error corrections or reformulations before they are further processed. An automatic classification of edits facilitates the approaches building upon revision history data and might help to increase the precision of such systems. This holds not only for the approaches outlined by Nelken and Yamangil, but also for most of the applications presented in the remainder of this section. We roughly divided them into the domains of spelling error correction and paraphrasing.

#### 5.2.2.1 Spelling Error Correction

Zesch [43] extracts a corpus of real-word spelling errors (malapropisms), which can only be detected by evaluating the context they appear in. For example, in the sentence, "That is the very defect of the matter, sir" (from Shakespeare's "The Merchant of Venice"), *defect* is confused with *effect*. This type of error is generally not detected by conventional spelling correctors. In the past, training and/or test data for this type of error has mostly been created artificially, e.g. by automatically replacing words with similar words from a dictionary. Zesch's approach is an attempt to generate a corpus of naturally occurring malapropisms. Therefore, the author extracts pairs of sentences with minimal changes from consecutive Wikipedia revisions. To determine such sentence pairs, a restrictive filter is applied on each pair of phrases in adjacent revisions, ruling out pairs that are either equal or exceeding

```
<modif id="142" wp_before_rev_id="1842309" wp_after_rev_id="1842337"
    wp_comment="Statistiques">
  <before>
    Taux de croissance de la <m num_words="1">pop.</m>: 0,24% (en 2001)
  </before>
  <after>
    Taux de croissance de la <m num_words="1">population</m>: 0,24% (en 2001)
  </after>
</modif>
```

**Fig. 5.3** A slightly truncated entry from the WiCoPaCo, the coded edit in the </m>-tag is highlighted

a small threshold in their character length difference. The sentences are annotated with part-of-speech tags and lemmata. Further filters ensure that the sentences differ in just one token, which may not be a number or a case change. The edit distance between the old and the new token must be below a threshold. With regard to the semantic level, edits involving misspelled words (the edit must not be an error that can be detected using a conventional spelling corrector), stopwords, named entities (the new token may not be a named entity) and semantically motivated changes (direct semantic relations are determined using WordNet) are filtered out. The resulting dataset was manually corrected to remove cases of vandalism and examples that could not be ruled out by the above described filter mechanism because they did not provide enough context. It has been generated from five million English and German revisions and contains altogether 686 error pairs.[8] The author used his data to compare statistical and knowledge-based approaches for detecting real-word spelling errors. Through this analysis he shows that artificial datasets tend to overestimate the performance of statistical approaches while underestimating the results of knowledge-based ones. This way, he proves the usefulness of the corpus of naturally occurring real-word errors created from the Wikipedia revision history, because it offers a more realistic scenario for the task of evaluating real-word spelling error correction.

In another application working with spelling errors, Max and Wisniewski [18] present the Wikipedia Correction and Paraphrase Corpus (WiCoPaCo). Different to the aforementioned approach, they analyze various types of edits. Their data originates from article revisions in the French Wikipedia. Differences between modified paragraphs from adjacent revisions are determined using the longest common subsequence algorithm. Only edits with a maximum of seven changed words are kept. Edits which exclusively add or delete tokens are not considered. Further filters rule out edits changing more than a certain number of words, changes that only affect punctuation and bot edits. *Bots* are automatic scripts operating in Wikipedia to carry out repetitive tasks, mostly for maintenance. The remaining data is tokenized and markup is removed. The actual edit is aligned in the context of the paragraphs of $r_v$ (denoted by </before>) and $r_{v-1}$ (</after>), see Fig. 5.3 for an example. The resulting corpus consists of 408,816 edits (v2.0), coded in an XML

[8]Freely accessible at http://code.google.com/p/dkpro-spelling-asl/.

format as shown in Fig. 5.3. This format stores, together with the textual data, meta data such as the user comment (denoted by `wp_comment`). To build a corpus of spelling errors, the authors filter out 74,100 real-word errors and 72,493 non-word errors using a rule-based approach. Among all edits affecting only a single word, they apply two rules. First, a `hunspell` correction system detects for both $r_{v-1}$ and $r_v$ whether the modified word $w_v$ or $w_{v-1}$ is in the dictionary. This way, they find:

- Non-word corrections ($w_{v-1}$ is erroneous, $w_v$ is correct),
- Real-word errors and paraphrases (both $w_v$ and $w_{v-1}$ are correct)
- And proper noun or foreign word edits, spam and wrong error corrections ($w_v$ is erroneous).

The second rule distinguishes between real-word errors and paraphrases. Therefore, a maximum character edit distance of three between $w_v$ and $w_{v-1}$ is allowed for spelling corrections, assuming that most of the spelling error corrections change three or less characters. Additionally, for the non-word corrections, edits with a character edit distance greater than five are ruled out. The authors justify this step with the need to filter out spam. They published the resulting data as a freely available corpus, the WiCoPaCo.[9] To evaluate the spelling error subset of WiCoPaCo, the authors randomly split the data into training and test set. They create candidate sets based on both `hunspell` rules and error correction patterns from their corpus. The later comprises two lists:

- A list of words built by applying the most frequent error correction scripts (e.g. e → é) extracted from their corpus to misspelled words and
- A list with all corrections of misspelled words from the training set.

For evaluation, they count the number of candidate sets containing the correct word, using the training set to build the candidate sets. The results show that the combined approach improves over a system based solely on `hunspell.` Improvement is particularly high for real-word errors. This is in line with the findings by Zesch [43] who also pointed out the importance of naturally occurring real-word error datasets. However, Max and Wisniewski do not test their approach on different data than the WiCoPaCo corpus.

After manual inspection of the WiCoPaCo data, Max and Wisniewski also developed a classification system to categorize edits in Wikipedia. Their system separates changes which preserve the meaning from those that alter the meaning. The former are further divided into edits modifying the spelling (such as spelling errors) and edits modifying the wording (e.g. paraphrases). Edits altering the meaning are divided into spam and valid meaning changes (such as simplifications). Based on the paraphrases in their corpus, the authors analyzed the probabilities of transformations of POS sequences (e.g. DET ADJ NOM → DET NOM). As a possible application, they propose employing these probabilities to assess the grammaticality of paraphrases when several candidates exist. The quantitative analysis and classification of paraphrases in WiCoPaCo is subject to future work.

---

[9] See http://wicopaco.limsi.fr/.

#### 5.2.2.2 Paraphrasing

Yatskar et al. [40] present an unsupervised method to extract lexical simplifications in the Simple English Wikipedia. They do not aim at simplifying entire sentences, but words or expressions, e.g. when "annually" is replaced by the simpler version "every year". In order to obtain a training corpus, they extract sentence pairs from adjacent revisions. Alignment of sentences is carried out based on the cosine similarity measure utilizing TF-IDF scores [22]. To calculate the latter, sentences are treated as documents and adjacent revisions as the document collection. From aligned sentences, the longest differing segments are calculated (*edits*) and changes longer than five words are filtered out. The authors introduce two different approaches to extract simplifications.

In the first approach (*edit model*), probabilities for edits to be simplifications derive from a model of different edits that are performed in the Simple and the Complex English Wikipedia. Based on edits in the Simple Wikipedia, the probability for an edit to be a simplification is calculated. On the opposite, the Complex English Wikipedia is used to filter out non-simplifications. To do so, the authors make the simplifying assumption that all edits in the Complex Wikipedia correspond to what they call "fixes", i.e. spam removal or corrections of grammar or factual content. Furthermore, they assume that vandalism does not exist and that the probability of a fix operation in the Simple Wikipedia is proportional to the probability of a fix operation in the Complex English Wikipedia. In their second approach (*meta data model*), Yatskar et al. use revision meta data to detect simplifications, namely the revision comments. They inspect all revisions containing the string "simpl" in their comment. Among the detected revisions, all possible edits are ranked by an association metric (Pointwise Mutual Information).

In a preliminary evaluation, the top 100 sentence pairs from each approach and a random selection from a user-generated list[10] have manually been annotated by native and non-native speakers of English as being a simplification or not. The inter-annotator agreement among the three annotators is sufficient with $\kappa = 0.69$. The authors used baselines returning the most frequent edits and random edits from the Simple English Wikipedia; both yielded a precision of 0.17. For the meta data method, a precision of 0.66 is reported, the edit model approach achieved a precision of 0.77, whereas the user-generated list had the highest precision with 0.86. With regard to recall, the authors report that the edit model generated 1,079 pairs and the meta data model 2,970 pairs, of which 62 and 71 % respectively, were not included in the user-generated list. The annotated datasets have been published and are freely available.[11]

In a similar approach, Woodsend and Lapata [38] additionally use syntactic information for a data-driven model of sentence simplification. Like Yatskar et al.,

[10]The Simple Wikipedia author `Specerk` offers a list of transformation pairs: http://simple.wikipedia.org/w/index.php?title=User:Spencerk/list_of_straight-up_substitutables.

[11]See http://www.cs.cornell.edu/home/llee/data/simple/.

they obtain their training data from the Simple and the Complex English Wikipedia. Two methods to create parallel corpora of simple and complex sentences are applied: first, the authors align sentences from Simple and Complex English Wikipedia articles (article corpus), and second, they align sentences from adjacent revisions in the Simple Wikipedia (revision corpus). In both corpora, the markup is removed. In the article corpus, the authors align parallel articles via the interwiki (language) links between the Simple and the Complex Wikipedia. Sentence alignment in parallel articles is established using TF-IDF scores to measure sentence similarity [22]. In the revision corpus, they select suitable revisions according to comment keywords, e.g. "simple", "clarification" or "grammar". Appropriate revisions $r_v$ are compared to $r_{v-1}$ followed by calculating the diff to find modified sections. Within those, the corresponding sentences are aligned via a word-based diff, resulting in 14,831 paired sentences. The aligned sentences are syntactically parsed. The parsed sentence pairs are used to train a Quasi-synchronous grammar (QG, similar to the content-based method of Zanzotti and Pennacchiotti [41], cf. below). Given a syntax tree $T_1$, the QG generates monolingual translations $T_2$ of this tree. Nodes in $T_2$ are aligned to one or more nodes in $T_1$. Alignment between direct parent nodes takes place when more than one child node (lexical nodes, i.e. words) are aligned. This way, a set of lexical and syntactic simplification rules as well as sentence splitting rules are generated, yielding transformations such as the following, which splits a sentence:

> John Smith walked his dog and afterwards met Mary. →
> John Smith walked his dog. He met Mary later.

Woodsend and Lapata solve the problem of finding the optimal QG transformations to simplify source sentences with an integer linear programming approach. In short, they use an objective function which guides the transformation towards a simpler language of the output, e.g. a lower number of syllables per word or of words per sentence. The authors evaluate their approach based on human judgments and readability measures. Human judgments include an evaluation of the output sentence with respect to the readability (whether is was easier to read than the input sentence), the grammaticality and the preservation of the meaning. The models are tested on the dataset used in Zhu et al. [45], who also align sentences from the Simple and the Complex English Wikipedia. With regard to the calculated readability measures, both the model trained on the revision corpus and the model trained on the article corpus do not outperform a baseline relying on the user-generated list previously used by Yatskar et al. [40] (cf. footnote 10). Considering the human judgments, the model trained on the revision corpus outperforms the article corpus model in all of the evaluation aspects. This result supports our assumption that the incorporation of revision history data not only helps to increase the amount of training data but also improves the performance of certain NLP applications.

Zanzotti and Pennacchiotti [41] apply semi-supervised machine learning for the task of Recognizing Textual Entailment (RTE) pairs from the Wikipedia revision history. They describe four essential properties of a textual entailment dataset and why data coming from Wikipedia's revision history is appropriate for this, i.e. the data is

- *Not artificial*, as it is extracted from authentic Wikipedia texts
- *Balanced*, i.e. equal in number of positive entailment pairs (when new information is added to the old content or old content is paraphrased) and negative entailment pairs (when the new information contradicts the old content or the entailment is reverse); this is roughly the case for their data as shown in the following
- *Not biased with respect to lexical overlap*, i.e. the lexical overlap of positive and negative entailment pairs should be balanced, this is mostly true for the Wikipedia revision data, as usually only a few words are changed both for positive and for negative entailment pairs
- *Homogeneous to existing RTE corpora* with respect to the entailment pairs contained in these corpora, this is roughly the case for their data as shown in the following.

Their approach to separate positive lexical entailment candidates from negative ones is based on co-training. Co-training is designed to learn from labeled data $L$ and unlabeled data $U$ and has to access the corpus in two different and independent views. Two different classifiers, each of them working with features from one of the two views, are trained on copies of $L$, defined as $L_1$ and $L_2$. These classifiers are used to classify data from $U$, resulting in different classifications $U_1$ and $U_2$. Finally, the best-classified examples in $U_1$ are added to $L_2$, resp. $U_2$ to $L_1$. This procedure is iteratively repeated until a stopping condition is met. As for the two views, the authors suggest a *content-based view* (features based on the textual difference of two revisions) and a *comment-based view* (features based on the comment of $r_v$). The features in the content-based view rely on syntactic transformations. A feature will be activated, if the syntactic transformation rule associated with that feature unifies with the syntax tree representations of a pair of sentences. The authors do not specify in detail how these pairs are generated. In the comment view, features are based on a bag-of-words model. The latter is calculated from the comment words, which have been filtered with respect to the stop words.

For the evaluation of their approach, Zanzotto and Pennacchiotti randomly selected 3,000 instances of positive and negative entailment pairs from 40,000 English Wikipedia pages (*wiki_unlabeled* dataset). Additionally, they manually annotate 2,000 entailment pairs. The inter-annotator agreement on a smaller development corpus of 200 examples is $\kappa = 0.60$. After removing vandalism and spelling corrections they obtained 945 positive and 669 negative entailment pairs (*wiki* dataset). The datasets are freely available.[12] The authors compared the *wiki* dataset to other corpora from RTE Challenges, namely the datasets from the RTE-1, RTE-2 and RTE-3 challenges [10]. To evaluate the quality of the *wiki* dataset, they split it into equally sized development, training and test set. The classification of

[12]See http://art.uniroma2.it/zanzotto/.

**Table 5.2** Trustworthiness and article quality assessment approaches based on the Wikipedia revision history

| Reference | Type of revision features | Criteria for evaluation | Language |
|---|---|---|---|
| [42] | Author reputation, edit type | Featured, cleanup, other | English |
| [5] | Author score, edit size | Featured, articles for deletion | Italian |
| [37] | Quantitative surface features | Featured, non-featured | English |
| [33] | Semantic convergence | Good, non-good | English |
| [11] | Revision cycle patterns | Featured, good, B-, C-class, start, stub | English |

positive and negative entailment pairs is carried out by a Support Vector Machine trained on the features from the content-based view. The authors report an accuracy of 0.71 for that approach when applied to the *wiki* data, compared to 0.61 for the RTE-2 dataset. Combining *wiki* data with the RTE challenge datasets for training did not show significant decrease or increase of accuracy. Therefore, the authors conclude that the *wiki* dataset is homogeneous to the RTE datasets. To evaluate the co-training approach, they use RTE-2 as labeled set and *wiki_unlabeled* as unlabeled set. RTE-2 does not allow for the comment-based view. Hence, the comment-view classifier is not activated until the first training examples are added from the content-based classifier. Performance is reported to become stable after several iterations with approximately 40 unlabeled examples and accuracy around 0.61. The authors conclude that their semi-supervised approach successfully serves to expand existing RTE datasets with data extracted from Wikipedia.

Having discussed example NLP applications based on the Wikipedia revision history data, we now focus on how revision information can be used to assess article quality.

### 5.2.3 Article Trustworthiness, Quality and Evolution

The revision history of a page in Wikipedia contains information about how, when and by whom an article has been edited. This property has been used to automatically assess the quality of an article. In this context, quality in Wikipedia is related to trustworthiness. However, the trustworthiness of an article and its quality are not necessarily the same. Rather, trustworthiness can be seen as a means of successfully communicating text quality to users. In the context of Wikipedia, trustworthiness is often related with the skills and expert knowledge of the author of a revision, whereas quality is rather measured in terms of the textual content itself. Certainly, this distinction is not always made, and different studies use the terms differently and sometimes interchangeable. In the following, we present several studies that make use of the Wikipedia revision history to analyze article quality and trustworthiness. An overview of the approaches can be found in Table 5.2.

Zeng et al. [42] were one of the first to develop and evaluate a model of article trustworthiness based on revision histories. Their model is based on author reputation, edit type features and the trustworthiness of the previous revision. As for the edit type features, the number of deleted and/or inserted words is measured. The reputation of authors is approximated by their editing privileges. Certain actions in Wikipedia, e.g. blocking other users, can be carried out only by privileged users. Furthermore, registered authors can be distinguished from unregistered users and blocked users. The authors apply a Dynamic Bayesian network depending on these features to estimate the trustworthiness of a revision based on a sequence of previous states, i.e. revisions. To account for uncertainty in the trustworthiness of authors and in the edit type features, beta probability distributions for the trustworthiness values of the network are assumed. The trustworthiness of $r_v$ is equal to the trustworthiness of $r_{v-1}$ plus the inserted trustworthy content minus the deleted trustworthy content, i.e. incorrectly removed portions of text. The amount of trustworthy and untrustworthy content is determined by an author's reputation. To evaluate the model, the authors built a corpus of internally reviewed articles, altogether containing 40,450 revisions. Wikipedia has an internal review system which labels articles that meet certain predefined quality[13] criteria, e.g. they should be comprehensive, contain images where appropriate, etc. The highest rating of an article is *featured*. However, distinguished articles not yet fulfilling all criteria to be featured can be also labeled as *good*. On the contrary, articles tagged for *cleanup*, do not meet the necessary quality standards as defined in the Wikipedia Manual of Style.[14] To evaluate their model, Zeng et al. calculate an article's mean trust distribution, an indicator of the trustworthiness of its latest revision, based on the above Bayesian network. They find that featured articles have the highest average of mean trust distributions, while cleanup articles show the lowest values. The authors carry out a manual inspection of changes in average trustworthiness values throughout the history of an article, showing that these changes correspond to major edit types like insertions or deletions of large quantities of text. The model is thus able to reproduce a realistic picture of the trustworthiness of articles based on their revision history.

Cusinato et al. [5] have a similar view on article quality in Wikipedia, as proposed in their system called QuWi. The system is based on an approach originally developed for quality assessment in peer reviewed scholarly publishing introduced by Mizzaro [21]. This model assigns quality and steadiness scores to articles, authors and readers. Scores for each of them are updated when a reader judges a paper, based on the following assumptions: the scores of authors are bound to and updated with the judgment scores of their articles, weighted with the article's steadiness. The weight of an article judgment depends on the score of the reader rating it. Readers' scores are bound to and updated with the appropriateness of their judgments, based on their agreement with the average rating of the articles they judged. Steadiness for articles, authors and readers increases with every

---

[13] http://en.wikipedia.org/wiki/WP:FA_Criteria

[14] http://en.wikipedia.org/wiki/WP:Manual_of_Style

corresponding judgment made. To adapt this model to Wikipedia, Cusinato et al. use the following adjustments:

(a) Articles have more than one author, hence, judgments have to be based on single contributions, i.e. edits between adjacent revisions.
(b) Edits cannot be rated directly, hence, the readers' judgment on an edit is measured implicitly by analyzing the next edit on the article, i.e. the next revision. In other words, the author of $r_v$ automatically becomes the reader of $r_{v-1}$.

Modifications express negative votes, while unmodified content is considered to be positive. The score of a contribution is calculated based on the ratio between modified and unmodified text (i.e. the reader's judgment), weighted by the score of the reader. Based on contribution scores, author and reader scores are calculated as explained above, derived from the approach by Mizzaro [21]. Finally, article scores are assigned based on the scores of the words contained in the article, weighted by the word length. Word scores are calculated based on the author's and (previous) readers' scores, each of them averaged by their steadiness scores.

The authors tested their system on 19,917 articles from the Science category of the June 2007 snapshot from the Italian Wikipedia. They ran the score calculation on the entire set of revisions and recorded article scores at six equally distributed timestamps, including the latest ones. As expected, average article scores increase over time. Featured articles and articles proposed for deletion were used to evaluate the calculated scores. The average score (ranging between 0 and 1) of the 19 featured articles contained in their corpus is 0.88, significantly higher than the total average 0.42, whereas 75 articles for deletion have an average score of 0.27. This work demonstrates an interesting way to apply Wikipedia revision information to an existing model of quality assessment and has been shown to work successfully on a small part of the Italian Wikipedia revision history. The approach could further be improved by accounting for bot edits and vandalism.

Wilkinson and Huberman [37] analyze correlations between quantitative revision features and article quality. Based on the number of revisions made in all Wikipedia articles they develop a model of article growth. In this model, the number of revisions in a given timeframe is on average proportional to the number of previous changes (i.e. revisions) made to the article. As a result, older articles are edited more often, or, as the authors put it: "edits beget edits". They verify their assumption with an empirical analysis over all page revisions in the English Wikipedia between January 2001 and November 2006, except for redirect and disambiguation pages and revisions by bots. Furthermore, they explore correlations between article quality and editing. Therefore, the authors analyze age- and topic-normalized featured and non-featured articles according to their number of revisions and their number of distinct authors. Their findings show that featured articles have a statistically significant higher number of revisions and distinct authors when compared to non-featured articles. To account for the cooperation among authors, they do the same for Talk pages (see Sect. 5.3), resulting in an even more significant difference between featured and non-featured articles. The authors conclude that high-quality articles in Wikipedia can be distinguished from other articles by the larger numbers of article edits, Talk pages edits and distinct authors.

#### 5.2.3.1 Analysis of Article Lifecycles

Using the revision count as a proxy for article quality seems to yield interesting results. However, it must be considered that featured articles in Wikipedia receive special attention because of their status. Therefore, the following approaches go further and explicitly treat articles as constructs going through different phases or stages of maturity, i.e. they study the evolution. The revision history is the only source of information for this purpose.

Thomas and Sheth [33] introduce a notion of article stability, which they call Semantic Convergence. They assume an article to be mature (i.e. trustworthy), when it is semantically stable. Semantic stability is defined in terms of semantic distance in a TF-IDF vector space representation of revision milestones. The TF-IDF space is calculated over all words occurring in an article's entire revision history. A revision milestone is defined as a combination of all revisions in 1 week, word counts for milestones are calculated as medians. This way, the authors aim to balance different editing frequencies for individual articles. They test their hypothesis on a dataset of 1,393 articles labeled as good and 968 non-labeled articles with a revision history consisting of at least 50 revisions. For evaluation, they measure the pairwise cosine distance between adjacent revision milestones and the distance between every revision milestone and the final revision. They show that articles generally move towards a stable state, i.e. that the semantic distance between revision milestones drops with time. When it comes to predicting the maturity for a single article at a given point of time, their measure does not prove to be reliable. However, knowledge about the past of an article helps to detect its present state, because articles which have already undergone stable revision milestones are less likely to change. Good and non-good articles did not show a significant difference in terms of their stability. Hence, if an article is labeled as good, it does not necessarily mean that its content is stable.

Han et al. [11] use a Hidden Markov Model to analyze the history of a Wikipedia article as a sequence of states. They define a number of states an article usually passes before reaching a convergence state. The states in their Markov model are as follows: building structure, contributing text, discussing text, contributing structure and text, discussing structure and text/content agreement. The observation variables used to determine the Markov states are calculated between a pair of consecutive revisions and are divided into:

- Update type (insertion, deletion, modification),
- Content type (structure, content, format) and
- Granularity type (extent of the edit).

Sequences or series of sequences of states are combined to form so called Revision Cycle Patterns. The authors aim to find correlations between human evaluated quality classes and revision cycle patterns to automatically assess the quality of an article. Therefore, they test their model on a corpus containing articles which have been labeled according to the Wikipedia internal quality grading scheme[15] as either featured, A-class, good, B- and C-class as well as start and stub-class. They create a model based on the following steps. First, Revision Cycle Patterns for each quality class in the corpus are extracted. Recurring sequences of states are detected via frequent items mining. Second, these are clustered to discover the dominant patterns and third, clusters of cycle patterns are related with quality labels. With this method, the percentage of correctly classified articles is between 0.98 for featured articles and 0.85 for the stub class. The authors report that their approach outperforms the results in Dalip et al. [6], who work on the same task and data, but without using features based on revision history data. Thus, the revision history based features turn out to be helpful for this task.

A combination of the revision-related features with language features regarding style, structure or readability as presented in [6] is an emerging topic. To the best of our knowledge, no effort has yet been made to incorporate all of the available revision-based information with plain text language features to assess article quality. This indicates a promising direction for future work on quality assessment and trustworthiness. Furthermore, as already mentioned, a clear definition of quality and trustworthiness in Wikipedia has not been established yet. The above outlined studies all have slightly different concepts of quality and trustworthiness. The evaluation methods are almost exclusively based on the human assigned Wikipedia-internal quality labels as explained above. This is a shortcoming, as the criteria for these ratings can change over time, and the quality assessment process may not be reproducible for external raters. A broader analysis of article quality which goes beyond user-assigned labels, together with a comprehensive definition of text quality, is thus required.

### *5.2.4 Vandalism Detection*

This subsection explains the usage of revision history data to detect spam or vandalistic edits in Wikipedia. Vandalism is a major problem in Wikipedia, since anybody can edit most of its content. About 6 to 7 % of all revisions in the English Wikipedia are estimated to be vandalized [3,26]. In short, vandalism or spam is "any addition, removal, or change of content in a deliberate attempt to compromise the integrity of Wikipedia".[16] Vandalistic additions, removals or changes to an article can only be detected using revision history data, because at least two revisions need to be compared: a trustworthy, not vandalized revision $r_{v-1}$ and a possibly

[15] WikiProject article quality grading scheme: http://en.wikipedia.org/wiki/Wikipedia:Version_1.0_Editorial_Team/Assessment.

[16] From http://en.wikipedia.org/w/index.php?title=Wikipedia:Vandalism&oldid=489137966. The same page also offers a list of frequent types of vandalism.

**Table 5.3** Vandalism detection approaches and the features they use. For each of them, the best classification results on two corpora are given. Please note that these numbers are not entirely comparable due to the use of different classifiers (given in brackets) and different training and test sets

| Reference | Basic feature types[a] | WEBIS-VC07 | PAN-WVC 10 |
|---|---|---|---|
| [28] | T, L, M, R | 0.85 $F_1$ (Log. Regr.) | – |
| [4] | Statistical language model | 0.90 $F_1$ (J48 Boost) | – |
| [36] | T, L/S, M, R | 0.95 $F_1$ (Log. Regr. Boost) | 0.85 $F_1$ (Log. Regr. Boost) |
| [1] | T, L, M, R | – | 0.98 AUC (Rand. Forest) |
| [12] | T, L, M, R | – | 0.97 AUC (Rand. Forest) |

[a] *T* Textual, *L* Language, *S* Syntax, *M* Meta data, *R* Reputation

vandalized revision $r_v$. Malicious edits are supposed to be reverted as quickly as possible by other users, which in practice seems to work quite well. Different median survival times for vandalized revisions are quoted, ranging from less than 3 min [34] to 11.3 min [13], depending on the type of vandalism.

Wikipedia has a revert system which serves to undo unwanted edits (cf. Fig. 5.2 *d*) and particularly, vandalistic edits. A small number of automatic bots watch changes and revert obvious vandalism. At the time of writing, `ClueBot NG` was the main anti-vandalism bot in the English Wikipedia.[17] In contrast to many of the previous rule-based anti-vandalism bots, its detection algorithm is based on machine learning techniques.

The International Competition on Wikipedia Vandalism Detection is a good starting point for work on vandalism detection in Wikipedia. It evaluates vandalism detection based on the PAN Wikipedia vandalism corpus (WVC) 10 and 11.[18] Each of these corpora contains around 30,000 edits of Wikipedia articles labeled as *regular* or *vandalism*.

State-of-the-art approaches formulate Wikipedia vandalism detection as a machine learning task. Malicious edits have to be separated from regular ones based on different features. For an overview of the approaches, see Table 5.3. Vandalism detection approaches can be classified according to their adoption of features. We categorize the features as proposed in Adler et al. [1]:

- Textual Features: language independent
- Language Features: language specific
- Meta Data Features: author, comment and timestamp
- Reputation Features: author and article reputation

[17] Cf. a list of Anti-vandalism bots compiled by the author `Emijrp`: http://en.wikipedia.org/w/index.php?title=User:Emijrp/Anti-vandalism_bot_census&oldid=482285684.

[18] See http://www.webis.de/research/corpora/pan-wvc-10 and http://www.uni-weimar.de/cms/medien/webis/research/corpora/pan-wvc-11.html.

**Table 5.4** Examples of features for vandalism detection as used in [1]

| Textual | Language | Meta data | Reputation |
|---|---|---|---|
| Ratio digits to characters | Freq. of vulgarisms | Comment length | Author behavior hist. |
| Ratio upper/lowercase characters | Freq. of pronouns | Author is registered | Author geogr. region |
| Length of longest token | Freq. of typos | Time of day | Reputation for article |

Although most of the presented studies use this kind of distinction between different types of features, there is no absolute agreement on how to categorize them. Likewise, some works might include an *author-is-registered* feature to meta data, while others consider such a feature as author reputation. Textual features involve language-independent characteristics of an edit, such as the number of changed characters, upper- to lowercase ratio and similar. To analyze (natural) language features, language-related knowledge is required. This knowledge comes from language-specific word lists or dictionaries (of swearwords etc.), and/or further processing with NLP tools (POS-tagger, parser etc.). Meta data features refer to information coming from a revision's meta data, like the comment or the time of its creation. Reputation features need detailed information about the author or edited article, e.g. the number of an author's past edits or the article's topical category. Examples of such features can be found in Table 5.4.

Potthast et al. [28] are among the first to present several important features for vandalism detection. They used the so called WEBIS-VC07 corpus which contains 940 pairs of adjacent revisions. Among them, the authors manually labeled 301 vandalistic revisions. Their features are mainly based on textual differences between revisions, including char- and word-based ones as well as dictionaries to detect vulgar words. They use some features based on meta-data. Among these, the *edits-per-user* feature shows the highest recall. The authors cross-validate a classifier based on Logistic Regression on their dataset, comparing the classifier's result to the baseline performance of rule-based anti-vandalism bots. They report 0.83 precision at 0.77 recall using a classifier based on logistic regression, outperforming the baseline primarily with respect to the recall. It should be noted, however, that anti-vandalism bots are designed to have high precision aiming to avoid reverting false positives, i.e. revisions that have not been vandalized. Insofar, vandalism detection approaches might focus on high precision rather than high recall.

Chin et al. [4] do not use any meta information in their vandalism classification approach. Their features come from a statistical language model, which assigns probabilities to the occurrence of word sequences. To account for repeated vandalism, they not only compare a revision to the immediately preceding one, but also to other preceding ones. Their test corpus consists of the entire revision history for two large and often vandalized articles. Instead of labeling huge amounts of data by hand, they apply an active learning approach, where classifiers are built iteratively, starting with the labeled data from the WEBIS-VC07 corpus [28]. After each iteration, only the 50 most probable vandalism edits are manually labeled. An evaluation using a Boosting classifier with J48 Decision Trees based on their features resulted in 0.90 $F_1$ on this corpus. The increase over the approach in Potthast et al. [28] is primarily due to a higher precision. Additionally, Chin et al. classify types of vandalism and edits in general. Consequently, they not only label the presence or absence of vandalism but also its type, e.g. Graffiti (inserting irrelevant or profane text), Large-scale Editing (inserting or replacing text with a large amount of malicious text) or Misinformation (replacing existing text with false information). Their analysis shows that Graffiti accounts for more than half of all vandalistic edits. The authors used a Logistic Regression classifier and Support Vector Machines as models for their active learning approach. After three to four iterations, the Logistic Regression classifier yielded best results with 0.81 average precision. An error analysis shows that the wiki markup and unknown words (e.g. template names) cause the language model to fail. The model considers unknown strings as out-of-vocabulary and hence assigns them a high probability of being vandalism. Furthermore, separating reverts from vandalized revisions turns out to be difficult. This could be addressed by including meta data like the comment into their features.

Wang et al. [36] focus on syntactic and semantic features in their approach of vandalism classification. They distinguish lexically ill-formed, syntactically ill-formed and ill-intentioned types of vandalism, aiming to account for cases of vandalistic edits or spam which are harder to recognize. This happens when authors try to hide bad intentions by inserting well-formed off-topic comments, biased opinions or (implicit) advertisement. To detect this type of edits, a classifier needs to have access to a wider range of language features. Therefore, the authors apply what they call shallow syntactic and semantic modeling. Their features are based on the title of an article and the diff text between adjacent revisions. Since this data contains sparse information for training a classifier, they use it as a query to various web search engines to build article-specific models. The top-ranked retrieved results are labeled with POS-tags. The authors use n-grams of POS-tags alone as well as n-grams of POS-tags and words to model a syntactic and a semantic representation of each revision. Their classifier also uses meta data (the comment) and reputation (number of revisions per author) features along with lexical ones (e.g. vulgarism and web slang). The experiments are conducted with the WEBIS-VC07 and the PAN-WVC 10, both split into training and test set. Classification is done by a Logistic Model Trees classifier and a Logistic Regression classifier with Boosting. With the WEBIS-VC07 corpus and the Logistic Regression Boosting classifier using all of their features they report the highest $F_1$ of 0.95. This is an improvement of around 15 % compared to the results in [28]. On the PAN-WVC 10 dataset, they obtain a maximum $F_1$-score of 0.85 with almost equal recall and precision using the Logistic Regression Boosting classifier. When compared to a system based on meta data, reputation and lexical features solely, the shallow syntactic and semantic features introduced an improvement of around 3 %.

Adler et al. [1] present a combination of previous approaches, resulting in a system based on meta data, textual and natural language features. Table 5.4 lists several of the features they use. Together with new results based on a meta-classifier which combines previously applied features, they provide an overview of existing work on vandalism detection and an extensive list of classification features. Furthermore, they distinguish between immediate and historic vandalism. The latter refers to vandalized revisions in an article's revision history. Some of their features can only be applied to historic vandalism, as they refer to subsequent revisions. The experiments are performed on the PAN-WVC 10 corpus. A ten-fold cross-validation with a Random Forest classifier shows that the usage of features for historic vandalism increases performance from 0.969 to 0.976 AUC (area under the ROC curve). This is due to features like *next-comment-indicates-revert* and *time-until-next-edit*. As a possible reason for the efficiency of the latter feature, the authors state that frequently edited pages are more likely to be vandalized, without explicitly giving a source for that assumption. They conclude that the gain from using language features should not be overestimated, based on the fact that the calculation of language-specific features is generally more time-consuming than the calculation of textual, meta data or reputation features.

Javanmardi et al. [12] classify their features in a similar way, i.e. into meta data, textual and language model features and present their detailed descriptions. The last category is based on the Kullback-Leibler distance between two unigram language models. As in [1], they use the PAN-WVC 10 corpus in their experiments and divide it into training and test data. A Random Forest classifier performed best, yielding a maximum AUC value of 0.955 on the test data, and 0.974 with a three-fold cross-validation on the training set. Javanmardi et al. experimented with the same corpus as [1]. However, their results are not comparable, as [1] employed cross-validation and Javanmardi et al. did not. The evaluation of different groups of features shows that their language model and meta data features are less significant than textual and reputation features. This partly contradicts the findings of Adler et al. [1], who find reputation features to be less important than other meta data features such as time or comment length. Javanmardi et al. explain this with differences in their definition of features. Furthermore, the classification approaches are different: Adler et al. use a meta-classifier combining different classifiers that have been developed for certain features, whereas Javanmardi et al. use one classifier trained and tested with different groups of features. To identify redundant individual features, the authors use a Logistic Regression approach. Among the most important individual features, they identify *Insert-Special-Words* (insertions of vulgarism, spam, sex etc. words) and a feature related to the author reputation.

We conclude that absolute agreement on the importance of different features for vandalism detection does not exist. In any case, vandalism detection depends on methods and algorithms developed to calculate the difference between adjacent revisions. Textual and language features use this kind of information, and revision meta data information has also proved to be helpful. Author reputation is in some cases bound to an edit history for authors, which also demands for a precalculation based on the revision histories. As already pointed out earlier in our survey, structured and systematic access to the type of edits performed in each revision, could also help the generation of features for vandalism detection. We consider this as a reference for future research.

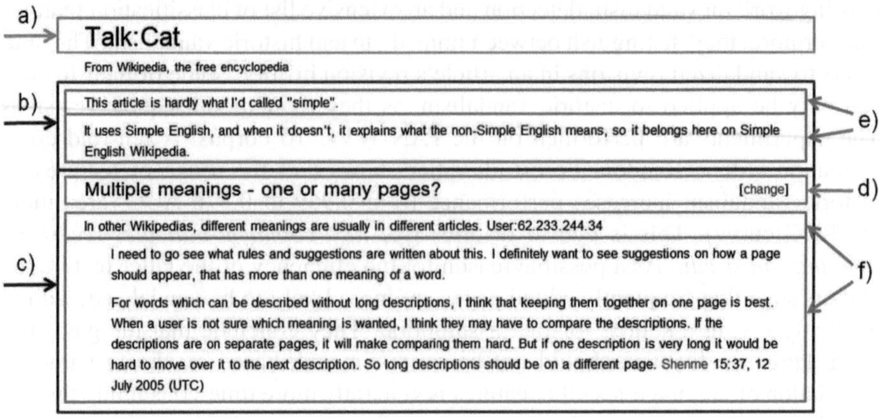


**Fig. 5.4** Structure of a Talk page: (*a*) Talk page title, (*b*) untitled discussion topic, (*c*) titled discussion topic, (*d*) topic title, (*e*) unsigned turns, (*f*) signed turns

## 5.3 Discussions in Wikipedia

So far, we have shown that the revision history of Wikipedia is a valuable resource for many NLP applications. By regarding the whole evolution of an article rather than just its latest version, it is possible to leverage the dynamic properties of Wikipedia. The article revision history reflects a product-centered view of the collaborative writing process. In order to fully understand collaborative writing and, in turn, collaboratively constructed resources, is is necessary to include the writing process itself into the equation.

In joint writing, authors have to externalize processes that are not otherwise made explicit, like the planning and the organization of the text. Traditionally, these processes could only be observed indirectly by conducting interviews with the authors. In Wikipedia, however, coordination and planning efforts can be observed on the article Talk pages on which the Wikipedians discuss the further development of the articles (see Fig. 5.4). These discussion spaces are a unique resource for analyzing the processes involved in collaborative writing.

Technically speaking, a Talk page is a normal wiki page located in one of the Talk namespaces (see Table 5.5). Similar to a web forum, they are divided into discussions (or topics) and contributions (or turns). What distinguishes wiki discussions from a regular web forum, however, is the lack of a fixed, rigid thread structure. There are no dedicated formatting devices for structuring the Talk pages besides the regular wiki markup.

Each Talk page is implicitly connected to an article by its page name—e.g., the Talk page Talk:Germany corresponds to the article Germany. It is, however, not possible to establish explicit connections between individual discussions on the page and the section of the article that is being discussed. Each namespace in Wikipedia

**Table 5.5** Wikipedia namespaces and functional Talk page classes

| Basic namespaces | Talk namespaces | Functional class |
|---|---|---|
| Main | Talk | Article |
| User | User talk | User |
| Wikipedia | Wikipedia talk | Meta |
| MediaWiki | MediaWiki talk | Meta |
| Help | Help talk | Meta |
| File | File talk | Item |
| Template | Template talk | Item |
| Category | Category talk | Item |
| Portal | Portal talk | Item |
| Book | Book talk | Item |

has a corresponding Talk namespace resulting in a total of ten different types of Talk pages (Table 5.5) which can be categorized into four functional classes:

- **Article Talk pages** are mainly used for the coordination and planning of articles.
- **User Talk pages** are used as the main communication channel and social networking platform for the Wikipedians.
- **Meta Talk pages** serve as a platform for policy making and technical support.
- **Item-specific Talk pages** are dedicated to the discussion of individual media items (e.g., pictures) or structural devices (e.g., categories and templates).

The users are asked to structure their contributions using paragraphs and indentation. One *turn* may consist of one or more paragraphs, but no paragraph may span over several turns. Turns that reply to another contribution are supposed to be indented to simulate a thread structure. We call this *soft threading* as opposed to *explicit threading* in web forums.

Users are furthermore encouraged to append signatures to their contributions to indicate the end of a turn (see Fig. 5.5). There are extensive policies[19] that govern the usage and format of signatures. They usually should contain the username of the author and the time and date of the contribution. However, users' signatures do not adhere to a uniform format, which makes reliable parsing of user signatures a complex task. Moreover, less than 70 % of all users explicitly sign their posts [35]. In some cases, depending on the setup of an individual Talk page, automatic scripts—so-called "bots"—take over whenever an unsigned comment is posted to a Talk page and add the missing signature. While this is helpful for signature-based discourse segmentation, it is misleading when it comes to author identification (see Fig. 5.5, signature 5.6).

Due to the lack of discussion-specific markup, contribution boundaries are not always clear-cut. They may even change over time, for instance if users insert

[19]http://en.wikipedia.org/wiki/WP:SIGNATURE

The Rambling Man (talk) 18:20, 27 February 2012 (UTC) (5.1)

–66.53.136.85 21:41, 2004 Aug 3 (UTC) (5.2)

– Taku (5.3)

– Preceding unsigned comment added by 121.54.2.122 (talk) 05:33, 10 February 2012 (UTC) (5.4)

–SineBot (talk) 08:43, 31 August 2009 (UTC) (5.5)

Imzadi 1979 > 09:20, 20 May 2011 (UTC) (5.6)

– ♫Greatorangepumpkin♫ T 14:14, 17 December 2010 (UTC) (5.7)

**Fig. 5.5** Examples for user signatures on Talk pages: (5.1) Standard signature with username, link to user Talk page and timestamp (5.2) Signature of an anonymous user (5.3) Simple signature without timestamp (5.4,5.5) Bot-generated signatures (5.6,5.7) Signatures using colors and special unicode characters as design elements

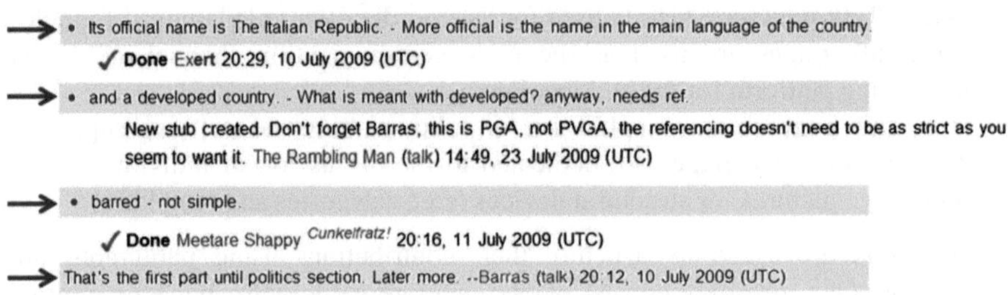


**Fig. 5.6** Inserted comments within user turn

their own comments into existing contributions of other users, which results in non-linear discussions (see Fig. 5.6). This makes automatic processing of Talk pages a challenging task and demands a substantial amount of preprocessing.

There are ongoing attempts to improve the usability of the discussion spaces with extensions for explicit threading[20] and visual editing.[21] However, these enhancements have been tested in only selected small Wikimedia projects and have not yet been deployed to the larger wikis.

In order to prevent individual Talk pages from becoming too long and disorganized, individual discussions can be moved to a discussion archive.[22] Discussion archives are marked with an "Archive" suffix and usually numbered consecutively. The oldest discussion archive page for the article "Germany", for example, is named Talk:Germany/Archive_1. There are two possible procedures for archiving a Talk

[20]http://www.mediawiki.org/wiki/Extension:LiquidThreads

[21]http://www.mediawiki.org/wiki/Visual_editor

[22]http://en.wikipedia.org/wiki/WP:ARCHIVE

page: the *cut-and-paste procedure* and the *move procedure*. While it is not possible to determine directly which method has been used to create an archive, the choice has important implications for page processing. The cut-and-paste procedure copies the text from an existing Talk page to a newly created archive page. All revisions of this Talk page remain in the revision history of the original page. The move procedure renames (i.e., moves) an existing Talk page and adds the numbered archive suffix to its page title. Afterwards, a new Talk page is created that is then used as the new active Talk space. Archives created with the latter procedure maintain their own revision history, which simplifies the revision-based processing of these pages.

Even though there is no discussion-specific markup to structure Talk pages, so-called *templates* can be used to better organize the discussions. In their simplest form, templates are small wiki pages that can be embedded in another page using a shortcut. These templates are commonly used to embed info banners and predefined messages into the lead section of articles and Talk pages or to mark important decisions in a discussion. For example, the template

{{*consensus*|*It was decided to remove the section*}}

is replaced with

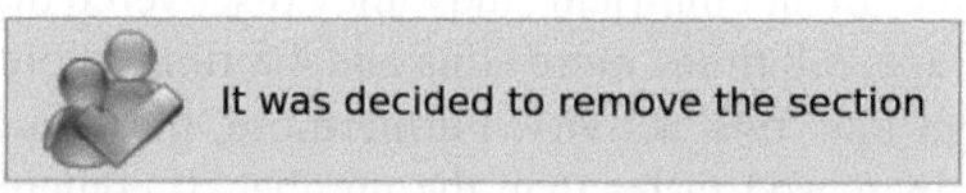


to highlight the consensus of a discussion. Depending on the individual template, the embedded content is either transcluded (i.e., inserted into the page on runtime but not in the source code), or substituted (i.e., inserted directly in the source code). While the latter approach is easier to process, most templates follow the transclusion method.

A specific subset of templates is used as a tagset for labeling articles and Talk pages. By adding the template {{*controversial*}} to a Talk page, an information banner is placed in the lead section of the Talk page and the associated article is tagged as controversial. A complete overview of Talk space specific templates can be found on the corresponding Wikipedia policy pages.[23] The cleanup and flaw markers are especially helpful criteria for filtering articles and Talk pages for corpus creation or further analysis.

The remainder of this section will discuss the merits of Talk pages as a resource for NLP and present a selection of qualitative and quantitative studies of discussions in Wikipedia. An overview can be found in Table 5.6.

[23]http://en.wikipedia.org/wiki/WP:TTALK

**Table 5.6** Qualitative and quantitative analyses of Wikipedia Talk pages

| Reference | Focus | Corpus size | Wikipedia | Tagset |
|---|---|---|---|---|
| [35] | Coordination | 25 TP | English | 11 |
| [29, 30] | Coordination | 100 TP | English | 15 |
| [13] | Coordination, Conflict | – | English | – |
| [9] | Coordination, Information quality | 100 TP | Simple English | 17 |
| [32] | Information Quality | 60 TP | English | 12 |
| [24] | Authority claims | 30 D | English | 6 |
| [2] | Authority claims, Alignment moves | 47 TP | English | 6[a], 8[b] |
| [15] | User interaction | – | English | – |
| [17] | User interaction | – | Venetian | – |

*TP* Talk pages, *D* Discussions

[a] Authority claims

[b] Alignment moves (3 positive, 5 negative)

## *5.3.1 Work Coordination and Conflict Resolution*

Viégas et al. [35] were among the first to draw attention to Wikipedia Talk pages as an important resource. In an empirical study, they discovered that articles with Talk pages have, on average, 5.8 times more edits and 4.8 times more participating users than articles without any Talk activity. Furthermore, they found that the number of new Talk pages increased faster than the number of content pages. In order to better understand how the rapidly increasing number of Talk pages are used by Wikipedians, they performed a qualitative analysis of selected discussions. The authors manually annotated 25 "purposefully chosen"[24] Talk pages with a set of 11 labels in order to analyze the aim and purpose of each user contribution. Each turn was tagged with one of the following labels:

- *request for editing coordination*
- *request for information*
- *reference to vandalism*
- *reference to Wikipedia guidelines*
- *reference to internal Wikipedia resources*
- *off-topic remark*
- *poll*
- *request for peer review*
- *information boxes*
- *images*
- *other*

The first two categories, requests for coordination (58.8 %) and information (10.2 %), were most frequently found in the the analyzed discussions, followed by off-topic remarks (8.5 %), guideline references (7.9 %), and references to internal

[24] According to [35], "[t]he sample was chosen to include a variety of controversial and non-controversial topics and span a spectrum from hard science to pop culture."

resources (5.4 %). This shows that Talk pages are not used just for the "retroactive resolution of disputes", as the authors hypothesized in their preliminary work [34]; rather, they are used for proactive coordination and planning of the editorial work.

Schneider et al. [29, 30] pick up on the findings of Viégas et al. and manually analyze 100 Talk pages with an extended annotation schema. In order to obtain a representative sample for their study, they define five article categories to choose the Talk pages from: *most-edited articles*, *most-viewed articles*, *controversial articles*, *featured articles*, and a *random set of articles*. In addition to the 11 labels established in [35], Schneider et al. classify the user contributions as

- *references to sources outside Wikipedia*
- *references to reverts, removed material or controversial edits*
- *references to edits the discussant made*
- *requests for help with another article*

The authors evaluated the annotations from each category separately and found that the most frequent labels differ between the five classes. Characteristic peaks in the class distribution could be found for the "reverts" label, which is a strong indicator for discussions of controversial articles. Interestingly, the controversial articles did not have an above-average discussion activity, which was initially expected due to a high demand of coordination. The labels "off-topic", "info-boxes", and "info-requests" peak in the random category, which are apt to contain shorter Talk pages than the average items from the other classes. In accordance with [35], coordination requests are the most frequent labels in all article categories, running in the 50–70 % range. The observed distribution patterns alone are not discriminative enough for identifying the type of article a Talk page belongs to, but they nevertheless serve as valuable features for Talk page analysis.

Furthermore, the labels can be used to filter or highlight specific contributions in a long Talk page to improve the usability of the Talk platform. In [30], the authors perform a user study in which they evaluate a system that allows discussants to manually tag their contribution with one of the labels. Most of the 11 participants in the study perceived this as a significant improvement in the usability of the Talk page, which they initially regarded as confusing. Given enough training data, this classification tasks can be tackled automatically using machine learning algorithms.

In a large-scale quantitative analysis, Kittur et al. [13] confirm earlier findings by [35] and demonstrate that the amount of work on content pages in Wikipedia is decreasing while the *indirect work* is on the rise. They define indirect work as "excess work in the system that does not directly lead to new article content." Besides the efforts for work coordination, indirect work comprises the resolution of conflicts in the growing community of Wikipedians. In order to automatically identify conflict hot spots or even to prevent future disputes, the authors developed a model of conflict on the article level and demonstrate that a machine learning algorithm can predict the amount of conflict in an article with high accuracy. In contrast to the works discussed above, Kittur et al. do not employ a hand-crafted coding schema to generate a manually annotated corpus; rather, they extract the *"controversial"* tags that have been assigned to articles with disputed content by

**Table 5.7** Page-level features proposed in [13]

| Feature | Page |
|---|---|
| Revisions[a] | Article[4], Talk[1], Article/Talk |
| Page length | Article, Talk, Article/Talk |
| Unique editors[a] | Article[5], Talk, Article/Talk |
| Unique editors[a]/Revisions[a] | Article, Talk[3] |
| Links from other articles[a] | Article, Talk |
| Links to other articles[a] | Article, Talk |
| Anonymous edits[a,b] | Article[7], Talk[6] |
| Administrator edits[a,b] | Article, Talk |
| Minor edits[a,b] | Article, Talk[2] |
| Reverts[a,c] | Article |

[a] Raw counts
[b] Percentage
[c] By unique editors
[1–7] Feature utility rank

Wikipedia editors. This human-labeled conflict data is obtained from a full Wikipedia dump with all page revisions (*revision dump*) using the Hadoop[25] framework for distributed processing. The authors define a measure called *Controversial Revision Count* (CRC) as "the number of revisions in which the 'controversial' tag was applied to the article". These scores are used as a proxy for the amount of conflict in a specific article and are predicted by a Support Vector Machine regression algorithm from raw data. The model is trained on all articles that are marked as controversial in their latest revision and evaluated by means of five-fold cross validation. As features, the authors define a set of page- level metrics based on both articles and talk pages (see Table 5.7). They evaluated the usefulness of each feature, which is indicated by the individual ranks as superscript numbers in the table.

The authors report that the model was able to account for almost 90 % of the variation in the CRC scores ($R^2 = 0.897$). They furthermore validate their model in a user study by having Wikipedia administrators evaluate the classification results on 28 manually selected articles that have not been tagged as controversial. The results of this study showed that the CRC model generalizes well to articles that have never been tagged as controversial. This opens up future applications like identifying controversial articles before a critical point is reached.

### *5.3.2 Information Quality*

The information quality (IQ) of collaboratively constructed knowledge resources is one of their most controversially disputed aspects. These resources break with the traditional paradigm of editorial quality assurance usually found in expert-mediated knowledge bases and allow anyone to view and edit the information

[25] http://hadoop.apache.org/

at their discretion. As collaboratively constructed resources become increasingly important, it is a vital necessity to measure their quality to ensure the reliability and thus the trustworthiness of their content. In Sect. 5.2.3, we already discussed how the revision history has been used to assess the article quality and trustworthiness using Wikipedia internal quality categories as frames of references. The assumption is that any article similar to a known high quality article with respect to the features defined by the individual assessment approach is again of high quality. While this approach is useful for evaluating different types features as to how they are able to quantify quality related aspects, the notion of quality itself is somewhat limited, since it provides no insight into the concrete problems a given article suffers from. In order to assess the information quality and potential quality problems of an article, a more fine grained concept of quality is needed. In Wikipedia, the information in Talk pages contains valuable insights into the readers' judgments of articles and comments about their potential deficiencies. Consequently, an analysis of these Talk pages with respect to article information quality is a good starting point for establishing a fine grained quality assessment model for Wikipedia.

From an information scientific perspective, Stvilia et al. [32] raise the question of how quality issues in Wikipedia are discussed by the community and how the open and unstructured discussions on Talk pages can be an integral part of a successful quality assurance process. The authors manually analyze 60 discussion pages in order to identify which types of IQ problems have been discussed by the community. They determine 12 IQ problems along with a set of related causal factors for each problem and actions that have been suggested by the community to tackle them.

For instance, IQ problems in the quality dimension *complexity* may be caused by low readability or complex language and might be tackled by replacing, rewriting, simplifying, moving, or summarizing the problematic article content. They furthermore identify trade-offs among these quality dimensions of which the discussants on Talk pages are largely aware. For example, an improvement in the dimension *completeness* might result in a deterioration in the *complexity* dimension. This model of IQ problems is useful for NLP applications in two ways: (1) as a frame of reference for automatic quality assessment of collaboratively created content, and (2) for automatically improving its quality using NLP techniques. In order to measure quality automatically, it is important to define what quality is and how it can be measured. Here, the proposed model is a sound basis for grounding any language processing approach to quality assessment with the users' understanding of quality. In order to use automatically calculated scores to improve article quality automatically, it is necessary to identify which actions can be taken to increase the quality scores in each dimension. This is also provided by the model proposed in [32].

Ferschke et al. [9] take the next step towards an automatic analysis of the discussions in Wikipedia. Inspired by the aforementioned IQ model, they develop an annotation schema for the discourse analysis of Talk pages aimed at the coordination effort for article improvement (see Table 5.8). With 17 labels in four categories, the schema captures article criticism and explicit user actions aimed at resolving IQ problems as well as the flow of information and the attitude of the discussants

**Table 5.8** Annotation schema for the discourse analysis of Wikipedia Talk pages proposed in [9]

| Article criticism | Explicit performative | Information content | Interpersonal |
|---|---|---|---|
| Missing content | Suggestion | Information providing | Positive (+) |
| Incorrect content | Reference to resource | Information seeking | Partially +/− |
| Unsuitable content | Commitment to action | Information correcting | Negative (−) |
| Structural problem | Report of action | | |
| Stylistic problem | | | |
| Objectivity issues | | | |
| Other | | | |

towards each other. The authors create a corpus of 100 Talk pages from the Simple English Wikipedia which they automatically segmented into individual discussions and turns by using the revision history for identifying turn boundaries and for attributing the correct authors without relying on user signatures. They manually label the corpus using their annotation schema and report a chance-corrected inter-annotator agreement between two raters of $\kappa = 0.67$ over all labels. In order to automatically label the turns in unseen Talk pages, the authors use the annotated corpus as training data for a set of machine learning algorithms and train individual classifiers for each label. They combine the best performing classification models into a classification pipeline which they use to label untagged discussions. They report an overall classification performance of $F_1 = 0.82$ evaluated on ten-fold cross-validation. The automatic classification of turns in Wikipedia Talk pages is a necessary prerequisite to investigating the relations between article discussions and article edits, which, in turn, is an important step towards understanding the processes of collaboration in large-scale wikis. Moreover, it enables practical applications that help to bring the content of Talk pages to the attention of article readers.

### *5.3.3 Authority and Social Alignment*

Information quality discussions in Wikipedia can have a big impact on articles. They usually aim at keeping articles in line with Wikipedia's guidelines for quality, neutrality and notability. If such a discussion is not grounded on authoritative facts but rather on subjective opinions of individual users, a dispute about content removal, for example, may lead to the unjustified removal of valuable information. Wikipedia Talk pages are, for the most part, pseudonymous discussion spaces and most of the discussants do not know each other personally. This raises the question how the users of Talk pages decide which claim or statement in a discussion can be trusted and whether an interlocutor is reliable and qualified.

Oxley et al. [24] analyze how users establish credibility on Talk pages. They define six categories of *authority claims* with which users account for their reliability and trustfulness (see Table 5.9). Based on this classification, Bender et al. [2] created a corpus of social acts in Wikipedia Talk pages (AAWD). In addition to authority claims, the authors define a second annotation layer to capture *alignment moves*—i.e., expressions of solidarity or signs of disagreement

**Table 5.9** Authority claims proposed in [2, 24]

| Claim type | Based on |
|---|---|
| Credentials | Education, work experience |
| Experiential | Personal involvement in an event |
| Institutional[a] | Position within the organizational structure |
| Forum | Policies, norms, rules of behavior (in Wikipedia) |
| External | Outside authority or resource |
| Social Expectations | Beliefs, intentions, expectations of social groups (outside of Wikipedia) |

[a] Not encoded in the AAWD corpus

among the discussants. At least two annotators labeled each of the 5,636 turns extracted from 47 randomly sampled Talk pages from the English Wikipedia. The authors report an overall inter-annotator agreement of $\kappa = 0.59$ for authority claims and $\kappa = 0.50$ for alignment moves.

Marin et al. [16] use the AAWD corpus to perform machine learning experiments targeted at automatically detecting authority claims of the *forum* type (cf. Table 5.9) in unseen discussions. They particularly focus on exploring strategies for extracting lexical features from sparse data. Instead of relying only on $n$-gram features, which are prone to overfitting when used with sparse data, they employ knowledge-assisted methods to extract meaningful lexical features. They extract word lists from Wikipedia policy pages to capture policy-related vocabulary and from the articles associated with the Talk pages to capture vocabulary related to editor discussions. Furthermore, they manually create six word lists related to the labels in the annotation schema. Finally, they augment their features with syntactic context gained from parse trees in order to incorporate a higher level linguistic context and to avoid the explosion of the lexical feature space that is often a side effect of higher level $n$-grams. Based on these features, the authors train a maximum entropy classifier to decide for each sentence whether it contains a forum claim or not.[26] The decision is then propagated to the turn level if the turn contains at least one forum claim. The authors report an $F_1$-score for the evaluation set of 0.66.

Besides being a potential resource for social studies and online communication research, the AAWD corpus and approaches to automatic classification of social acts can be used to identify controversial discussions and online trolls.[27]

### 5.3.4 User Interaction

It is not only the content of Talk pages which has been the focus of recent research, but also the social network of the users who participate in the discussions. Laniado et al. [15] create Wikipedia discussion networks from Talk pages in order to capture

[26] The corpus was split into training set (67 %), development set (17 %) and test set (16 %).

[27] A *troll* is a participant in online discussions with the primary goal of posting disruptive, off-topic messages or provoking emotional responses.

structural patterns of interaction. They extract the thread structure from all article and user Talk pages in the English Wikipedia and create tree structures of the discussions. For this, they rely on user signatures and turn indentation. The authors consider only registered users, since IP addresses are not unique identifiers for the discussants. In the directed article reply graph, a user node A is connected to a node B if A has ever written a reply to any contribution from B on any article Talk page. They furthermore create two graphs based on User Talk pages which cover the interactions in the personal discussion spaces in a similar manner.

The authors analyze the directed degree assortativity of the extracted graphs. In the article discussion network, they found that users who reply to many different users tend to interact mostly with inexperienced Wikipedians while users who receive messages from many users tend to interact mainly with each other. They furthermore analyzed the discussion trees for each individual article, which revealed characteristic patterns for individual semantic fields. This suggests that tree representations of discussions are a good basis for metrics for characterizing different types of Talk pages while the analysis of User Talk pages might be a good foundation for identifying social roles by comparing the different discussion fingerprints of the users.

A different aspect of the social network analysis in Wikipedia is examined by Massa [17]. He aims at reliably extracting social networks from User Talk pages. Similarly to [15], he creates a directed graph of user interactions. The interaction strength between two users is furthermore quantified by weighted edges with weights derived from the number of messages exchanged by the users. The study is based on networks extracted from the Venetian Wikipedia. Massa employs two approaches to extract the graphs automatically, one based on parsing user signatures and the other based on the revision history. He compares the results with a manually created gold standard and found that the revision based approach produces more reliable results than the signature approach, which suffers from the extreme variability of the signatures. However, history based processing often resulted in higher weights of the edges, because several edits of a contribution are counted as individual messages. A history-based algorithm similar to the one used by Ferschke et al. [9] could account for this problem. Massa furthermore identifies several factors that impede the network extraction, like noise in form of bot messages and vandalism, inconsistently used usernames, and unsigned messages. While these insights might be a good basis for future work on network extraction tasks, they are limited by the small Venetian Wikipedia on which the study is based. Talk pages in larger Wikipedias are much longer, more complex and are apt to contain pitfalls not recognized by this work.

## 5.4 Tools and Resources

In the following, we describe tools for processing revisions and discussions from Wikipedia as well as corpora which offer this content in a structured form.

**Table 5.10** Tools for accessing Wikipedia articles, revisions and Talk pages

| Reference and name | Type of data | API | License |
|---|---|---|---|
| MediaWiki API | Pages and revisions | web service | – |
| JWPL [44] | Pages (incl. Talk) | Java | LGPL |
| Wikipedia Revision Toolkit [8] | Revisions | Java | LGPL |
| Wikipedia Miner [20] | Articles | Java | GPL |
| WikiXRay [31] | Quantitative statistics | Python, R | GPL |

### *5.4.1 Tools for Accessing Wikipedia Articles, Revisions and Talk Pages*

We give an overview of tools for accessing and processing Wikipedia articles, revisions, discussions or statistics about them (see overview in Table 5.10). All of them are freely available, some are open-source. This list does *not* include special purpose scripts[28] provided by Wikipedia users or individual projects hosted on the Wikimedia Toolserver (see below).

The MediaWiki API[29] provides direct access to MediaWiki databases including Wikipedia. It can be accessed via a web service[30] or various client code wrappers.[31] Many bots and Toolserver utilities use this facility to get the data they need and to edit pages. The MediaWiki API supports various actions like *query*, *block* or *edit* and output formats such as JSON, PHP or XML. As it works directly on the MediaWiki databases, the API provides real time access to Wikipedia. This is a discriminative feature comparing it to any other API that is working on static dumps (cf. Sect. 5.4.2).

The Java Wikipedia Library (JWPL) [44] offers a Java-based programming interface for accessing all information in different language versions of Wikipedia in a structured manner. It includes a MediaWiki markup parser for in-depth analysis of page contents. JWPL works with a database in the background, the content of the database comes from a dump, i.e. a static snapshot of a Wikipedia version. JWPL offers methods to access and process properties like in- and outlinks, templates, categories, page text—parsed and plain—and other features. The *Data Machine* is responsible for generating the JWPL database from raw dumps. Depending on what data are needed, different dumps can be used, either including or excluding the Talk page namespace.

The Wikipedia Revision Toolkit [8] expands JWPL with the ability to access Wikipedia's revision history. To this end, it is divided into two tools, the

[28]A compilation of these can be found under http://en.wikipedia.org/wiki/WP:WikiProject_User_scripts/Scripts

[29]http://www.mediawiki.org/wiki/API

[30]http://en.wikipedia.org/w/api.php

[31]http://www.mediawiki.org/wiki/API:Client_code

*TimeMachine* and the *RevisionMachine*. The TimeMachine is capable of restoring any past state of the encyclopedia, including a user-defined interval of past versions of the pages. The RevisionMachine provides access to the entire revision history of all Wikipedia articles. It stores revisions in a compressed form, keeping only differences between adjacent revisions. The Revision Toolkit additionally provides an API for accessing Wikipedia revisions along with meta data like the comment, timestamp and information about the user who made the revision.

Wikipedia Miner [20] offers a Java-based toolkit to access and process different types of information contained in Wikipedia articles. Similar to JWPL, it has an API for structured access to basic information of an article. Categories, links, redirects and the article text, plain or as MediaWiki markup, can also be accessed as Java classes. It runs a preprocessed Java Berkeley database in the background to store the information contained in Wikipedia. Wikipedia Miner has a focus on concepts and semantic relations within Wikipedia. It is able to detect and sense-disambiguate Wikipedia topics in documents, i.e. it can be used to wikify plain text. Furthermore, the framework compares terms and concepts in Wikipedia, calculating their semantic relatedness or related concepts based on structural article properties (e.g. in-links) or machine learning. In contrast to JWPL, it cannot be used to access and process the revision history of an article. The capability of its parser is rather limited, e.g. no templates or infoboxes can be processed.

WikiXRay [31] is a collection of Python and GNU R scripts for the quantitative analysis of Wikipedia data. It parses plain Wikimedia dumps and imports the extracted data into a database. This database is used to provide general quantitative statistics about editors, pages and revisions.

Finally, the Wikimedia Toolserver[32] is a hosting platform for tools dedicated to processing Wikimedia data. The tools and scripts on the Toolserver are mainly developed by Wikipedia editors and researchers for Wiki maintenance and analysis. The unique advantage of running software on the Toolserver is the direct access to data from mirrored Wikimedia databases. The databases offer more information than the downloadable data dumps and are always kept up-to-date. However, computing resources are limited, so that the Toolserver is not an appropriate platform for running applications that demand much processing power.

### 5.4.2 Resources Based on Data from Wikipedia's Article and Talk Pages

This paragraph assembles a list of corpora containing either data directly exported from Wikipedia pages, revisions or discussion pages, or data that has been extracted and annotated from one of these sources for different tasks. Rather than being exhaustive, this list is meant to give a short overview of existing data collections

[32]http://toolserver.org/

**Table 5.11** Resources based on Wikipedia articles and Talk pages

| Resource | Based on | Annotations | Format | License |
|---|---|---|---|---|
| WVC [26,27] | Revisions | Vandalism | CSV | CC |
| WiCoPaCo [18] | Revisions | Spelling errors and paraphrases | XML | GFDL |
| [40] | Revisions | Lexical simplifications | CSV | – |
| [41] | Revisions | Textual entailment | XML, TXT | – |
| [43] | Revisions | Real-word spelling errors | TXT | CC |
| SEWD corpus [9] | Discussions | Dialog acts | XMI, MMAX | CC |
| AAWD corpus [2] | Discussions | Social acts | XTDF | – |

that have been introduced in the course of this chapter and are freely available. The resources we presented can roughly be divided into corpora produced from the article revisions and from Talk pages. Table 5.11 provides an overview.

The main source of raw data from Wikipedia is usually one of the so called Wikimedia *dumps*.[33] These dumps are snapshots of different content from Wikimedia Wiki projects, usually stored in large compressed XML and SQL files, with various releases throughout a year. The XML dumps store text including MediaWiki markup and metadata, separated by namespace. The main *page data* dumps are usually divided into three sets: *pages-articles* contains current versions of pages excluding Talk- and User-pages, *pages-meta-current* contains current page versions including Talk- and user-pages and *pages-meta-history* contains all revisions of all pages. Besides the tools mentioned in Sect. 5.4.1, there are various programs[34] available to handle Wikipedia dumps, all of them being limited to downloading or importing the dumps into databases for further processing.

## 5.5 Conclusion

For the last several years, the importance of Wikipedia in academic research has been continuously growing. In particular, NLP researchers increasingly find it to be a valuable resource to analyze the process of collaborative writing. To demonstrate this, we focused on the dynamic aspects of Wikipedia—that is, on the fact that its content is constantly changing. The massive amount of data that is generated by storing each edit to any page in Wikipedia offers numerous possibilities to create task-specific corpora, such as training data for tasks such as spelling error detection and information quality assessment. Although the number of studies on this kind of data has increased, to the best of our knowledge, there is no comprehensive introduction to existing applications in this field. In this survey, we therefore sought

[33] http://dumps.wikimedia.org/

[34] http://meta.wikimedia.org/wiki/Data_dumps#Tools

to analyze and compare methods for analyzing the process of collaborative writing based on Wikipedia's revision history and its discussion pages.

Section 5.2 described the concept of page revisions in Wikipedia. After defining the necessary terms, we explained various approaches generating training data for NLP tasks from the semistructured revision history data. Most of these approaches are either related to spelling error detection and correction or paraphrasing. A common way to process revision data is to calculate the changes (i.e., edits) between adjacent revisions and subsequently select suitable edit examples for further processing. The latter can be done by applying filters either on the raw edit data or after post-processing the contained lexical or syntactical information. Furthermore, approaches differ in whether they keep or ignore wiki markup such as links and headlines. Another series of approaches using revision history data aims to assess article quality. We distinguished different studies by the type of revision features they employ and by their definition of article quality. Revision features include quantitative properties like raw edit counts, but also various concepts of stability of article contents. Article quality criteria are mostly based on the Wikipedia internal review system. While article quality is an important factor for the reliability of the encyclopedic content in Wikipedia, vandalism is a serious problem to address. Vandalism detection is the task of distinguishing between valid and malicious edits in Wikipedia. It thus naturally uses revision history data, as the decision whether an edit is vandalistic or not is most likely based on the analysis of the changes between one revision and another. Advanced vandalism detection algorithms are based on machine learning and thus utilize a wide range of features. Typical vandalism features are based on changes in the article and/or on meta data. We compared both types of approaches.

Collaboration in a large text collection like Wikipedia is a demanding task and therefore needs coordination. Wikipedia offers a space for discussion among the authors, the so-called Talk pages. Whereas information from the revision history in Wikipedia is mainly used to support specific NLP applications (e.g. by augmenting the amount of training data), Wikipedia discussions are mostly analyzed to find out more about the process of collaboration in Wikipedia. Section 5.3 introduced the concept of Talk pages in Wikipedia and explained the challenges related to their processing. We analyzed various types of quantitative and qualitative NLP studies of Wikipedia discussions. A number of them focus on the utility of Talk pages for coordination and conflict resolution among the authors of an article. We introduced approaches using labels to categorize the purpose of contributions in discussions. They agree in the finding that coordination requests are the most frequent type of contributions. Another approach uses a machine learning model which is, amongst others, based on Talk page features to identify highly controversial articles. As an extension to the work discussed in Sect. 5.2.3, we reported on approaches analyzing the information quality of Wikipedia contents based on discussion page properties. We presented two studies with a focus on quality assessment based on a qualitative analysis of Talk page contributions. Both of them developed a model of information quality useful to NLP applications. We then turned to the social aspects of the discussion pages in Wikipedia. First, we introduced a corpus of so called social acts in Talk pages, along with various studies based on authority and social alignment among Wikipedia authors. Second, we explained two approaches investigating a network of user interaction in Wikipedia based on the thread structure of Talk pages.

A summary of tools and corpora for accessing and processing collaboratively constructed discourse in Wikipedia is presented in Sect. 5.4. We explained different methods and tools to access Wikipedia revisions and/or Talk pages. Additionally, we gave a summary of the freely accessible corpora presented in Sects. 5.2 and 5.3.

We have discussed approaches exploiting Wikipedia's revision history and its Talk pages. However, to understand the process of collaborative writing in Wikipedia even better, edit history information and discussion page contents should be brought together in future work. For example, it would be necessary to establish links between coordination and conflict resolution efforts on Talk pages and edits on the article. The resulting correlations could possibly answer a couple of very interesting questions with regard to the relevance or success of Wikipedia discussions. For example, it might be interesting to analyze the numbers of topics discussed on Talk pages which have actually been addressed by edits on the article itself. We think that this is a promising direction for future investigations based on the findings we presented in this survey.

**Acknowledgements** This work has been supported by the Volkswagen Foundation as part of the Lichtenberg-Professorship Program under grant No. I/82806, and by the Hessian research excellence program "Landes-Offensive zur Entwicklung Wissenschaftlich-ökonomischer Exzellenz" (LOEWE) as part of the research center "Digital Humanities". We thank the anonymous reviewers for their valuable comments.

# References

1. Adler BT, Alfaro L, Mola-Velasco SM, Rosso P, West AG (2011) Wikipedia vandalism detection: combining natural language, metadata, and reputation features. In: Gelbukh A (ed) Computational linguistics and intelligent text processing. Lecture notes in computer science. Springer, Berlin, pp 277–288
2. Bender EM, Morgan JT, Oxley M, Zachry M, Hutchinson B, Marin A, Zhang B, Ostendorf M (2011) Annotating social acts: authority claims and alignment moves in Wikipedia talk pages. In: Proceedings of the workshop on language in social media, Portland, OR, USA, pp 48–57
3. Buriol LS, Castillo C, Donato D, Leonardi S, Millozzi S (2006) Temporal analysis of the Wikigraph. In: Proceedings of the 2006 IEEE/WIC/ACM international conference on web intelligence, Hong Kong, China, pp 45–51
4. Chin SC, Street WN, Srinivasan P, Eichmann D (2010) Detecting Wikipedia vandalism with active learning and statistical language models. In: Proceedings of the 4th workshop on information credibility, Hyderabad, India
5. Cusinato A, Della Mea V, Di Salvatore F, Mizzaro S (2009) QuWi: quality control in Wikipedia. In: Proceedings of the 3rd workshop on information credibility on the web. ACM, Madrid, pp 27–34
6. Dalip DH, Gonçalves MA, Cristo M, Calado P (2009) Automatic quality assessment of content created collaboratively by web communities. In: Proceedings of the joint international conference on digital libraries, Austin, TX, USA, pp 295–304

7. Emigh W, Herring SC (2005) Collaborative authoring on the web: a genre analysis of online encyclopedias. In: Proceedings of the 38th annual Hawaii international conference on system sciences, Waikoloa, Big Island, HI, USA
8. Ferschke O, Zesch T, Gurevych I (2011) Wikipedia revision toolkit: efficiently accessing Wikipedia's edit history. In: Proceedings of the 49th annual meeting of the association for computational linguistics: human language technologies. System demonstrations, Portland, OR
9. Ferschke O, Gurevych I, Chebotar Y (2012) Behind the article: recognizing dialog acts in Wikipedia talk pages. In: Proceedings of the 13th conference of the European chapter of the association for computational linguistics, Avignon, France
10. Giampiccolo D, Trang Dang H, Magnini B, Dagan I, Cabrio E, Dolan B (2007) The third PASCAL recognizing textual entailment challenge. In: Proceedings of the ACLPASCAL workshop on textual entailment and paraphrasing, Prague, Czech Republic, pp 1–9
11. Han J, Wang C, Jiang D (2011) Probabilistic quality assessment based on article's revision history. In: Proceedings of the 22nd international conference on database and expert systems applications, Toulouse, France, pp 574–588
12. Javanmardi S, McDonald DW, Lopes CV (2011) Vandalism detection in Wikipedia: a high-performing, feature-rich model and its reduction through lasso. In: Proceedings of the 7th international symposium on Wikis and open collaboration, Mountain View, CA, USA, pp 82–90
13. Kittur A, Suh B, Pendleton B, Chi EH (2007) He says, she says: conflict and coordination in Wikipedia. In: Proceedings of the SIGCHI conference on human factors in computing systems, San Jose, CA, USA, pp 453–462
14. Knight K, Marcu D (2000) Statistics-based summarization—step one: sentence compression. In: Proceedings of the seventeenth national conference on artificial intelligence and twelfth conference on innovative applications of artificial intelligence, Austin, TX, USA, pp 703–710
15. Laniado D, Tasso R, Kaltenbrunner A, Milano P, Volkovich Y (2011) When the Wikipedians talk: network and tree structure of Wikipedia discussion pages. In: Proceedings of the 5th international conference on weblogs and social media, Barcelona, Spain, pp 177–184
16. Marin A, Zhang B, Ostendorf M (2011) Detecting forum authority claims in online discussions. In: Proceedings of the workshop on languages in social media, Portland, OR, USA, pp 39–47
17. Massa P (2011) Social Networks of Wikipedia. In: Proceedings of the 22nd ACM conference on hypertext and hypermedia, Eindhoven, Netherlands, pp 221–230
18. Max A, Wisniewski G (2010) Mining naturally-occurring corrections and paraphrases from Wikipedia's revision history. In: Proceedings of the 7th conference on international language resources and evaluation, Valletta, Malta
19. Medelyan O, Milne D, Legg C, Witten IH (2009) Mining meaning from Wikipedia. Int J Human Comput Stud 67(9):716–754
20. Milne D, Witten IH (2009) An open-source toolkit for mining Wikipedia. In: Proceedings of the New Zealand computer science research student conference, Auckland, New Zealand
21. Mizzaro S (2003) Quality control in scholarly publishing: a new proposal. J Am Soc Inf Sci Technol 54(11):989–1005
22. Nelken R, Shieber SM (2006) Towards robust context-sensitive sentence alignment for monolingual corpora. In: Proceedings of the 11th conference of the European chapter of the association for computational linguistics, Trento, Italy
23. Nelken R, Yamangil E (2008) Mining Wikipedia's article revision history for training computational linguistics algorithms. In: Proceedings of the 1st AAAI workshop on Wikipedia and artificial intelligence, Chicago, IL, USA
24. Oxley M, Morgan JT, Hutchinson B (2010) "What I Know Is...": establishing credibility on Wikipedia talk pages. In: Proceedings of the 6th international symposium on wikis and open collaboration, Gdańsk, Poland, pp 2–3
25. Posner IR, Baecker RM (1992) How people write together. In: Proceedings of the 25th Hawaii international conference on system sciences, Wailea, Maui, HI, USA, pp 127–138

26. Potthast M (2010) Crowdsourcing a Wikipedia vandalism corpus. In: Proceedings of the 33rd annual international ACM SIGIR conference on research and development on information retrieval, Geneva
27. Potthast M, Holfeld T (2011) Overview of the 2nd international competition on Wikipedia vandalism detection. In: Notebook papers of CLEF 2011 labs and workshops, Amsterdam, Netherlands
28. Potthast M, Stein B, Gerling R (2008) Automatic vandalism detection in Wikipedia. In: Proceedings of the 30th European conference on advances in information retrieval, Glasgow, Scotland, UK, pp 663–668
29. Schneider J, Passant A, Breslin JG (2010) A content analysis: how Wikipedia talk pages are used. In: Proceedings of the 2nd international conference of web science, Raleigh, NC, USA, pp 1–7
30. Schneider J, Passant A, Breslin JG (2011) Understanding and improving Wikipedia article discussion spaces. In: Proceedings of the 2011 ACM symposium on applied computing, Taichung, Taiwan, pp 808–813
31. Soto J (2009) Wikipedia: a quantitative analysis. Ph.D. thesis, Universidad Rey Juan Carlos, Madrid
32. Stvilia B, Twidale MB, Smith LC, Gasser L (2008) Information quality work organization in Wikipedia. J Am Soc Inf Sci Technol 59(6):983–1001
33. Thomas C, Sheth AP (2007) Semantic convergence of Wikipedia articles. In: Proceedings of the IEEE/WIC/ACM international conference on web intelligence, Washington, DC, USA, pp 600–606
34. Viégas FB, Wattenberg M, Dave K (2004) Studying cooperation and conflict between authors with history flow visualizations. In: Proceedings of the SIGCHI conference on human factors in computing systems, Vienna, Austria, pp 575–582
35. Viégas FB, Wattenberg M, Kriss J, Ham F (2007) Talk before you type: coordination in Wikipedia. In: Proceedings of the 40th annual Hawaii international conference on system sciences, Big Island, HI, USA, pp 78–78
36. Wang WY, McKeown KR (2010) Got you!: automatic vandalism detection in Wikipedia with web-based shallow syntactic-semantic modeling. In: Proceedings of the 23rd international conference on computational linguistics, Beijing, China, pp 1146–1154
37. Wilkinson DM, Huberman BA (2007) Cooperation and quality in Wikipedia. In: Proceedings of the 2007 international symposium on wikis, Montreal, Canada, pp 157–164
38. Woodsend K, Lapata M (2011) Learning to Simplify Sentences with quasi-synchronous grammar and integer programming. In: Proceedings of the conference on empirical methods in natural language processing, Edinburgh, Scotland, UK, pp 409–420
39. Yamangil E, Nelken R (2008) Mining Wikipedia revision histories for improving sentence compression. In: Proceedings of the 46th annual meeting of the association for computational linguistics: human language technologies. Short papers, association for computational linguistics, Columbus, OH, USA, pp 137–140
40. Yatskar M, Pang B, Danescu-Niculescu-Mizil C, Lee L (2010) For the sake of simplicity: unsupervised extraction of lexical simplifications from Wikipedia. In: Proceedings of the 2010 annual conference of the North American chapter of the association for computational Linguistics, Los Angeles, CA, USA, pp 365–368
41. Zanzotto FM, Pennacchiotti M (2010) Expanding textual entailment corpora from Wikipedia using co-training. In: Proceedings of the 2nd COLING-workshop on the people's web meets NLP: collaboratively constructed semantic resources, Beijing, China
42. Zeng H, Alhossaini MA, Ding L, Fikes R, McGuinness DL (2006) Computing trust from revision history. In: Proceedings of the 2006 international conference on privacy, security and trust, Markham, Ontario, Canada, pp 1–10
43. Zesch T (2012) Measuring contextual fitness using error contexts extracted from the Wikipedia revision history. In: Proceedings of the 13th conference of the European chapter of the association for computational linguistics, Avignon, France

44. Zesch T, Müller C, Gurevych I (2008) Extracting lexical semantic knowledge from Wikipedia and wiktionary. In: Proceedings of the 6th international conference on language resources and evaluation, Marrakech, Morocco
45. Zhu Z, Bernhard D, Gurevych I (2010) A monolingual tree-based translation model for sentence simplification. In: Proceedings of the 23rd international conference on computational linguistics, Beijing, China, pp 1353–1361
46. Zobel J, Dart P (1996) Phonetic string matching: lessons from information retrieval. In: Proceedings of the 19th annual international ACM SIGIR conference on research and development in information retrieval, Zurich, Switzerland, pp 166–172

# Chapter 6
# ConceptNet 5: A Large Semantic Network for Relational Knowledge

**Robyn Speer and Catherine Havasi**

**Abstract** ConceptNet is a knowledge representation project, providing a large semantic graph that describes general human knowledge and how it is expressed in natural language. Here we present the latest iteration, ConceptNet 5, with a focus on its fundamental design decisions and ways to interoperate with it.

## 6.1 Introduction

The wisdom of crowds can be found all over the Web. Some of the most significant recent advances in collecting the world's knowledge appear in resources such as Wikipedia and Wiktionary, which are written for people by large numbers of people, yet converge on a structure that can be made understandable by computers. Meanwhile, "games with a purpose" collect large quantities of specific knowledge while simply providing entertainment in return. Both are knowledge sources that can provide a wealth of information to computers about how people use and understand language, as long as it can be compiled into a useful and scalable representation.

ConceptNet is a project that creates such a representation of crowd-sourced knowledge, providing a large semantic graph that describes general human knowledge and how it is expressed in natural language. The scope of ConceptNet includes words and common phrases in any written human language. It provides a large set of background knowledge that a computer application working with natural language text should know.

These words and phrases are related through an open domain of predicates, such as *IsA* or *UsedFor*, describing not just how words are related by their lexical

R. Speer (✉) · C. Havasi (✉)
MIT Media Lab, 20 Ames St., Cambridge, MA 02142, USA
e-mail: rspeer@arborelia.net; havasi@media.mit.edu

I. Gurevych and J. Kim (eds.), *The People's Web Meets NLP*, Theory and Applications of Natural Language Processing, DOI 10.1007/978-3-642-35085-6_6,

**Table 6.1** The most common interlingual relations in ConceptNet, with example sentence frames in English and their number of collected edges

| Relation | # edges | Sentence pattern |
|---|---|---|
| IsA | 7,956,303 | *NP* is a kind of *NP*. |
| PartOf | 536,648 | *NP* is part of *NP*. |
| AtLocation | 535,278 | Somewhere *NP* can be is *NP*. |
| RelatedTo | 319,471 | *NP* is related to *NP*. |
| HasProperty | 303,921 | *NP* is *AP*. |
| UsedFor | 254,563 | *NP* is used for *VP*. |
| DerivedFrom | 242,853 | *TERM* is derived from *TERM*. |
| Causes | 233,727 | The effect of *VP* is *NP*\|*VP*. |
| CapableOf | 167,405 | *NP* can *VP*. |
| MotivatedByGoal | 173,111 | You would *VP* because you want *VP*. |
| HasSubevent | 154,214 | One of the things you do when you *VP* is *NP*\|*VP*. |
| Desires | 95,779 | *NP* wants to *VP*. |
| HasPrerequisite | 69,474 | *NP*\|*VP* requires *NP*\|*VP*. |
| HasA | 56,691 | *NP* has *NP*. |
| CausesDesire | 51,338 | *NP* makes you want to *VP*. |
| MadeOf | 43,278 | *NP* is made of *NP*. |
| DefinedAs | 39,406 | *NP* is defined as *NP*. |
| HasFirstSubevent | 35,242 | The first thing you do when you *VP* is *NP*\|*VP*. |
| ReceivesAction | 24,609 | *NP* can be *VP*. |
| LocatedNear | 12,679 | You are likely to find *NP* near *NP*. |
| SimilarTo | 11,635 | *NP* is like *NP*. |
| SymbolOf | 11,302 | *NP* represents *NP*. |
| HasLastSubevent | 8,689 | The last thing you do when you *VP* is *NP*\|*VP*. |
| CreatedBy | 1,979 | You make *NP* by *VP*. |

definitions, but also how they are related through common knowledge. We will refer to these as *relations*. The most common ones appear in Table 6.1.

For example, ConceptNet's knowledge about "jazz" includes not just the properties that define it, such as *IsA*(jazz, genre of music); it also includes incidental facts such as

- *AtLocation*(jazz, new orleans)
- *UsedFor*(saxophone, jazz), and
- *Plays percussion in*(jazz drummer, jazz).

A cluster of related concepts and the ConceptNet relations that connect them is visualized in Fig. 6.1.

ConceptNet originated as a representation for the knowledge collected by the Open Mind Common Sense project [21], which uses a long-running interactive Web site to collect new statements from visitors to the site, and asks them target questions about statements it thinks may be true. Later releases included knowledge from similar websites in other languages, such as Portuguese and Dutch, and collaborations with online word games that automatically collect general knowledge, yielding further knowledge in English, Japanese, and Chinese.

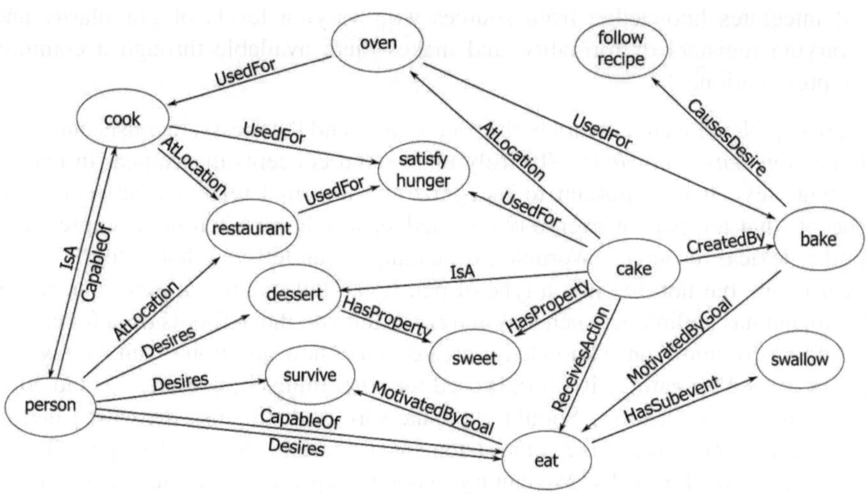


**Fig. 6.1** A high-level view of the knowledge ConceptNet has about a cluster of related concepts

ConceptNet gives a foundation of real-world knowledge to a variety of AI projects and applications. Previous versions of ConceptNet [11] have been used, for example, to build a system for analyzing the emotional content of text [6], to create a dialog system for improving software specifications [14], to recognize activities of daily living [24], to visualize topics and trends in a corpus of unstructured text [23], and to create public information displays by reading text about people and projects from a knowledge base [12].

There are similar ongoing projects that collect crowd-sourced knowledge from similar sources. BabelNet [18] and MENTA [7], for example, create a large, structured, multilingual taxonomy from a combination of Wikipedia's structured knowledge and WordNet. WikiNet [17] also mines structured knowledge from Wikipedia, while a project by Blanco et al. [4] creates a ConceptNet-like representation whose input primarily comes from unstructured machine reading. ConceptNet's niche is defined by its representational decisions, which are particularly suited for some kinds of text understanding applications:

- Its concepts are connected to, and defined by, natural language words and phrases that can also be found in free text.
- It includes not just definitions and lexical relationships, but also the "common sense" associations that ordinary people make among these concepts. Its sources range in formality from dictionaries to online games.
- It puts more emphasis on collecting information about common words than about named entities (which is why, for example, it collects more from Wiktionary than it does from Wikipedia).
- The concepts are not limited to a single language; they can be from any written language.

- It integrates knowledge from sources with varying levels of granularity and varying registers of formality, and makes them available through a common representation.

ConceptNet aims to contain both specific facts and the messy, inconsistent world of *common sense knowledge*. To truly understand concepts that appear in natural language text, it is important to recognize the informal relations between these concepts that are part of everyday knowledge, which are often under-represented in other lexical resources. WordNet, for example, can tell you that a dog is a type of carnivore, but not that it is a type of pet. It can tell you that a fork is an eating utensil, but has no link between *fork* and *eat* to tell you that a fork is used for eating.

Adding common sense knowledge creates many new questions. Can we say that "a fork is used for eating" if a fork is used for other things besides eating, and other things are used for eating? Should we make sure to distinguish the eating utensil from the branching of a path? Is the statement still true in cultures that typically use chopsticks instead of forks? We can try to collect representations that answer these questions, while pragmatically accepting that much of the content of a common sense knowledge base will leave them unresolved.

### *6.1.1 Motivation for ConceptNet 5*

In comparison to previous versions of ConceptNet, the new goals of ConceptNet 5 include:

- Incorporating knowledge from other crowd-sourced resources with their own communities and editing processes, particularly data mined from Wiktionary and Wikipedia.
- Adding links to other resources such as DBPedia [2], Freebase [5], and WordNet [10].
- Supporting machine-reading tools such as ReVerb [9], which extracts relational knowledge from Web pages.
- Finding translations between concepts represented in different natural languages.

ConceptNet 5 is intended to grow freely and absorb knowledge from many sources, with contributions from many different projects. We aim to allow different projects to contribute data that can easily be merged into ConceptNet 5 without the difficulty of aligning large databases.

Combining all these knowledge sources in a useful way requires processes for normalizing and aligning their different representations, while avoiding information loss. It also requires a system for comparing the reliability of the collected knowledge, as such knowledge can come from a variety of processes, sometimes involving unreliable sources (such as players of online games) and sometimes involving unreliable processes (parsers and transformations between representations).

In a sense, while ConceptNet 4 and earlier versions collected facts, ConceptNet 5 also at a higher level collects *sources* of facts. This greatly expands its domain, makes it interoperable with many other public knowledge resources, and makes it applicable to a wider variety of text-understanding applications.

## 6.2 Knowledge in ConceptNet 5

ConceptNet expresses *concepts*, which are words and phrases that can be extracted from natural language text; we call them "concepts" instead of terms to account for the fact that they can be more or less specific than a typical term. ConceptNet also contains *assertions* of the ways that these concepts relate to each other. These assertions can come from a wide variety of sources that create *justifications* for them. The current sources of knowledge in ConceptNet 5 are:

- The Open Mind Common Sense website (http://openmind.media.mit.edu), which collects common-sense knowledge mostly in English, but has more recently supported other languages.
- Sister projects to OMCS in Portuguese [1] and Dutch [8].
- The multilingual data, including translations between assertions, collected by the GlobalMind project, a spin-off of OMCS.
- "Games with a purpose" that collect common knowledge, including Verbosity [26] in English, *nadya.jp* in Japanese, and the "pet game" [16] on the popular Taiwanese bulletin board PTT, collecting Chinese knowledge in traditional script.
- A new process that scans the English Wiktionary (a Wikimedia project at en.wiktionary.org that defines words in many languages in English). In addition to extracting structured knowledge such as synonyms and translations, it also extracts some slightly-unstructured knowledge. For example, it extracts additional translations from the English-language glosses of words in other languages. The process is similar to that of UKPL [27] but targets ConceptNet's representation.
- WordNet 3.0 [10], including cross-references to its RDF definition at http://semanticweb.cs.vu.nl/lod/wn30/ [25].
- The semantic connections between Wikipedia articles represented in DBPedia [2], with cross-references to the corresponding DBPedia resources. DBPedia contains a number of collections, in different languages, representing relationships with different levels of specificity. So far we use only the English collection, and only use links that translate to our standard relations "IsA", "PartOf", and "AtLocation".
- Relational statements mined from Wikipedia's free text using ReVerb [9], run through a filter we designed to keep only the statements that are going to be most useful to represent in ConceptNet. We discarded statements whose ReVerb scores were too low, and those that contained uninformative terms such as "this".

Adding knowledge from other free projects such as WordNet does more than just increase the coverage of ConceptNet; it also allows us to align entries in ConceptNet with those in WordNet, and refer to those alignments without having to derive them again. This is an important aspect of the Linked Data movement: different projects collect data in different forms, but it is best when there is a clear way to map from one to the other. When the data is linked, ConceptNet enhances the power of WordNet and vice versa.

Adding data from Wiktionary was key in unifying the data collected in many different languages. In ConceptNet 4, each language was a separate connected component; now all the languages of ConceptNet are highly interlinked.

ConceptNet 5 is growing as we find new sources and new ways to integrate their knowledge. As of April 2012, it contains 12.5 million edges, representing about 8.7 million assertions connecting 3.9 million concepts. 2.78 million of the concepts appear in more than one edge. Its most represented language is English, where 11.5 million of the edges contain at least one English concept. The next most represented languages are Chinese (900,000 edges), Portuguese (228,000 edges), Japanese (130,000 edges), French (106,000 edges), Russian (93,700 edges), Spanish (92,400 edges), Dutch (90,000 edges), German (86,500 edges), and Korean (71,400 edges). The well-represented languages largely represent languages for which multilingual collaborations with Open Mind Common Sense exist, with an extra boost for languages that are well-represented in Wiktionary.

Additional sources that may be added include the plan-oriented knowledge in Honda's Open Mind Indoor Common Sense [13], connections to knowledge in Freebase [5], ontological connections to SUMO and MILO [19], and new processes that scan well-structured Wiktionaries in other target languages, such as Japanese and German.

### 6.2.1 Representation

ConceptNet 5 is conceptually represented as a hypergraph. Its assertions can be seen as edges that connect its nodes, which are concepts (words and phrases), via relations that are also nodes. These assertions, however, can be *justified* by other assertions, knowledge sources, or processes. The predicates that label them can be one of a set of interlingual relations, such as *IsA* or *UsedFor*, or they can be automatically-extracted relations that are specific to a language, such as *is_known_for*, or underspecified prepositional relations such as *is_on*.

The values of the predicates – referred to hereafter as the *relation* of each assertion – are represented using concept nodes as well. The structure of edges surrounding two assertions appears in Fig. 6.2. The most common interlingual relations we identify in ConceptNet appear in Table 6.1.

One way to represent a hypergraph is to reify all edges as nodes, with lower-level relationships such as "$x$ is the first argument of $y$" becoming the new edges. We experimented with representations of reified hypergraphs, but found that the result

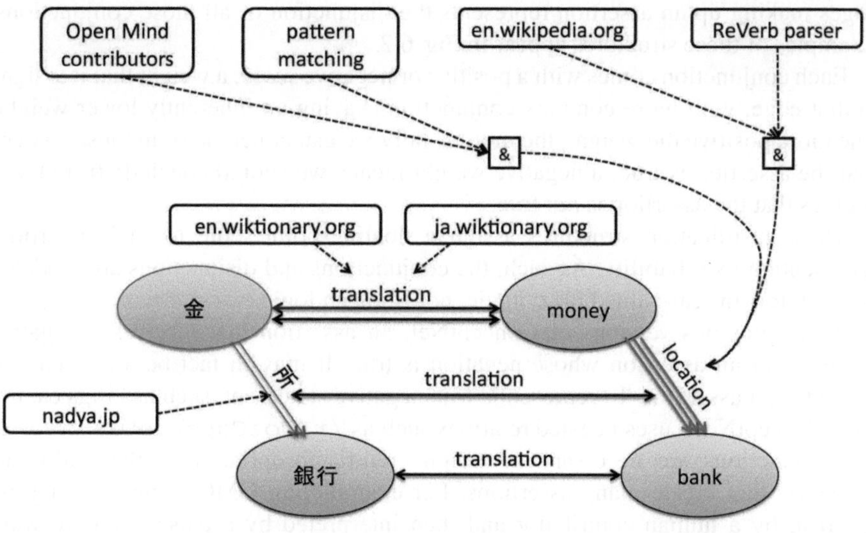


**Fig. 6.2** An example of two assertions in ConceptNet 5, and the edges they involve. *Rounded rectangles* and *dotted edges* represent knowledge sources; *solid edges* are grouped together into assertions

was exceptionally difficult to query as the database grew. Asking simple questions such as "What are the parts of a car?" in a hypergraph is a complex, multi-step query, and we found no mature database system that could perform all the queries we needed efficiently.

Instead, we store almost all of the relevant information about an edge as properties on that edge. Each assertion is still reified with a unique ID, but that ID is only referred to within the assertion or in higher-level assertions about that assertion, such as translations.

In particular, an edge in ConceptNet 5 is an *instance* of an assertion, as learned from some knowledge source. The same assertion might be represented by a large bundle of edges, when we learn it in many different ways; these all have the same assertion ID, along with algorithmically-generated unique edge IDs that we can use to deduplicate data later.

A hypergraph can be represented in a standard graph format such as RDF, but only by reifying all of its edges. It is straightforward to export an RDF version of ConceptNet 5 that conveys the same information, but the overhead created by reifying everything would make it a poor choice for a native representation.

### 6.2.2 *Assertion Scores*

The sources that justify each assertion form a structure that can be seen as a disjunction of conjunctions. Each edge – that is, each instance of an assertion – indicates a combination of sources that produced that edge, while the bundle of edges making up an assertion represents the disjunction of all those conjunctions. Examples of these structures appear in Fig. 6.2.

Each conjunction comes with a positive or negative *score*, a weight that it assigns to that edge, with more complex conjunctions having an inherently lower weight. The more positive the weight, the more solidly we can conclude from these sources that the assertion is true; a negative weight means we should conclude from these sources that the assertion is *not* true.

These justification structures assign a floating-point score to each assertion, representing its reliability. As such, the conjunctions and disjunctions are modeled on operators in real-valued fuzzy logic, not Boolean logic.

As in previous versions of ConceptNet, an assertion that receives a negative score is not an assertion whose negation is true. It may in fact be a nonsensical or irrelevant assertion. To represent a true negative statement, such as "Pigs cannot fly", ConceptNet 5 uses negated relations such as `/r/NotCapableOf`.

Conjunctions are necessary to assign credit appropriately to the multi-part processes that create many assertions. For example, an OMCS sentence may be typed in by a human contributor and then interpreted by a parser, and we want the ability to examine the collected data and determine whether the human is a reliable data source as well as whether the parser is. As another example, relations mined from Wikipedia using ReVerb depend on both the reliability of Wikipedia and of ReVerb.

### *6.2.3 Granularity*

The different knowledge sources that feed ConceptNet 5 represent concepts at different levels of granularity, especially in that they can be ambiguous or disambiguated. Concepts are often ambiguous when we acquire them from natural-language text. Other concepts are explicitly disambiguated by a resource such as WordNet or Wiktionary. ConceptNet 5 contains, for example, the ambiguous node /c/en/jazz. A source such as Wiktionary might define it as a noun, yielding the more specific concept /c/en/jazz/n, and it may even distinguish the word sense from other possible senses, yielding /c/en/jazz/n/musical_art_form.

These URLs do not represent the same node, but the nodes they represent are highly related. This indicates that when we add a way to query ConceptNet 5, described in Sect. 6.3.1, we need to structure the index so that a query for /c/en/jazz also matches /c/en/jazz/n/musical_art_form.

### *6.2.4 Normalizing and Aligning Concepts*

ConceptNet deals with natural-language data, but it should not store the assertion that "a cat is an animal" in a completely different way than "cats are animals".

Therefore, we represent each concept using *normalized* versions of the concept's text. The process for creating a normalized concept differs by language. Some examples are:

- *Running*, in English: /c/en/run
- *Rennen*, in Dutch: /c/nl/renn
- *Run (baseball)*, a disambiguated English word: `/c/en/run/n/baseball`

ConceptNet 5 uses our custom Python package called *metanl*[1] for lemmatization (reducing words to a root form) and other kinds of normalization. *metanl* provides a straightforward Python interface to our preferred stemmers and lemmatizers in many different languages.

The normalization process in English is an extension of WordNet's Morphy algorithm as provided by NLTK [3], plus removal of a very small number of stopwords, and a transformation that undoes CamelCase on knowledge sources that write their multiple-word concepts that way. In Japanese, we use the commonly-used MeCab algorithm for splitting words and reducing the words to a dictionary form [15]. In many European languages, we use the Snowball stemmer for that language [20] to remove stop words and reduce inflected words to a common stem.

Normalization inherently involves discarding information, but since ConceptNet 3, we have ensured that this information is stored with the assertion and not truly discarded. Every edge that forms every assertion is annotated with how it was expressed in natural language. That information is important in some applications such as generating natural-language questions to ask, as the AnalogySpace system [22] does with ConceptNet data; it is also very important so that if we change the normalization process one day, the original data is not lost and there is a clear way to determine which new concepts correspond to which old concepts.

### 6.2.5 URIs and Namespaces

An important aspect of the representation used by ConceptNet 5 is that it is free from arbitrarily-assigned IDs, such as sequential row numbers in a relational database. Every node and edge has a URI, which contains all the information necessary to identify it uniquely and no more.

Concepts (normalized terms) are the fundamental unit of representation in ConceptNet 5. Each concept is represented by a URI that identifies that it is a concept, what language it is in, its normalized text, and possibly its part of speech and disambiguation. A concept URI looks like `/c/en/run/n/basement`.

[1] http://github.com/commonsense/metanl

The predicates that relate concepts can be multilingual relations such as `/r/IsA`: this represents the "is-a" or "hypernym" relation that will be expressed in different ways, especially when the text is in different languages.

Processes that read free text, such as ReVerb, will produce relations that come from natural language and cannot be aligned in any known way with our multilingual relations. In this case, the relation is in fact another concept, with a specified language and a normalized form. In the text "A bassist performs in a jazz trio", the relation is /c/en/perform_in.

The fact that interlingual relations and language-specific concepts can be interchanged in this way is one reason we need to distinguish them with the namespaces /r/ and /c/. The namespaces are as short as possible so as to not waste memory and disk space; they appear millions of times in ConceptNet.

There is a namespace /s/ for data sources that justify an edge. These contain, for example, information extraction rules such as /s/rule/reverb, human contributors such as /s/contributor/omcs/rspeer, and curated projects such as /s/wordnet/3.0.

An assertion URI contains all the information necessary to reconstruct that assertion. For example, the assertion that "jazz is a kind of music" has the URI /a/[/r/IsA/,/c/en/jazz/,/c/en/music/]. By using the special path components /[/ and /]/, we can express arbitrary tree structures within the URI, so that the representation can even represent assertions about assertions without ambiguity. The advantage of this is that if multiple branches of ConceptNet are developed in multiple places, we can later merge them simply by taking the union of the edges. If they acquire the same fact, they will assign it the same ID.

Assertions will be represented multiple times by multiple edges. Edge IDs also take into account all the information that uniquely identifies the edge. There is no need to represent this information in a way from which its parts can be reconstructed; doing so would create very long edge IDs that would repeat the majority of the data contained in the edge. Instead, edge IDs are the hexadecimal SHA-1 hash of all the unique components, separated by spaces. These IDs can be queried to get an arbitrary subset of edges, which is very useful for evaluation.

### *6.2.6 Graph Statistics*

A simple transformation of ConceptNet 5 allows us to consider it as a simple undirected graph. We consider there to be an edge between the two arguments of every assertion. We add an implicit edge from every disambiguated concept to its ambiguous form, and from every reified assertion to its two arguments: for example,

The resulting graph[2] has 9,611,524 distinct edges among 3,930,196 nodes.

[2]Statistics apply to the May 1, 2012 release.

The largest connected component contains 3,675,400 nodes. The second largest component, with 727 nodes, contains all the instances of /c/en/olympic_result from DBPedia, such as /c/en/belgium_at_1972_winter_olympics.

ConceptNet 5 is not overwhelmed with dangling edges; the 2-core of ConceptNet 5 (the maximal subgraph in which every node has degree ≥2) contains 8,286,862 edges among 2,512,028 nodes.

## 6.3 Storing and Accessing ConceptNet Data

As ConceptNet grows larger and is used for more purposes, it has been increasingly important to separate the data from the interface to that data. A significant problem with ConceptNet 3, for example, was that the only way to access it was through the same Django database models that created it.

ConceptNet 5 fully separates the data from the interface. The data in ConceptNet 5 is a flat list of edges, available in JSON or as tab-separated values. A flat file is in fact the most useful format for many applications:

- Many statistics about ConceptNet can be compiled by iterating over the full list of data, which neither a database nor a graph structure is optimized for.
- A subset of the information in each line of the flat file is the appropriate input for many machine learning tools.
- A flat file can be easily converted to different formats using widely-available tools.
- It is extremely easy to merge flat files. It is sometimes sufficient simply to put them in the same directory and iterate over both. If deduplication is needed, one can use highly optimized tools to sort the lines and make them unique.

However, a flat file is not particularly efficient for querying. A question such as "What are the parts of a car?" involves a very small proportion of the data, which could only be found in a flat file by iterating over the entire thing. Thus, we build indexes *on top of* ConceptNet 5.

### *6.3.1 Indexes*

Currently, we index ConceptNet 5 with a combination of Apache Solr and MongoDB. We provide access to them through a REST API, as well as transformations of the data that a downstream user can import into a local Solr index or MongoDB database. The Solr index seems to be the most useful and scalable, and its distributed queries make it simple to distribute it between sites, so it is the primary index that we currently use. For example, we can maintain the main index while our collaborators in Taiwan maintain a separate index, including up-to-date information they have collected, and now a single API query can reach both.

Using the Solr server, we can efficiently index all edges by all lemmas (normalized words) they contain and prefixes of any URIs they involve. A search for `rel:/r/PartOf` and `end:/c/en/wheel OR end:/c/en/wheel/*` will find all edges describing the parts of a wheel, automatically ordered by the absolute value of their score. The Solr index would not make sense as a primary way to store the ConceptNet data, but it allows very efficient searches for many kinds of queries a downstream user would want to perform.

The flat file of ConceptNet 5 contains 7.5 GB of text. A Solr index performs best when it can keep all its data, plus overhead for indexing, in memory instead of swapping it to disk. This is a large memory requirement for a single computer. However, when we shard the index across two *m1.large* instances on Amazon EC2, each with 7.5 GB of RAM, the data and index fit in memory. This is sufficient to respond to queries on any of the indexed fields in 100–500 ms.

### *6.3.2 Downloading*

ConceptNet's usefulness as a knowledge platform depends on its data being freely available under a minimally restrictive license, and not (for example) tied up in agreements to use the data only for research purposes. ConceptNet 5 can be downloaded or accessed through a Web API at its web site, http://conceptnet5.media.mit.edu, and may be redistributed or reused under a choice of two Creative Commons licenses.

The flat files containing ConceptNet 5 data are available at: http://conceptnet5.media.mit.edu/downloads/

Python code for working with this data, transforming it, and building indexes from it is maintained on GitHub in the "conceptnet5" project: https://github.com/commonsense/conceptnet5.

## 6.4 Evaluation

To evaluate the current content of ConceptNet, we put up a website for 48 h that showed a random sample of the edges in ConceptNet. It showed the natural language form of the text (which was machine-generated in the cases where the original data was not in natural language) and asked people to classify the statement as "Generally true", "Somewhat true", "I don't know", "Unhelpful or vague", "Generally false", and "This is garbled nonsense". People were invited to participate via e-mail and social media. They were shown 25 results at a time. We got 81 responses that evaluated a total of 1,888 statements, or 1,193 if "Don't know" results are discarded.

All participants were English speakers, so we filtered out statements whose surface text was not in English. Statements that translate another language to English were left in, but participants were not required to look them up, so in many cases they answered "Don't know".

**Table 6.2** The breakdown of responses to an evaluation of random statements in ConceptNet 5

| Dataset | False | Nonsense | Vague | Don't know | Sometimes | True | Total |
|---|---|---|---|---|---|---|---|
| Existing ConceptNet | 34 | 50 | 15 | 19 | 117 | 300 | 535 |
| WordNet | 4 | 17 | 0 | 11 | 9 | 35 | 76 |
| Wiktionary, English-only | 2 | 5 | 3 | 9 | 6 | 10 | 35 |
| Wiktionary, translations | 4 | 6 | 2 | 233 | 8 | 51 | 304 |
| DBPedia | 10 | 36 | 9 | 389 | 41 | 238 | 723 |
| Verbosity | 10 | 41 | 7 | 2 | 32 | 51 | 143 |
| ReVerb | 2 | 15 | 15 | 19 | 3 | 5 | 59 |
| GlobalMind translations | 0 | 0 | 0 | 4 | 0 | 0 | 4 |
| Negative edges | 4 | 2 | 0 | 0 | 1 | 2 | 9 |

We have grouped the results by *dataset*, distinguishing edges that come from fundamentally different sources. The datasets are:

- **Existing ConceptNet**: statements previously collected by Common Sense Computing projects, which can be found in ConceptNet 4.
- **WordNet**: connections from WordNet 3.0.
- **Wiktionary, English-only**: monolingual information from the English Wiktionary, such as synonyms, antonyms, and derived words.
- **Wiktionary, translations**: translations in Wiktionary from some other language to English. As these are numerous compared to other sources, we kept only 50 % of them.
- **DBPedia**: Triples from DBPedia's `instance_types_en` dataset. As these are numerous compared to other sources, we kept only 25 % of them.
- **Verbosity**: Statements collected from players of Verbosity on gwap.com.
- **ReVerb**: Filtered statements extracted from ReVerb parses of a corpus of Wikipedia's front-paged articles.
- **GlobalMind translations**: translations of entire assertions between languages.

We also separated out **negative edges**, those which previous contributors to ConceptNet have rated as not true, confirming that most of them are rated similarly now.

The breakdown of results appears in Table 6.2. Their relative proportions, excluding the "Don't know" responses, are graphed in Fig. 6.3.

We can see that people often answered "Don't know" when faced with very specific knowledge, which is to be expected when presenting expert knowledge to arbitrary people.

All the examples of higher-level assertions that translate assertions between languages were rated as "Don't know". A more complete evaluation could be performed in the future with the help of bilingual participants who could evaluate translations.

The processes of extracting translations from Wiktionary and triples from DBPedia performed very well, while the ReVerb data – faced with the hardest task, extracting knowledge from free text – did poorly. The few negative-score edges were mostly rated as false, as expected, though 3 out of 9 of the respondents to them

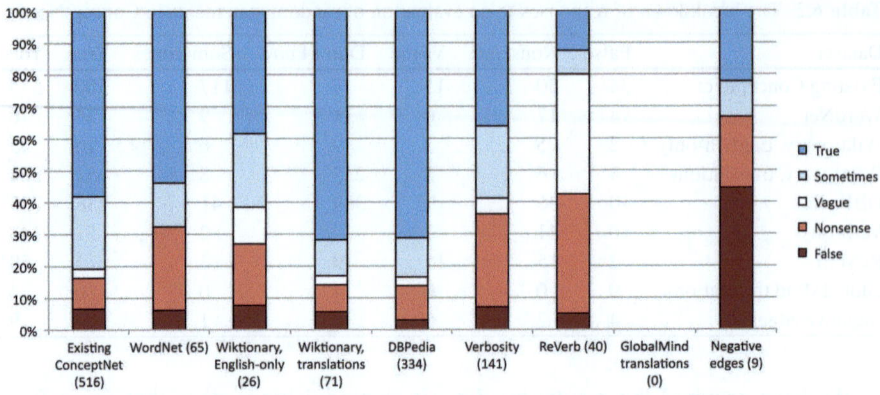


**Fig. 6.3** The relative proportions of responses people gave about each dataset

disagreed. The core data in ConceptNet was evaluated nearly exactly the same as data mined from Wiktionary.

Interestingly, existing ConceptNet data was rated better than WordNet data, which was often rated as "nonsense"; perhaps the average WordNet edge is an assertion so obscure that a human evaluator will not even recognize it as making sense, or perhaps our own process of generating artificial English-language glosses of the WordNet edges is at fault.

A typical example of a WordNet edge rated "nonsense" is: *white (flesh of any of a number of slender food fishes especially of Atlantic coasts of North America)* is part of *white (any of several food fishes of North American coastal waters)*. When describing obscure senses of the word "white", this is actually a highly specific true statement, but our evaluator did not decipher it. Other statements rated "nonsense" include statements that reflect an unintuitive taxonomy, such as *illinois class battleship* is a *product* from DBPedia.

These should be distinguished from errors in which the lack of context is an inherent problem with the statement. When a process such as ReVerb extracts a statement without the context that makes it make sense, such as "*Critics* have seen *Jake*", evaluators correctly mark it as nonsense.

We present the results as they are, keeping in mind that a future evaluation should be designed to provide evaluators with more of the context they need to make accurate judgments. The presence of awkwardly-worded statements with absolutely no context had a negative effect on the evaluation of all sources, but was particularly penalizing to highly-specific statements such as those in WordNet.

### *6.4.1 Next Steps*

The overall accuracy of approximately 79 % across all sources is sufficient for many purposes but motivates future work on verifying and cleaning up the data.

The ConceptNet data presents many starting points for machine learning, which could help to both refine the ConceptNet data and to create new resources from it. The ConceptNet 5 Web API[3] currently supports using dimensionality reduction, as in [22], to list similar concepts to a query. Useful future tasks include automatically learning from and refining the assertion scores to learn which sources and combinations of sources provide the most reliable information, aligning the most similar word senses within a language and across different languages, and recognizing paraphrases and nearly-equivalent statements that support one another.

## References

1. Anacleto J, Lieberman H, Tsutsumi M, Neris V, Carvalho A, Espinosa J, Zem-Mascarenhas S (2006) Can common sense uncover cultural differences in computer applications? In: Proceedings of IFIP world computer conference, Santiago, Chile
2. Auer S, Bizer C, Kobilarov G, Lehmann J, Cyganiak R, Ives Z (2007) DBpedia: a nucleus for a web of open data. In: Aberer K, Choi KS, Noy N, Allemang D, Lee KI, Nixon L, Golbeck J, Mika P, Maynard D, Mizoguchi R, Schreiber G, Cudré-Mauroux P (eds) The semantic web. Lecture notes in computer science, vol 4825. Springer, Berlin/Heidelberg, pp 722–735
3. Bird S, Klein E, Loper E (2009) Natural language processing with Python. O'Reilly Media, Beijing
4. Blanco E, Cankaya HC, Moldovan D (2011) Commonsense knowledge extraction using concepts properties. In: Proceedings of the 24th Florida artificial intelligence research society conference (FLAIRS-24), Palm Beach, FL, USA, pp 222–227
5. Bollacker K, Evans C, Paritosh P, Sturge T, Taylor J (2008) Freebase: a collaboratively created graph database for structuring human knowledge. In: Proceedings of the 2008 ACM SIGMOD international conference on management of data, SIGMOD '08. ACM, New York, pp 1247–1250. doi:http://doi.acm.org/10.1145/1376616.1376746
6. Cambria E, Hussain A, Havasi C, Eckl C (2010) SenticSpace: visualizing opinions and sentiments in a multi-dimensional vector space. In: Knowledge-based and intelligent information and engineering systems. Springer, Heidelberg, pp 385–393
7. de Melo G, Weikum G (2010) Menta: inducing multilingual taxonomies from Wikipedia. In: Proceedings of the 19th ACM international conference on information and knowledge management, CIKM '10. ACM, New York, pp 1099–1108. doi:http://doi.acm.org/10.1145/1871437.1871577
8. Eckhardt N (2008) A kid's open mind common sense. PhD thesis, Tilburg University, http://ilk.uvt.nl/downloads/pub/papers/GV_thesis_NienkeFINAL.pdf
9. Etzioni O, Banko M, Soderland S, Weld DS (2008) Open information extraction from the web. Commun ACM 51:68–74. doi:http://doi.acm.org/10.1145/1409360.1409378
10. Fellbaum C (1998) WordNet: an electronic lexical database. MIT, Cambridge, MA
11. Havasi C, Speer R, Alonso J (2007) ConceptNet 3: a flexible, multilingual semantic network for common sense knowledge. In: recent advances in natural language processing, Borovets, Bulgaria, pp 27–29. http://web.mit.edu/~rspeer/www/research/cnet3.pdf
12. Havasi C, Borovoy R, Kizelshteyn B, Ypodimatopoulos P, Ferguson J, Holtzman H, Lippman A, Schultz D, Blackshaw M, Elliott G, Ng C (2011) The glass infrastructure: using common sense to create a dynamic, place-based social information system. In: Proceedings of 2011 conference on innovative applications of artificial intelligence. AAAI, San Francisco

[3]https://github.com/commonsense/conceptnet5/wiki/API

13. Kochenderfer MJ (2004) Common sense data acquisition for indoor mobile robots. In: Proceedings of the nineteenth national conference on artificial intelligence (AAAI-04), San Jose, California, at the San Jose Convention Center, July 25–29, 2004, pp 605–610
14. Korner S, Brumm T (2009) Resi – a natural language specification improver. In: IEEE international conference on semantic computing, 2009, ICSC '09, pp 1–8. doi:10.1109/ICSC.2009.47
15. Kudo T, Yamamoto K, Matsumoto Y (2004) Applying conditional random fields to Japanese morphological analysis. In: Lin D, Wu D (eds) Proceedings of EMNLP 2004. Association for Computational Linguistics, Barcelona, pp 230–237
16. Kuo YL, Lee JC, Chiang KY, Wang R, Shen E, Chan CW, Hsu JYJ (2009) Community-based game design: experiments on social games for commonsense data collection. In: Proceedings of the ACM SIGKDD workshop on human computation, HCOMP '09. ACM, New York, pp 15–22, doi:http://doi.acm.org/10.1145/1600150.1600154
17. Nastase V, Strube M, Boerschinger B, Zirn C, Elghafari A (2010) Wikinet: a very large scale multi-lingual concept network. In: LREC, Valletta, Malta, May 17–23, 2010
18. Navigli R, Ponzetto SP (2010) Babelnet: building a very large multilingual semantic network. In: Proceedings of the 48th annual meeting of the association for computational linguistics, ACL '10. Association for Computational Linguistics, Stroudsburg, pp 216–225. http://dl.acm.org/citation.cfm?id=1858681.1858704
19. Niles I, Pease A (2001) Towards a standard upper ontology. In: Proceedings of the international conference on formal ontology in information systems – volume 2001, FOIS '01. ACM, New York, pp 2–9. doi:http://doi.acm.org/10.1145/505168.505170
20. Porter MF (2001) Snowball: a language for stemming algorithms. Published online at http://snowball.tartarus.org/texts/introduction.html. Accessed 24 Oct 2011
21. Singh P, Lin T, Mueller ET, Lim G, Perkins T, Zhu WL (2002) Open Mind Common Sense: knowledge acquisition from the general public. In: On the move to meaningful internet systems, 2002 – DOA/CoopIS/ODBASE 2002 Confederated international conferences DOA, CoopIS and ODBASE 2002. Springer, London, pp 1223–1237. http://www.media.mit.edu/~push/ODBASE2002.pdf
22. Speer R, Havasi C, Lieberman H (2008) AnalogySpace: reducing the dimensionality of common sense knowledge. In: Proceedings of AAAI, Chicago, Illinois, July 13-17, 2008
23. Speer R, Havasi C, Treadway N, Lieberman H (2010) Finding your way in a multi-dimensional semantic space with Luminoso. In: Proceedings of the 15th international conference on intelligent user interfaces, Hong Kong, China, February 7–10, 2010
24. Ullberg J, Coradeschi S, Pecora F (2010) On-line ADL recognition with prior knowledge. In: Proceeding of the 2010 conference on STAIRS 2010: proceedings of the fifth Starting AI Researchers' symposium. IOS, Amsterdam, pp 354–366. http://dl.acm.org/citation.cfm?id=1940526.1940556
25. van Assem M, Isaac A, von Ossenbruggen J (2010) Wordnet 3.0 in RDF. Published online at http://semanticweb.cs.vu.nl/lod/wn30/. Accessed 24 Oct 2011
26. von Ahn L, Kedia M, Blum M (2006) Verbosity: a game for collecting common-sense facts. In: Proceedings of the SIGCHI conference on human factors in computing systems, CHI '06. ACM, New York, pp 75–78. doi:http://doi.acm.org/10.1145/1124772.1124784
27. Zesch T, Müller C, Gurevych I (2008) Extracting lexical semantic knowledge from Wikipedia and Wiktionary. In: Proceedings of the 6th conference on language resources and evaluation (LREC). http://elara.tk.informatik.tu-darmstadt.de/publications/2008/lrec08_camera_ready.pdf

# Chapter 7
# An Overview of BabelNet and its API for Multilingual Language Processing

**Roberto Navigli and Simone Paolo Ponzetto**

**Abstract** In this chapter we present BabelNet, a very large multilingual semantic network. We first describe the two-stage approach used to build it, namely: (a) the automatic integration of lexicographic information from WordNet with encyclopedic knowledge from Wikipedia; (b) the combination of Wikipedia's manually-edited translations with the output of a state-of-the-art machine translation system. Next, we present in detail statistics about the current version of BabelNet, which consists of a very large semantic network with lexicalizations for six languages (Catalan, English, French, German, Italian and Spanish). The figures all indicate that, thanks to our methodology, we are able to effectively create a knowledge repository containing wide-coverage lexical knowledge for many different languages. Finally, we present an overview of the Application Programming Interface (API) which enables easy programmatic access to all levels of information encoded in BabelNet.

## 7.1 Introduction

The availability of wide-coverage machine readable knowledge is an old, yet unsolved problem in Artificial Intelligence [53].[1] In the field of Natural Language Processing (NLP), in particular, many applications have been shown to benefit from the availability of lexical knowledge at different levels, including, among others, text summarization [30], named entity disambiguation [9], Question Answering [14,21], text categorization [12, 41, 58], coreference resolution [49, 51] and plagiarism detection [6], to name a few. Crucially, the availability of wide-coverage lexical

---

[1] See also [10] for a discussion from a machine learning perspective.

R. Navigli · S.P. Ponzetto (✉)
Dipartimento di Informatica, Sapienza University of Rome, Rome, Italy
e-mail: navigli@di.uniroma1.it; ponzetto@di.uniroma1.it

I. Gurevych and J. Kim (eds.), *The People's Web Meets NLP*, Theory and Applications of Natural Language Processing, DOI 10.1007/978-3-642-35085-6_7,

knowledge repositories has been shown to have a positive impact also on a core language understanding task such as Word Sense Disambiguation [33, 34], where rich and high-quality knowledge benefits both knowledge-rich systems [35, 48] and supervised classifiers [42, 60].

Seminal efforts addressed these problems by manually creating knowledge repositories, which led to full-fledged computational lexicons and ontologies, such as WordNet [11] and Cyc [20]. However, building such resources requires dozens of years, an effort which must be repeated for each new language, as shown by manually built multilingual knowledge repositories like EuroWordNet [57], MultiWordNet [46], BalkaNet [56], and the Multilingual Central Repository [2], among others. As a result, resources for non-English languages often have much poorer coverage and a clear bias exists towards conducting research in resource-rich languages such as English.

Historically, the limited coverage and high costs associated with manually created resources led to a great deal of work on acquiring structured knowledge automatically with minimal supervision [8]: however, while recent contributions in information extraction have successfully exploited large amounts of textual data, most notably from the Web [5, 44, 45, 61, *inter alia*], their output is still not ontologized. Besides, while these methods are language-independent in nature, they have been applied almost exclusively to English, arguably because of the fact that this is the predominant language of the Web, and less data are available for resource-poor languages (thus leading to sparser data and poorer performance of the methods). Consequently, it is still not clear whether all these methods make it possible to acquire large amounts of machine-readable knowledge lexicalized in arbitrary languages, which, in turn, is expected to have a positive impact on multilingual applications. Recently, this problem has become ever more acute, due to the fact that recent trends in NLP show not only that the field is moving towards high-end tasks (such as recognizing textual entailment or sentiment analysis), but also that there is substantial interest to tackle these problems from a multilingual perspective – see, e.g., [22] and [24], *inter-alia*, for some recent proposals.

In the last few years, many researchers turned to collaborative resources like Wikipedia[2] and took advantage of its semi-structured content to develop methods which provide a middle ground between manual and fully automatic approaches [23]. Much work in the literature has been devoted to the extraction of structured information from Wikipedia, including the automatic acquisition of semantic relations [59], taxonomies [50], and full-fledged semantic networks [7, 16, 26, 29, 32, 54]. One major feature of Wikipedia is its richness of explicit and implicit semantic information, mostly about named entities (e.g., Apple as a company). However, its encyclopedic nature is also a major limit, in that it lacks full coverage for the lexicographic senses of a given lemma (e.g., the apple fruit and tree senses are merged in one single meaning).

[2]http://www.wikipedia.org

In this chapter, we propose a novel solution[3] to these problems – namely the acquisition of wide-coverage, fully-ontologized multilingual lexical knowledge about nouns and named entities in arbitrary languages – by means of BabelNet. At its core, BabelNet consists of a very large multilingual lexical knowledge base built on the basis of a two-tier methodology which: (a) automatically combines the largest available semantic lexicon of English (WordNet) with a wide-coverage collaboratively-edited encyclopedia (Wikipedia) on the basis of an unsupervised mapping algorithm; (b) complements human-edited translations from Wikipedia with those obtained from a state-of-the-art machine translation system applied to sense-labeled data from SemCor [28] and Wikipedia itself. Each of these phases is specifically targeted at overcoming the knowledge acquisition bottleneck and achieving wide coverage at the conceptual, relational and lexical level, respectively. Integrating WordNet with Wikipedia allows us, in fact, to get the best of both worlds both in terms of a very large repository of concepts and named entities, and a rich semantic network. This is because in this step not only do we complement fine-grained lexicographic information (mostly about nominal concepts) from WordNet with encyclopedic one (typically about named entities) from Wikipedia, but we also enrich the highly structured 'core' network of labeled semantic relations of the former with millions of unlabeled, topically associative relations from the latter. Next, since BabelNet is, at its core, built around a multilingual semi-structured resource such as Wikipedia, we are able to leverage two of its most distinguishing features – i.e., its multilinguality and large amounts of annotated linguistic data – to complement manual translations from human editors with Machine Translation.

The remainder of this chapter is organized as follows: in Sect. 7.2 we introduce the two resources which are used to build BabelNet, i.e., WordNet and Wikipedia. Section 7.3 presents an overview of our resource and its construction methodology. Section 7.4 provides statistics for its current version. In Sect. 7.5 we show how the different levels of information encoded in BabelNet can easily be accessed in a programmatic way on the basis of a Java API. We finally provide an overview of related work from the literature in Sect. 7.6, and conclude with final remarks and future work directions in Sect. 7.7.

## 7.2 Knowledge Resources

BabelNet aims at providing an "encyclopedic dictionary" by bringing WordNet and Wikipedia together. In the following we provide a brief overview of these two resources.

---

[3]This paper is based on [36] and [37]. We expand our previous work by giving in Sect. 7.4 statistics for the current version of BabelNet, as well as an overview of how to access it programmatically in Sect. 7.5.

**WordNet.** WordNet [11], a computational lexicon of the English language based on psycholinguistic principles, is by far the most popular lexical knowledge resource in the field of NLP. A concept in WordNet is represented as a synonym set (called *synset*), i.e. the set of words that share the same meaning. For instance, the concept wind is expressed by the following synset:

$$\{\ \mathsf{wind}^1_n,\ \mathsf{air\ current}^1_n,\ \mathsf{current\ of\ air}^1_n \},$$

where each word's subscript and superscript indicate its parts of speech (e.g. *n* stands for noun) and sense number, respectively. Words can be polysemous and therefore the same word, e.g., balloon, can appear into more than one synset, each identifying one of its different senses: for instance, the sense of balloon as aircraft ($\mathsf{balloon}^1_n$) or toy ($\mathsf{balloon}^2_n$). For each synset, WordNet provides a textual definition, or gloss. For example, the gloss of the synset identifying the aircraft meaning of balloon is: "large tough nonrigid bag filled with gas or heated air".

Synsets are related to each other by means of *lexical* and *semantic relations*. Lexical relations identify connections between words like, for instance, antonyms (i.e., being the opposites of each other, cf. strong vs. weak). Semantic relations, instead, apply to all synonyms of a synset (since they share the same meaning): examples of such relations include hypernymy (expressing concept generalization, e.g., $\mathsf{balloon}^1_n$ *is-a* $\mathsf{lighter\text{-}than\text{-}air\ craft}^1_n$), hyponymy (expressing concept specialization, e.g., $\mathsf{hot-air\ balloon}^1_n$ *is-a* $\mathsf{balloon}^1_n$) and meronymy (capturing *part-of* relations: e.g., $\mathsf{gasbag}^2_n$ is a meronym of $\mathsf{balloon}^1_n$).

In addition to the standard WordNet relations, in this work we also consider *gloss* relations. Given a synset $S$ and its set of disambiguated gloss words $gloss(S) = \{\ \mathsf{s}_1, \ldots, \mathsf{s}_k\ \}$,[4] we define a semantic gloss relation between $S$ and each synset $S_i$ containing a sense $s_i \in gloss(S)$, $i = 1, \ldots, k$. For instance, the disambiguated gloss for $\mathsf{balloon}^1_n$ contains, among others, senses like $\mathsf{air}^1_n$ and $\mathsf{gas}^2_n$, so $S$ – i.e., $\mathsf{balloon}^1_n$ – is related to both of the latter synsets via the gloss relation.

**Wikipedia.** Our second resource, Wikipedia, is a multilingual Web-based encyclopedia. It is a collaborative open source medium edited by volunteers to provide a very large domain-independent repository of encyclopedic knowledge. Each article in Wikipedia is represented as a page (henceforth, Wikipage) and presents the knowledge about a specific concept (e.g. BALLOON (AIRCRAFT))[5] or named entity (e.g. MONTGOLFIER BROTHERS).[6] The title of a Wikipage (e.g. BALLOON (AIRCRAFT)) is composed of the lemma of the concept defined (e.g., balloon)

[4] Sense disambiguated glosses are distributed by the Princeton WordNet project at http://wordnet.princeton.edu/glosstag.shtml.

[5] Throughout this chapter, we use Sans Serif for words, SMALL CAPS for Wikipedia pages and CAPITALS for Wikipedia categories.

[6] Throughout the paper, unless otherwise stated, we use the general term *concept* to denote either a concept or a named entity.

plus an optional label in parentheses which specifies its meaning if the lemma is ambiguous (e.g., aircraft).

The text in Wikipedia is partially structured. Apart from articles often having tables and infoboxes (a special kind of table which summarizes the most important attributes of the entity referred to by a page), various relations exist between the pages themselves. These include:

(a) **Redirect pages**. These pages are used to forward to the Wikipage containing the actual information about a concept of interest. This is used to point alternative or highly related expressions for a concept to the same encyclopedic entry, and thus models loose *synonymy*. For instance, CHARLIÈRE and GONDOLA (BALLOON) both redirect to BALLOON (AIRCRAFT), whereas BALLOON (TOY) redirects to TOY BALLOON.
(b) **Disambiguation pages**. These pages collect links for a number of possible concepts an arbitrary expression could be referred to. This models *homonymy*, e.g., BALLOON (DISAMBIGUATION) links to both pages BALLOON (AIRCRAFT) and TOY BALLOON.
(c) **Internal links**. Wikipages typically contain hypertext linked to other Wikipages, which typically refer to highly related concepts. For instance, BALLOON (AIRCRAFT) links to AEROSTAT, GAS BALLOON, HYDROGEN, etc. whereas TOY BALLOON points to PUMP, BALLOON MAIL, and so on.
(d) **Inter-language links**. Wikipages also provide links to their counterparts (i.e. corresponding concepts) contained within wikipedias in other languages (e.g., the English Wikipage BALLOON (AIRCRAFT) links to the Italian PALLONE AEROSTATICO and German BALLON).
(e) **Categories**. Articles can be assigned to one or more categories, which are further categorized to provide a so-called "category tree", e.g. BALLOON (AIRCRAFT) is categorized under BALLOONING, which in turn is a subcategory of UNPOWERED AVIATION, and so on. In practice, this "tree" is not designed as a strict hierarchy, but rather as a multi-faceted categorization scheme to provide a thematic clustering of the encyclopedic entries.

Both WordNet and Wikipedia can be viewed as graphs. In the case of WordNet, nodes are synsets and edges lexical and semantic relations between synsets[7] whereas, in the case of Wikipedia, nodes are Wikipages and edges the hyperlinks between them (i.e., the above-mentioned *internal* links). By taking these two graphs and combining them, e.g. by merging nodes conveying the same meaning, we are able to produce an integrated resource, a task to which we now turn.

[7] Lexical relations link senses (e.g., $\text{dental}^1_a$ pertains-to $\text{tooth}^1_n$). However, relations between senses can be easily extended to the synsets which contain them, thus making all the relations connect synsets.

## 7.3 BabelNet

The construction of BabelNet has two key goals: first, integrating Wikipedia with WordNet automatically in an effective way; second, augmenting the resource with additional lexicalizations in many languages.

BabelNet encodes knowledge as a labeled directed graph $G = (V, E)$ where $V$ is the set of *nodes* – i.e., *concepts* such as $\text{balloon}_n^1$ or BALLOON (AIRCRAFT), and *named entities* like MONTGOLFIER BROTHERS – and $E \subseteq V \times R \times V$ is the set of *edges* connecting pairs of concepts (e.g., $\text{balloon}_n^1$ *is-a* $\text{lighter-than-air craft}_n^1$). Each edge is labeled with a *semantic relation* from $R$, i.e., $\{is-a,\ part-of,\ \ldots,\ \epsilon\}$, where $\epsilon$ denotes an unspecified semantic relation. Importantly, each node $v \in V$ contains a set of lexicalizations of the concept for different languages, e.g., { $\text{balloon}_{\text{EN}}$, $\text{Ballon}_{\text{DE}}$, $\text{aerostato}_{\text{ES}}$, ..., $\text{montgolfière}_{\text{FR}}$ }. We call such multilingually lexicalized concepts *Babel synsets*. Concepts and relations in BabelNet are harvested from the largest available semantic lexicon of English, WordNet, and a wide-coverage collaboratively-edited encyclopedia, Wikipedia (both introduced in Sect. 7.2). In order to build the BabelNet graph, we collect at different stages:

(a) From WordNet, all available word senses (as *concepts*) and all the lexical and semantic pointers between synsets (as *relations*);
(b) From Wikipedia, all encyclopedic entries (i.e., Wikipages, as *concepts*) and semantically unspecified *relations* from hyperlinked text.

An overview of BabelNet is given in Fig. 7.1. WordNet and Wikipedia can overlap both in terms of concepts and relations: accordingly, in order to provide a *unified resource*, we merge the intersection of these two knowledge sources (i.e., their concepts in common) by establishing a mapping between Wikipages and WordNet senses. Next, to enable multilinguality, we collect the lexical realizations of the available concepts in different languages. Finally, we connect the multilingual Babel synsets by collecting all relations found in WordNet, as well as all wikipedias in the languages of interest. Thus, our methodology consists of three main steps:

(a) We **combine WordNet and Wikipedia** by automatically acquiring a mapping between WordNet senses and Wikipages. This avoids duplicate concepts and allows their inventories of concepts to complement each other.
(b) We **harvest multilingual lexicalizations** of the available concepts (i.e., Babel synsets) using (a) the human-generated translations provided by Wikipedia (the so-called *inter-language* links), as well as (b) a machine translation system to translate occurrences of the concepts within sense-tagged corpora.
(c) We **establish relations between Babel synsets** by collecting all relations found in WordNet, as well as all wikipedias in the languages of interest.

**Mapping Wikipedia to WordNet.** The first phase of our methodology aims at establishing links between Wikipages and WordNet senses. Formally, given the entire set of pages *Wikipages* and WordNet senses *WNSenses*, we want to acquire a mapping:

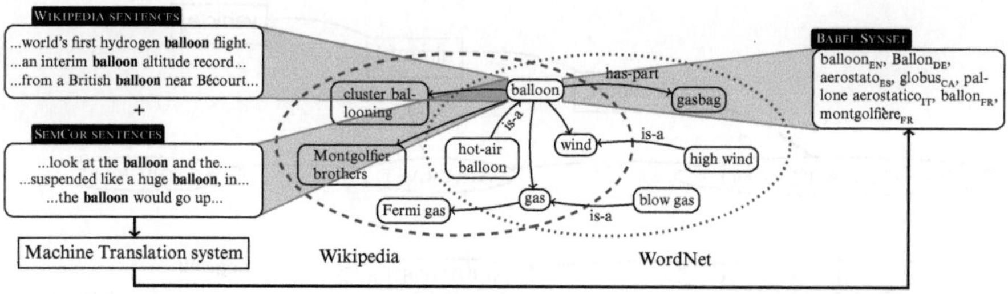


**Fig. 7.1** An illustrative overview of BabelNet

$$\mu : Wikipages \rightarrow WNSenses \cup \{\epsilon\}, \tag{7.1}$$

such that, for each Wikipage $w \in Wikipages$, we have:

$$\mu(w) = \begin{cases} s \in WNSenses(w) & \text{if a link can be established,} \\ \epsilon & \text{otherwise,} \end{cases} \tag{7.2}$$

where $WNSenses(w)$ is the set of senses of the *lemma* of $w$ in WordNet. For example, if our mapping methodology linked BALLOON (AIRCRAFT) to the corresponding WordNet sense $\text{balloon}_n^1$, we would have $\mu(\text{BALLOON (AIRCRAFT)}) = \text{balloon}_n^1$.

Our mapping algorithm, described in detail in [48], works by first leveraging resource-specific properties of our source and target resources, namely monosemy and redirections; next, given a Wikipedia page, it finds the WordNet sense that maximizes the probability of the sense providing an adequate corresponding concept for the page. In practice, resource mapping is viewed as a disambiguation process which pairs Wikipages and WordNet senses in such a way as to maximize the conditional probability of a WordNet sense given a Wikipedia page, on the basis of a bag-of-words disambiguation context.

**Translating Babel Synsets.** Once a mapping between English Wikipages and WordNet senses is established, BabelNet's multilingually lexicalized concepts are created as follows. Given a Wikipage $w$, and provided it is mapped to a sense $s$ (i.e., $\mu(w) = s$), we create a Babel synset $S \cup W$, where $S$ is the WordNet synset to which sense $s$ belongs, and $W$ includes: (i) $w$; (ii) the set of redirections to $w$; (iii) all its inter-language links (that is, translations of the Wikipage into other languages); (iv) the redirections to the inter-language links found in the Wikipedia of the target language. For instance, given that $\mu(\text{BALLOON (AIRCRAFT)}) = \text{balloon}_n^1$, the corresponding Babel synset is { $\text{balloon}_{\text{EN}}$, $\text{Ballon}_{\text{DE}}$, $\text{aerostato}_{\text{ES}}$, $\text{balón aerostático}_{\text{ES}}, \ldots,$ $\text{pallone aerostatico}_{\text{IT}}$ } (cf. Fig. 7.2). However, two issues arise: first, a concept might be covered only in one of the two resources (either WordNet or Wikipedia), meaning that no link can be established (e.g., FERMI GAS or $\text{gasbag}_n^1$ in Fig. 7.1); second, even if covered in both resources, the Wikipage for the concept

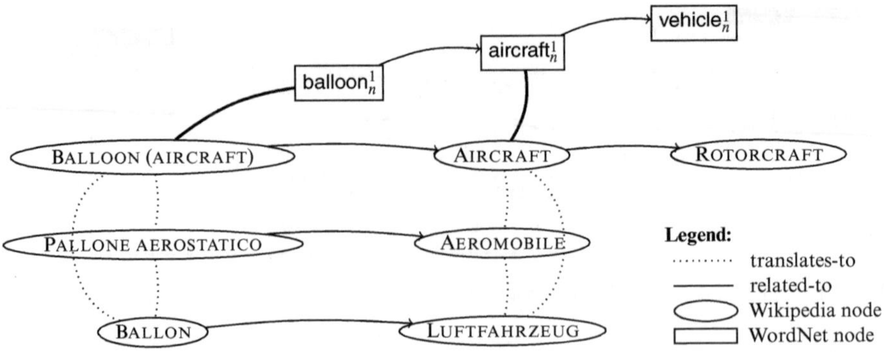


**Fig. 7.2** Translating Babel synsets based on Wikipedia's inter-language links. After mapping, Babel synsets integrating WordNet synsets and Wikipedia pages are straightforwardly translated by collecting the manual translations provided by editors as hyperlinks to wikipedias in languages other than English

might not provide any translation for the language of interest (e.g., the Catalan for BALLOON (AIRCRAFT) is missing in Wikipedia).

In order to address the above issues and thus guarantee high coverage for all languages we developed a methodology for translating senses in the Babel synset to missing languages. Given a WordNet word sense in our Babel synset of interest (e.g., balloon$^1_n$) we collect its occurrences in SemCor [28], a corpus of more than 200,000 words annotated with WordNet senses. We do the same for Wikipages by retrieving sentences in Wikipedia with links to the Wikipage of interest. By repeating this step for each English lexicalization in a Babel synset, we obtain a collection of sentences for the Babel synset (see left part of Fig. 7.1). Next, we apply state-of-the-art Machine Translation[8] and translate the set of sentences in all the languages of interest. Given a specific term in the initial Babel synset, we collect the set of its translations. We then identify the most frequent translation in each language and add it to the Babel synset. Note that translations are sense-specific, as the context in which a term occurs is provided to the translation system. For instance, in order to collect missing translations for BALLOON (AIRCRAFT) and its corresponding WordNet sense balloon$^1_n$, we collect from Wikipedia and translate sentences such as the following ones:

*Example 7.1.* Francois Pilatre de Rozier and François Laurent d'Arlandes flew in an aircraft lighter than air, a **[[Balloon (aircraft)|balloon]]**.

Similarly, from SemCor we collect and automatically translate, among others, the following sentence:

[8]We use the Google Translate API. An initial prototype used a statistical machine translation system based on Moses [18] and trained on Europarl [17]. However, we found such system unable to cope with many technical names, such as in the domains of sciences, literature, history, etc.

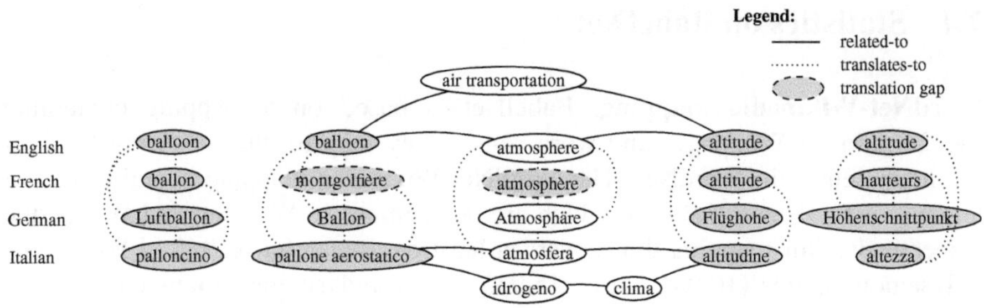


**Fig. 7.3** Translating Babel synsets based on a machine translation system. In order to fill lexical gaps (i.e. missing translations, typically for resource-poor languages), sense annotated data are collected from SemCor and Wikipedia, and their most frequent translations are included as network's lexicalizations

*Example 7.2.* Just like the **balloon**$^1_n$ would go up and you could sit all day and wish it would spring a leak or blow to hell up and burn and nothing like that would happen.

As a result, we can enrich the initial Babel synset with the following words: mongolfière$_{\text{FR}}$, globus$_{\text{CA}}$, globo$_{\text{ES}}$, mongolfiera$_{\text{IT}}$ (cf. Fig. 7.3). Note that we had no translation for Catalan and French in the first phase, because the inter-language link was not available, and we also obtain new lexicalizations for the Spanish and Italian languages.

**Harvesting semantic relations.** In order to provide a truly semantic network where concepts (i.e., Babel synsets, in our case) are connected by meaningful connections, we create edges expressing semantic relations between the synsets from WordNet and the wikipedias in the languages of interest. Given a Babel synset $s$, we first collect all relations found in WordNet for the corresponding WordNet synsets and senses it contains, if any. For instance, given the Babel synset containing the aircraft sense of balloon, we connect such synset with those Babel synsets containing the WordNet senses lighter-than-air craft$^1_n$, hot-air balloon$^1_n$, gasbag$^2_n$, etc. Similarly, we collect all relations from Wikipedia based on its internal hyperlink structure: for each Wikipage $w$ contained in $s$, we collect all Wikipedia links occurring in that page and for any such link from $w$ to $w'$ we establish an unspecified semantic relation $\epsilon$ between $s$ and the Babel synset $s'$ that contains $w'$ – that is, we link the Babel synset containing the Wikipage BALLOON (AIRCRAFT) to those containing AEROSTAT, GAS BALLOON, HYDROGEN, and so on. In order to harvest as many relevant relations as possible, we make use of *all* wikipedias in the available languages: that is, relations from wikipedias in languages other than English are also included. For instance, while the page BALLOON (AIRCRAFT) does not link directly to a highly related concept such as AIRSHIP, by pivoting on Italian (based on the inter-language links) we find that PALLONE AEROSTATICO links to DIRIGIBILE, so a link can be established between the two Babel synsets that contain both English and Italian lexicalizations.

## 7.4 Statistics on BabelNet

**WordNet-Wikipedia mapping.** BabelNet is based on a mapping containing 89,226 pairs of Wikipages and word senses they map to, thus covering 52 % of the noun senses in WordNet. The WordNet-Wikipedia mapping contains 72,572 lemmas, 10,031 and 26,398 of which are polysemous in WordNet and Wikipedia, respectively. Our mapping thus covers at least one sense for 62.9 % of WordNet's polysemous nouns (10,031 out of 15,935): these polysemous nouns can refer to 44,449 and 71,918 different senses in WordNet and Wikipedia, respectively, 13,241 and 16,233 of which are also found in the mapping.

**Lexicon.** BabelNet currently covers six languages, namely: English, Catalan, French, German, Italian and Spanish. The second column of Table 7.1 shows the number of lemmas for each language. In Table 7.2 we report instead the number of monosemous and polysemous words divided by part of speech. Given that we work with nominal synsets only, the numbers for verbs, adjectives and adverbs are the same as in WordNet 3.0. As for nouns, we observe a very large number of monosemous words (almost 23 million), but also a large amount of polysemous words (more than one million). Both numbers are considerably larger than in WordNet, because – as remarked above – words here denote both concepts (mainly from WordNet) and named entities (primarily from Wikipedia).

**Concepts.** BabelNet contains more than 3 million concepts, i.e., Babel synsets, and more than 26 million word senses (in any of the available languages). In Table 7.1 we report the number of synsets covered for each language (third column) and the number of word senses lexicalized in each language (fourth column). 72.3 % of the Babel synsets contain lexicalizations in all 6 languages and the overall number of word senses in English is much higher than those in the other languages (thanks to the high number of synonyms available in the English WordNet synsets and Wikipedia redirections). Each Babel synset contains 8.6 synonyms, i.e., word senses, on average in any language. The number of synonyms for each language ranges from 2.2 to 1.7 for English and Italian, respectively, with an average of 1.8 synonyms per language.

In Table 7.3 we show for each language the number of word senses obtained directly from WordNet, Wikipedia pages and redirections, as well as Wikipedia and WordNet translations (as a result of the translation process).

**Relations.** We now turn to relations in BabelNet. Relations come either from Wikipedia hyperlinks (in any of the covered languages) or WordNet. All our relations are semantic, in that they connect Babel synsets (rather than senses), however the relations obtained from Wikipedia are unlabeled. In Table 7.4 we show the number of lexico-semantic relations from WordNet, WordNet glosses and the six wikipedias used in our work. We can see that the major contribution comes from the English Wikipedia (50 million relations) and wikipedias in other languages (some million relations, depending on their size in terms of number of articles and links therein).

**Table 7.1** Number of lemmas, synsets and word senses in the six languages currently covered by BabelNet

| Language | Lemmas | Synsets | Word senses |
|---|---|---|---|
| English | 5,938,324 | 3,032,406 | 6,550,579 |
| Catalan | 3,518,079 | 2,214,781 | 3,777,700 |
| French | 3,754,079 | 2,285,458 | 4,091,456 |
| German | 3,602,447 | 2,270,159 | 3,910,485 |
| Italian | 3,498,948 | 2,268,188 | 3,773,384 |
| Spanish | 3,623,734 | 2,252,632 | 3,941,039 |
| Total | 23,935,611 | 3,032,406 | 26,044,643 |

**Table 7.2** Number of monosemous and polysemous words by part of speech (verbs, adjectives and adverbs are the same as in WordNet 3.0)

| POS | Monosemous words | Polysemous words |
|---|---|---|
| Noun | 22,763,265 | 1,134,857 |
| Verb | 6,277 | 5,252 |
| Adjective | 16,503 | 4,976 |
| Adverb | 3,748 | 733 |
| Total | 22,789,793 | 1,145,818 |

**Table 7.3** Composition of Babel synsets: number of synonyms from the English WordNet, Wikipedia pages and translations, as well as translations of WordNet's monosemous words and SemCor's sense annotations

| | | English | Catalan | French | German | Italian | Spanish | Total |
|---|---|---|---|---|---|---|---|---|
| English WordNet | | 206,978 | – | – | – | – | – | 206,978 |
| Wikipedia | pages | 2,955,552 | 123,101 | 524,897 | 506,892 | 404,153 | 349,375 | 4,863,970 |
| Wikipedia | redirections | 3,388,049 | 105,147 | 617,379 | 456,977 | 217,963 | 404,009 | 5,189,524 |
| Wikipedia | translations | – | 3,445,273 | 2,844,645 | 2,841,914 | 3,046,323 | 3,083,365 | 15,261,520 |
| WordNet | monosemous | – | 97,327 | 97,680 | 97,852 | 98,089 | 97,435 | 488,383 |
| WordNet | SemCor | – | 6,852 | 6,855 | 6,850 | 6,856 | 6,855 | 34,268 |
| Total | | 6,550,579 | 3,777,700 | 4,091,456 | 3,910,485 | 3,773,384 | 3,941,039 | 26,044,643 |

**Table 7.4** Number of lexico-semantic relations harvested from WordNet, WordNet glosses and the six wikipedias

| | English | Catalan | French | German | Italian | Spanish | Total |
|---|---|---|---|---|---|---|---|
| WordNet | 364,552 | – | – | – | – | – | 364,552 |
| WordNet glosses | 617,785 | – | – | – | – | – | 617,785 |
| Wikipedia | 50,104,884 | 978,006 | 5,613,873 | 5,940,612 | 3,602,395 | 3,411,612 | 69,651,382 |
| Total | 51,087,221 | 978,006 | 5,613,873 | 5,940,612 | 3,602,395 | 3,411,612 | 70,633,719 |

**Glosses.** Each Babel synset naturally comes with one or more glosses (possibly available in many languages). In fact, WordNet provides a textual definition for each English synset, while in Wikipedia a textual definition can be reliably obtained from

**Table 7.5** Glosses for the Babel synset referring to the concept of balloon as 'aerostat'

| | | |
|---|---|---|
| English | WordNet | Large tough nonrigid bag filled with gas or heated air. |
| English | Wikipedia | A balloon is a type of aerostat that remains aloft due to its buoyancy. |
| Catalan | | Un globus aerostàtic [1] és una aeronau aerostàtica no propulsada que es serveix del principi dels fluids d'Arquimedes per volar, entenent l'aire com un fluid. |
| French | | Un aérostat est un aéronef "plus léger que l'air", dont la sustentation est assurée par la poussée d'Archimède, contrairement à un aérodyne. |
| German | | Ein Ballon im heutigen Sprachgebrauch ist eine nicht selbsttragende, gasdichte Hülle, die mit Gas gefüllt ist und über keinen Eigenantrieb verfügt. |
| Italian | | Un pallone aerostatico è un tipo di aeromobile, un aerostato che si solleva da terra grazie al principio di Archimede. |
| Spanish | | Un globo aerostático es una aeronave aerostática no propulsada que se sirve del principio de los fluidos de Arquímedes para volar, entendiendo el aire como un fluido. |

the first sentence of each Wikipage.[9] Overall, BabelNet includes 4,683,031 glosses (2,985,243 of which are in English). In Table 7.5 we show the glosses for the Babel synset which refers to the concept of balloon as 'aerostat'.

## 7.5 Multilingual NLP in the Fast Lane with the BabelNet API

Similarly to WordNet, BabelNet consists, at its lowest level, of a plain text file. An excerpt of the entry for the Babel synset containing $\text{balloon}_n^1$ is shown in Fig. 7.4. The record contains (a) the synset's `id`; (b) the `region` of BabelNet where it lies (e.g., `WIKIWN` means at the intersection of WordNet and Wikipedia); (c) the corresponding (possibly empty) WordNet 3.0 synset `offset`; (d) the number of `senses` in all languages and their full listing; (e) the number of translation relations and their full listing; (f) the number of semantic `pointers` (i.e., relations) to other Babel synsets and their full listing. Senses encode information about their source – i.e., whether they come from WordNet (`WN`), Wikipedia pages (`WIKI`) or their redirections (`WIKIRED`), or are automatic translations (`WNTR` / `WIKITR`) – and about their language and lemma. In addition, translation relations between lexical items are represented as a mapping from source to target senses – e.g., `2_3,5,6,7`

[9]"The article should begin with a declarative sentence telling the nonspecialist reader what (or who) the subject is.", extracted from http://en.wikipedia.org/wiki/Wikipedia:Manual_of_Style/Lead_section#First_sentence. This simple, albeit powerful, heuristic has been successfully used in previous work to construct a corpus of definitional sentences [40].

```
bn:00008187n WIKIWN 02782093n 25 WN:EN:balloon WIKI:EN:Balloon_(aircraft)
                                 WIKI:DE:Ballon WNTR:ES:globo
                                 WIKI:IT:Pallone_aerostatico
                                 WIKI:ES:Aerostato WIKITR:CA:globus ...
                              13 1_4 2_3,5,6,7 ...
                             218 r bn:00006823n r bn:01071110n|FROM_IT
                                 @ bn:00051149n ...
```

**Fig. 7.4** The Babel synset for $\mathsf{balloon}_n^1$, i.e. its 'aircraft' sense (excerpt)

means that the second element in the list of senses (the English word balloon) translates into items #3 (German Ballon), #5 (Italian pallone aerostatico), #6 (Spanish aerostato), and #7 (Catalan globus). Finally, semantic relations are encoded using WordNet pointers and an additional symbol for Wikipedia relations (`r`), which can also specify the source of the relation (e.g., `FROM_IT` means that the relation was harvested from the Italian Wikipedia). In Fig. 7.4, the Babel synset inherits the WordNet hypernym (@) relation to $\mathsf{lighter\text{-}than\text{-}air\ craft}_n^1$ (offset `bn:00051149n`), as well as Wikipedia relations to the synsets of HYDROGEN (`bn:00006823n`) and VINCENZO LUNARDI (`bn:01071110n`, from Italian).

Information encoded in the text dump of BabelNet can be effectively accessed and automatically embedded within applications by means of a programmatic access. To this end, we developed a Java API, based on Apache Lucene[10] as backend, which indexes the textual dump and includes a variety of methods to access the the four main levels of information encoded in BabelNet, namely: (a) lexicographic (information about word senses), (b) encyclopedic (i.e. named entities), (c) conceptual (the semantic network made up of its concepts), (d) and multilingual level (information about word translations). Figure 7.5 shows a usage example of the BabelNet API. In the code snippet we start by querying the Babel synsets for the English word balloon (line 3). Next, we access different kinds of information for each synset: first, we print their id, source (WordNet, Wikipedia, or both), the corresponding, possibly empty, WordNet offsets, and 'main lemma' – namely, a compact string representation of the Babel synset consisting of its corresponding WordNet synset in stringified form, or the first non-redirection Wikipedia page found in it (lines 5–7). Then, we access and print the Italian word senses they contain (lines 8–10), and finally the synsets they are related to (lines 11–19). Thanks to carefully designed Java classes, we are able to accomplish all of this in about 20 lines of code.

[10]http://lucene.apache.org

```
BabelNet bn = BabelNet.getInstance();
System.out.println("SYNSETS WITH English word: \"balloon\"");
List<BabelSynset> synsets = bn.getSynsets(Language.EN, "balloon");
for (BabelSynset synset : synsets) {
  System.out.print("  =>(" + synset.getId() + ") SOURCE: " + synset.getSource() +
                   "; WN SYNSET: " + synset.getWordNetOffsets() + ";\n" +
                   "  MAIN LEMMA: " + synset.getMainLemma() + ";\n  SENSES (IT): { ");
  for (BabelSense sense : synset.getSenses(Language.IT))
    System.out.print(sense.toString()+" ");
  System.out.println("}\n  -----");
  Map<IPointer, List<BabelSynset>> relatedSynsets = synset.getRelatedMap();
  for (IPointer relationType : relatedSynsets.keySet()) {
    List<BabelSynset> relationSynsets = relatedSynsets.get(relationType);
    for (BabelSynset relationSynset : relationSynsets) {
      System.out.println("    EDGE " + relationType.getSymbol() +
                         " " + relationSynset.getId() +
                         " " + relationSynset.toString(Language.EN));
    }
  }
  System.out.println("  -----");
}
```

**Fig. 7.5** Sample BabelNet API usage. Thanks to carefully designed classes, all levels of information encoded in BabelNet can be accessed with a few lines of code

## 7.6 Related Work

BabelNet focuses around two key objectives, namely to automatically integrate lexicographic and encyclopedic knowledge from WordNet and Wikipedia effectively on the basis of an unsupervised algorithm, and to provide multilingual lexical information with wide coverage for all its languages. Accordingly, in the following we review previous work from the literature on the topics of lexical resource mapping and integration, and multilingual lexical knowledge acquisition.

### 7.6.1 Lexical Resource Mapping and Integration

The last years have seen a great deal of work on aligning and merging lexical semantic databases such as WordNet with collaboratively generated resources like Wikipedia and Wiktionary.[11] One of the first proposals to map Wikipedia to WordNet is presented by Ruiz-Casado et al. [52], who associate Wikipedia pages with their most similar WordNet synsets on the basis of the similarity between their respective bags-of-words representations (built using the pages' content and glosses, respectively). Later work on the automatic alignment and enrichment of GermaNet [19] with Wiktionary by Henrich et al. [15] exploits the structure of the semantic network to acquire sense descriptions when no glosses are available (as in the case of GermaNet): these sense descriptions are later used within a Lesk-like (i.e., word overlap based) algorithm [4] to automatically produce the mapping between the two resources.

In contrast to using bags-of-words, Suchanek et al. [54] build the YAGO ontology by relying instead on the so-called 'most frequent sense' heuristic typically used in Word Sense Disambiguation [33]. In YAGO, Wikipedia categories are linked to WordNet synsets on the basis of (a) a pre-processing step which approximates the label of categories consisting of complex noun phrases (e.g. AIRSHIP TECHNOLOGY) with their lexical head (i.e. the most important noun found in the label, e.g. technology); (b) a mapping which associates each category with (the synset containing) the sense of its label: in case more than one such synset exists (i.e., the label consists of a polysemous word), a link is established with the sense which is most frequent in SemCor [28], a sense-labeled corpus. This simple approach can, in turn, be improved in a variety of ways, e.g., by combining different sources of evidence like knowledge-based semantic similarity and distributional methods [55].

Supervised approaches to align Wikipedia to WordNet are explored by de Melo and Weikum in the construction of MENTA [26], a multilingual taxonomy. The proposed supervised model is built from a set of manually-labeled mappings, and uses a variety of features such as word-level information (term overlap between sets of synonyms and redirections, cosine similarity between the vector of glosses), as well as YAGO's most frequent sense heuristic (used here as a soft constraint, instead). An alternative method [43], instead, starts with bag-of-word representations for WordNet synsets and Wikipedia pages and computes their Personalized PageRank vector by running the PageRank algorithm over WordNet [1], while initializing the probabilities to the senses of the lemmatized words in the bag. These vectors are then fed into a vector similarity measure, e.g. cosine distance, to compute the strength of the mapping. Finally, the similarity scores are used, together with a set of manually labeled mappings, to learn in a supervised way the threshold for the minimum similarity that a sense pair must have to generate a mapping. Mapping heterogeneous resources on the basis of a supervised model is shown in these works

[11] http://www.wiktionary.org

to yield high-performing results: however, this approach has the downside that it cannot be applied to arbitrary resources where no manually-labeled mappings are available for training. This problem becomes even more acute as soon as the mapping method needs to be applied to other resources (e.g., Wiktionary [27]), or the mapping itself involves more than one resource [13]: in this case, in fact, the annotation effort must be repeated for each resource pair in turn.

A possible solution to the knowledge acquisition bottleneck suffered by supervised systems is to exploit structure within an unsupervised framework – e.g., since Wikipedia provides semi-structured content, its structured features can be exploited also for the mapping task. An approach of this kind is presented by Ponzetto and Navigli [47], who associate categories from WikiTaxonomy [50] with those synsets which have the highest degree of structural overlap (computed against WordNet's taxonomy). Using a graph-based technique is found to improve by a large margin over the most frequent sense heuristic: the mapping, in turn, can be used to restructure the Wikipedia taxonomy, in order to improve its quality by increasing its degree of alignment with the reference resource (i.e., WordNet).

### *7.6.2 Multilingual Lexical Knowledge Acquisition*

Similarly to the case of mapping lexical knowledge resources, much work has been done in recent years on the automatic acquisition of multilingual lexical knowledge bases. Important early work by de Melo and Weikum in [25] aims at developing a Universal WordNet (UWN) by extending a core taxonomic backbone provided by WordNet with additional information in languages other than English from existing wordnets, translation dictionaries, and parallel corpora. A later extension, named MENTA [26], consists of a large-scale taxonomy of named entities and their semantic classes built by integrating WordNet with Wikipedia. The common theme underlying the construction methodology of both UWN and MENTA is to collect large amounts of entities and translations from external resources, and integrate them within the clean taxonomic structure of WordNet (UWN), which is further enriched with additional concepts from Wikipedia on the basis of an automatic mapping (MENTA).

Nastase et al. develop WikiNet [32], a very large multilingual semantic network built by leveraging different elements of Wikipedia at the same time, including infoboxes, internal and inter-language links, and categories. The concept inventory of WikiNet is made out of all Wikipedia pages (entities, mostly) and categories (typically referring to classes). Relations between concepts are harvested on the basis of a link co-occurrence analysis of Wikipedia's markup text – i.e., a semantically unspecified relation is assumed to exist between pairs of articles hyperlinked within the same window – as well as from relations between categories generated using the heuristics developed in previous work by Ponzetto and Strube [50] and Nastase and Strube [31].

Multilingual lexical knowledge bases like UWN/MENTA and WikiNet are primarily concerned with conceptual knowledge. At the lexical level, in fact, these approaches essentially rely 'only' on Wikipedia's inter-language links (WikiNet), and additionally perform statistical inference, in order to produce a more coherent output (MENTA). BabelNet aims at providing a contribution which is complementary to these efforts by focusing, instead, on providing wide-coverage lexical knowledge for all languages. To this end, so-called 'translation gaps' (i.e., missing translations from Wikipedia) are filled using Statistical Machine Translation. The result is a very large lexical multilingual knowledge base, whose high-quality knowledge has been recently applied to a variety of monolingual and cross-lingual lexical semantic tasks like multilingual Word Sense Disambiguation [39] and semantic relatedness [38]. An approach complementary to BabelNet's main focus on multilingual wide coverage is presented in the construction of the UBY lexical knowledge base [13]. UBY aims, in fact, at achieving wide coverage by specifically targeting the alignment and integration of a wide range of resources in multiple languages, including manually assembled ones like WordNet, GermaNet and FrameNet [3], as well as collaboratively constructed resources such as Wikipedia and Wiktionary. In order to enable true resource interoperability, a special focus is given also to developing a standardized format for heterogeneous lexical knowledge bases built upon the ISO standard Lexical Markup Framework.

## 7.7 Conclusions

In this chapter we presented BabelNet, a wide-coverage multilingual knowledge resource obtained by means of an automatic construction methodology. Key to our approach is a two-tier methodology, namely: (a) an unsupervised method to produce a mapping between a multilingual encyclopedic knowledge repository (Wikipedia) and a computational lexicon of English (WordNet); (b) the use of a state-of-the-art machine translation system to collect a very large amount of multilingual concept lexicalizations, and complement Wikipedia's manually-edited translations.

BabelNet includes several million instances of semantic relations, mainly from Wikipedia (however, WordNet relations are labeled), and contains more than three million concepts (8.6 labels per concept on average). While BabelNet currently includes six languages, links to freely-available wordnets[12] can immediately be established by utilizing the English WordNet as an inter-language index. Indeed, BabelNet can be extended to virtually any language of interest and our translation method allows it to cope with any resource-poor language.

Our work opens up many exciting directions to extend and leverage BabelNet. As future work, we plan to enlarge our core network with other languages, including Eastern European, Arabic, and Asian languages. Our overarching vision is that

[12]http://www.globalwordnet.org

of a very large multilingual knowledge base which will enable knowledge-rich approaches for many different NLP applications on a multitude of languages. In fact, we have already taken the first steps in this direction by showing the beneficial effects of a multilingual approach to Word Sense Disambiguation [39] and computing semantic relatedness [38]. Accordingly, we plan to explore in the very near future the use of BabelNet for a wide spectrum of high-end multilingual NLP tasks, including discourse-level applications like cross-lingual summarization and question answering, and Information Retrieval tasks like knowledge-rich query translation and cross-lingual expansion.

## Downloads

BabelNet and its API are freely available at http://lcl.uniroma1.it/babelnet under a Creative Commons Attribution-Noncommercial-Share Alike License.

## Acknowledgments

The authors gratefully acknowledge the support of the ERC Starting Grant MultiJEDI No. 259234. Thanks go to Google for access to the University Research Program for Google Translate.

## References

1. Agirre E, Soroa A (2009) Personalizing PageRank for Word Sense Disambiguation. In: Proceedings of the 12th conference of the European chapter of the association for computational linguistics, Athens, Greece, 30 March–3 April 2009, pp 33–41
2. Atserias J, Villarejo L, Rigau G, Agirre E, Carroll J, Magnini B, Vossen P (2004) The MEANING multilingual central repository. In: Proceedings of the 2nd international global WordNet conference, Brno, Czech Republic, 20–23 Jan 2004, pp 80–210
3. Baker CF, Fillmore CJ, Lowe JB (1998) The Berkeley FrameNet project. In: Proceedings of the 17th international conference on computational linguistics and 36th annual meeting of the association for computational linguistics, Montréal, Québec, Canada, 10–14 Aug 1998
4. Banerjee S, Pedersen T (2003) Extended gloss overlap as a measure of semantic relatedness. In: Proceedings of the 18th international joint conference on artificial intelligence, Acapulco, Mexico, 9–15 Aug 2003, pp 805–810
5. Banko M, Cafarella MJ, Soderland S, Broadhead M, Etzioni O (2007) Open information extraction from the Web. In: Proceedings of the 20th international joint conference on artificial intelligence, Hyderabad, India, 6–12 Jan 2007, pp 2670–2676
6. Barrón-Cedeño A, Rosso P, Agirre E, Labaka G (2010) Plagiarism detection across distant language pairs. In: Proceedings of the 23rd international conference on computational linguistics, Beijing, China, 23–27 Aug 2010, pp 37–45

7. Bizer C, Lehmann J, Kobilarov G, Auer S, Becker C, Cyganiak R, Hellmann S (2009) DBpedia – a crystallization point for the web of data. J Web Semant 7(3):154–165
8. Buitelaar P, Cimiano P, Magnini B (eds) (2005) Ontology learning from text: methods, evaluation and applications. IOS, Amsterdam
9. Bunescu R, Paşca M (2006) Using encyclopedic knowledge for named entity disambiguation. In: Proceedings of the 11th conference of the European chapter of the association for computational linguistics, Trento, Italy, 3–7 Apr 2006, pp 9–16
10. Domingos P (2007) Toward knowledge-rich data mining. Data Min Knowl Disc 15(1):21–28
11. Fellbaum C (ed) (1998) WordNet: an electronic database. MIT, Cambridge, MA
12. Gabrilovich E, Markovitch S (2006) Overcoming the brittleness bottleneck using Wikipedia: enhancing text categorization with encyclopedic knowledge. In: Proceedings of the 21st national conference on artificial intelligence, Boston, MA, 16–20 July 2006, pp 1301–1306
13. Gurevych I, Eckle-Kohler J, Hartmann S, Matuschek M, Meyer CM, Wirth C (2012) UBY – a large-scale unified lexical-semantic resource based on LMF. In: Proceedings of the 13th conference of the European chapter of the association for computational linguistics, Avignon, France, 23–27 Apr 2012, pp 580–590
14. Harabagiu SM, Moldovan D, Paşca M, Mihalcea R, Surdeanu M, Bunescu R, Girju R, Rus V, Morarescu P (2000) FALCON: boosting knowledge for answer engines. In: Proceedings of the ninth text REtrieval conference, Gaithersburg, Maryland, 13–16 Nov 2000, pp 479–488
15. Henrich V, Hinrichs E, Vodolazova T (2011) Semi-automatic extension of GermaNet with sense definitions from Wiktionary. In: Proceedings of 5th language & technology conference, Poznań, Poland, 25–27 Nov 2011, pp 126–130
16. Hoffart J, Suchanek FM, Berberich K, Weikum G (2012) YAGO2: a spatially and temporally enhanced knowledge base from Wikipedia. Artif Intell. doi:10.1016/j.artint.2012.06.001
17. Koehn P (2005) Europarl: a parallel corpus for statistical machine translation. In: Proceedings of machine translation summit X, Phuket, Thailand, 2005, pp 79–86
18. Koehn P, Hoang H, Birch A, Callison-Burch C, Federico M, Bertoldi N, Cowan B, Shen W, Moran C, Zens R, Dyer C, Bojar O, Constantin A, Herbst E (2007) Moses: open source toolkit for statistical machine translation. In: Companion volume to the proceedings of the 45th annual meeting of the association for computational linguistics, Prague, Czech Republic, 23–30 June 2007, pp 177–180
19. Lemnitzer L, Kunze C (2002) GermaNet – representation, visualization, application. In: Proceedings of the 3rd international conference on language resources and evaluation, Las Palmas, Canary Islands, Spain, 29–31 May 2002, pp 1485–1491
20. Lenat DB (1995) Cyc: a large-scale investment in knowledge infrastructure. Commun ACM 38(11), pp 33–38
21. Lita LV, Hunt WA, Nyberg E (2004) Resource analysis for question answering. In: Companion volume to the proceedings of the 42nd annual meeting of the association for computational linguistics, Barcelona, Spain, 21–26 July 2004, pp 162–165
22. Lu B, Tan C, Cardie C, KB Tsou (2011) Joint bilingual sentiment classification with unlabeled parallel corpora. In: Proceedings of the 49th annual meeting of the association for computational linguistics, Portland, OR, 19–24 June 2011, pp 320–330
23. Medelyan O, Milne D, Legg C, Witten IH (2009) Mining meaning from Wikipedia. Int J Hum-Comput Stud 67(9):716–754. doi:10.1016/j.ijhcs.2009.05.004
24. Mehdad Y, Negri M, Federico M (2011) Using bilingual parallel corpora for cross-lingual textual entailment. In: Proceedings of the 49th annual meeting of the association for computational linguistics, Portland, OR, 19–24 June 2011, pp 1336–1345
25. de Melo G, Weikum G (2009) Towards a universal wordnet by learning from combined evidence. In: Proceedings of the eighteenth ACM conference on information and knowledge management, Hong Kong, China, 2–6 Nov 2009, pp 513–522
26. de Melo G, Weikum G (2010) MENTA: inducing multilingual taxonomies from Wikipedia. In: Proceedings of the nineteenth ACM conference on information and knowledge management, Toronto, Canada, 26–30 Oct 2010, pp 1099–1108

27. Meyer CM, Gurevych I (2011) What psycholinguists know about chemistry: aligning Wiktionary and WordNet for increased domain coverage. In: Proceedings of the 5th international joint conference on natural language processing, Chiang Mai, Thailand, 8–13 Nov 2011, pp 883–892
28. Miller GA, Leacock C, Tengi R, Bunker R (1993) A semantic concordance. In: Proceedings of the 3rd DARPA workshop on human language technology, Plainsboro, NJ, pp 303–308
29. Moro A, Navigli R (2012) WiSeNet: building a Wikipedia-based semantic network with ontologized relations. In: Proceedings of the twenty-first ACM conference on information and knowledge management, Maui, Hawaii, 29 Oct–2 Nov 2012
30. Nastase V (2008) Topic-driven multi-document summarization with encyclopedic knowledge and activation spreading. In: Proceedings of the conference on empirical methods in natural language processing, Waikiki, Honolulu, HI, 25–27 Oct 2008, pp 763–772
31. Nastase V, Strube M (2008) Decoding Wikipedia category names for knowledge acquisition. In: Proceedings of the 23rd conference on the advancement of artificial intelligence, Chicago, IL, 13–17 July 2008, pp 1219–1224
32. Nastase V, Strube M (2012) Transforming Wikipedia into a large scale multilingual concept network. Artif Intell. doi:10.1016/j.artint.2012.06.008
33. Navigli R (2009) Word Sense Disambiguation: a survey. ACM Comput Surv 41(2):1–69
34. Navigli R (2012) A quick tour of Word Sense Disambiguation, induction and related approaches. In: Bieliková M, Friedrich G, Gottlob G, Katzenbeisser S, Turán G (eds) SOFSEM 2012: theory and practice of computer science. Lecture notes in computer science, vol 7147. Springer, Heidelberg, pp 115–129
35. Navigli R, Lapata M (2010) An experimental study on graph connectivity for unsupervised Word Sense Disambiguation. IEEE Trans Pattern Anal Mach Intel 32(4):678–692
36. Navigli R, Ponzetto SP (2010) BabelNet: building a very large multilingual semantic network. In: Proceedings of the 48th annual meeting of the association for computational linguistics, Uppsala, Sweden, 11–16 July 2010, pp 216–225
37. Navigli R, Ponzetto SP (2012) BabelNet: the automatic construction, evaluation and application of a wide-coverage multilingual semantic network. Artif Intell. doi:10.1016/j.artint.2012.07.001
38. Navigli R, Ponzetto SP (2012) BabelRelate! A joint multilingual approach to computing semantic relatedness. In: Proceedings of the 26th conference on artificial intelligence, Toronto, ON, Canada, 22–26 July 2012, pp 108–114
39. Navigli R, Ponzetto SP (2012) Joining forces pays off: multilingual joint Word Sense Disambiguation. In: Proceedings of the 2012 joint conference on empirical methods in natural language processing and computational language learning, Jeju Island, South Korea, 12–14 July 2012, pp 1399–1410
40. Navigli R, Velardi P (2010) Learning Word-Class Lattices for definition and hypernym extraction. In: Proceedings of the 48th annual meeting of the association for computational linguistics, Uppsala, Sweden, 11–16 July 2010, pp 1318–1327
41. Navigli R, Faralli S, Soroa A, de Lacalle OL, Agirre E (2011) Two birds with one stone: learning semantic models for Text Categorization and Word Sense Disambiguation. In: Proceedings of the twentieth ACM conference on information and knowledge management, Glasgow, Scotland, UK, 24–28 Oct 2011, pp 2317–2320
42. Ng HT, Lee HB (1996) Integrating multiple knowledge sources to disambiguate word senses: an exemplar-based approach. In: Proceedings of the 34th annual meeting of the association for computational linguistics, Santa Cruz, CA, 24–27 June 1996, pp 40–47
43. Niemann E, Gurevych I (2011) The people's web meets linguistic knowledge: automatic sense alignment of Wikipedia and WordNet. In: Proceedings of the 9th international conference on computational semantics, Oxford, UK, pp 205–214
44. Paşca M (2007) Organizing and searching the World Wide Web of facts – Step two: Harnessing the wisdom of the crowds. In: Proceedings of the 16th world wide web conference, Banff, Canada, 8–12 May 2007, pp 101–110

45. Paşca M, Lin D, Bigham J, Lifchits A, Jain A (2006) Organizing and searching the world wide web of facts – step one: the one-million fact extraction challenge. In: Proceedings of the 21st national conference on artificial intelligence, Boston, MA, 16–20 July 2006, pp 1400–1405
46. Pianta E, Bentivogli L, Girardi C (2002) MultiWordNet: developing an aligned multilingual database. In: Proceedings of the 1st international global WordNet conference, Mysore, India, 21–25 Jan 2002, pp 21–25
47. Ponzetto SP, Navigli R (2009) Large-scale taxonomy mapping for restructuring and integrating Wikipedia. In: Proceedings of the 21st international joint conference on artificial intelligence, Pasadena, CA, 14–17 July 2009, pp 2083–2088
48. Ponzetto SP, Navigli R (2010) Knowledge-rich Word Sense Disambiguation rivaling supervised systems. In: Proceedings of the 48th annual meeting of the association for computational linguistics, Uppsala, Sweden, 11–16 July 2010, pp 1522–1531
49. Ponzetto SP, Strube M (2007) Knowledge derived from Wikipedia for computing semantic relatedness. J Artif Intell Res 30:181–212
50. Ponzetto SP, Strube M (2011) Taxonomy induction based on a collaboratively built knowledge repository. Artif Intell 175:1737–1756
51. Rahman A, Ng V (2011) Narrowing the modeling gap: a cluster-ranking approach to coreference resolution. J Artif Intell Res 40:469–521
52. Ruiz-Casado M, Alfonseca E, Castells P (2005) Automatic assignment of Wikipedia encyclopedic entries to WordNet synsets. In: Advances in web intelligence. Lecture notes in computer science, vol 3528. Springer, Berlin/New York, pp 380–386
53. Schubert LK (2006) Turing's dream and the knowledge challenge. In: Proceedings of the 21st national conference on artificial intelligence, Boston, MA, 16–20 July 2006, pp 1534–1538
54. Suchanek FM, Kasneci G, Weikum G (2008) Yago: a large ontology from Wikipedia and WordNet. J Web Semant 6(3):203–217
55. Toral A, Ferrández O, Agirre E, Muñoz R (2009) A study on linking Wikipedia categories to WordNet synsets using text similarity. In: Proceedings of the international conference on recent advances in natural language processing, Borovets, Bulgaria, 14–16 Sept 2009, pp 449–454
56. Tufiş D, Ion R, Ide N (2004) Fine-grained Word Sense Disambiguation based on parallel corpora, word alignment, word clustering, and aligned wordnets. In: Proceedings of the 20th international conference on computational linguistics, Geneva, Switzerland, 23–27 Aug 2004, pp 1312–1318
57. Vossen P (ed) (1998) EuroWordNet: a multilingual database with lexical semantic networks. Kluwer, Dordrecht
58. Wang P, Domeniconi C (2008) Building semantic kernels for text classification using Wikipedia. In: Proceedings of the 14th ACM SIGKDD international conference on knowledge discovery and data mining, Las Vegas, Nevada, 24–27 Aug 2008, pp 713–721
59. Wu F, Weld D (2008) Automatically refining the Wikipedia infobox ontology. In: Proceedings of the 17th world wide web conference, Beijing, China, 21–25 Apr 2008, pp 635–644
60. Yarowsky D, Florian R (2002) Evaluating sense disambiguation across diverse parameter spaces. Nat Lang Eng 9(4):293–310
61. Yates A, Etzioni O (2009) Unsupervised methods for determining object and relation synonyms on the web. J Artif Intell Res 34:255–296

# Chapter 8
# Hierarchical Organization of Collaboratively Constructed Content

**Jianxing Yu, Zheng-Jun Zha, and Tat-Seng Chua**

**Abstract** Huge collections of collaboratively constructed content (e.g. blogs, consumer reviews, etc.) are now available online. This content has become a valuable knowledge repository, which enables users to seek quality information. However, such content is often unorganized, leading to difficulty in information navigation and knowledge acquisition. This chapter focuses on discovering the structure of the content and organizing them accordingly, so as to facilitate users in understanding the knowledge inherent within the content. In particular, we employ one example of the collaboratively constructed content, i.e. consumer reviews on products, as a case study, and propose a domain-assisted approach to generate a hierarchical structure to organize the reviews. The hierarchy organizes product aspects as nodes following their parent-child relations. For each aspect, the reviews and corresponding opinions on this aspect are stored. Such hierarchy provides a well-visualized way to browse consumer reviews at different granularity to meet various users' needs, which can help to improve information dissemination and accessibility. We further apply the generated hierarchy to support the application of opinion Question Answering (opinion-QA) for products, which aims to generate appropriate answers for opinion questions about products. The experimental results on 11 popular products in 4 domains demonstrate the effectiveness of our approach.

## 8.1 Introduction

The rapid expansion of social media facilitates users to collaborate online on a variety of activities. These activities include discussing hot topics on blogs, expressing opinions on products via forum reviews, sharing videos on the media

J. Yu (✉) · Z.-J. Zha · T.-S. Chua
AS6, 13 Computing Drive, School of Computing, National University of Singapore, Singapore 117417, Singapore
e-mail: jianxing@comp.nus.edu.sg; zhazj@comp.nus.edu.sg; chuats@comp.nus.edu.sg

I. Gurevych and J. Kim (eds.), *The People's Web Meets NLP*, Theory and Applications of Natural Language Processing, DOI 10.1007/978-3-642-35085-6_8,

cites such as YouTube, asking and answering questions in community websites, etc. Huge collections of content are constructed by such collaborative activities and distributed on the Web. A recent study reports that the collaboratively constructed content is created at the rate of about 10 GB a day [49]. This content covers a wide range of media sources, such as question-answer service, digital video, blogging, podcasting, forum, review site, social networking, and wiki, etc. It also reflects most public collective knowledge, and has become a valuable information repository. For example, users can listen to the voice of customers from online content prior to purchasing products, accordingly advertisers can use the content to identify the potential customers and understand their needs for posing effective strategies on marketing. While the content is often unorganized, this leads to the difficulty in information navigation and knowledge acquisition. It is impractical for users to grasp the overview of the constructed content on a certain topic, and inefficient to browse specific details (e.g. sub-topics) in the content. Thus, there is a compelling need to organize the content and transform it into a useful knowledge structure, so as to enable users to understand the knowledge inherent within the content. Since the hierarchy can improve information dissemination and accessibility [9], we propose to generate a hierarchical structure for organizing the content.

In this chapter, we focus on consumer reviews on products, which is one example of the collaboratively constructed content. The reviews are utilized as a case study to illustrate the procedure of generating a hierarchical structure for organizing them. Generally, the reviews usually include consumer opinions on various products and some aspects of the products. An *aspect* here refers to a component or an attribute of a certain product. A sample review in Fig. 8.1 reveals positive opinions on the aspects such as "design," "interface" of the product *iPhone 3GS*. The opinionated information in these reviews is quite valuable for both consumers and firms. Consumers commonly seek public opinions from online consumer reviews prior to purchasing products, while many firms use online reviews as useful feedbacks in their product development, marketing, and consumer relationship management. In Fig. 8.2, we illustrate a sample of hierarchical organization of consumer reviews for product *iPhone 3G*. The hierarchy not only organizes all the product aspects and consumer opinions mentioned in the reviews, but also captures the parent-child relations among the aspects. With the hierarchy, users can easily grasp the overview of consumer reviews and browse the desired information, such as product aspects and consumer opinions. For example, users can find that 623 reviews, out of 9,245 reviews, are about the aspect "price", with 241 positive and 382 negative reviews.

Towards generating the hierarchy, we could refer to traditional methods in the domain of ontology learning, which first identify the concepts from text, then determine the parent-child relations among these concepts using either pattern-based or clustering-based methods [42]. Pattern-based methods usually suffer from inconsistency of the parent-child relations among concepts, while clustering-based methods often result in low accuracy [64]. Thus, by directly utilizing these methods to generate an aspect hierarchy from reviews, the resulting hierarchy is usually inaccurate, leading to unsatisfactory review organization. Moreover, the generated hierarchy may not be consistent with the information needs of the users which

**Fig. 8.1** Sample reviews on product *iPhone 3GS*

expect certain sub-topics to be present. On the other hand, domain knowledge of products is now available on the Web. This knowledge provides a broad structure that aims to answer the users' key information needs. For example, there are more than 248,474 product specifications in the forum website CNet.com [4]. These product specifications cover some product aspects and provide the coarse-grained parent-child relations among these aspects. Such domain knowledge is useful to help organize the product aspects into a hierarchy. While the initial hierarchy obtained from domain knowledge is good for broad structure of review organization, it is often too coarse and does not cover the specific aspects commented in the reviews well. Moreover, some aspects in the hierarchy may not be of interests to users in the reviews. In order to take advantages of the best of both worlds, we need to integrate initial domain knowledge structure, which reflects broad user interests in product, and distribution of reviews that indicates current interests and topics of concerns to users. Hence we need an approach to evolve the initial review hierarchy into one that reflects current users' opinions and interests.

Motivated by the above observations, we propose a domain-assisted approach to generate a review hierarchical organization by simultaneously exploiting the

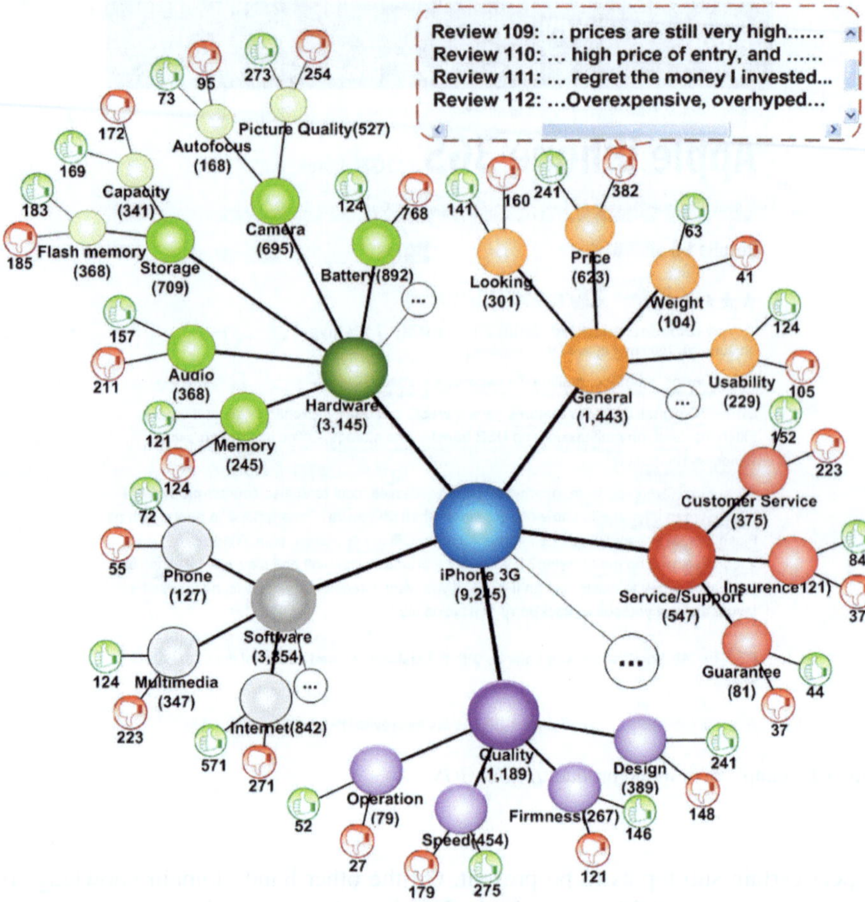


**Fig. 8.2** Sample hierarchical organization for product *iPhone 3G*

domain knowledge (e.g., the product specification) and consumer reviews. The framework of our approach is illustrated in Fig. 8.3. Given a collection of consumer reviews on a certain product, we first automatically acquire an initial aspect hierarchy from domain knowledge and identify the aspects in the reviews. We then develop a multi-criteria optimization approach to incrementally insert the identified aspects into appropriate positions of the initial hierarchy, and finally obtain a hierarchy that allocates all the aspects. The consumer reviews are then organized to their corresponding aspect nodes in the enhanced hierarchy. We further perform sentiment classification to determine consumer opinions on the aspects, and obtain the final hierarchical organization. Experiments are conducted on 11 popular products in 4 domains. There are 70,359 consumer reviews on these products totally. The dataset were crawled from multiple prevalent forum websites, such as CNet.com, Viewpoints.com, Reevoo.com and Pricegrabber.com etc. This dataset is

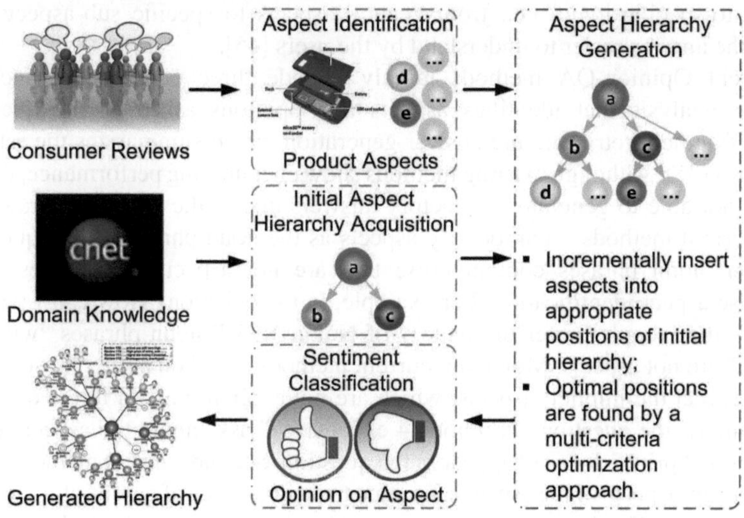


**Fig. 8.3** Overview of the hierarchical organization framework

released to facilitate future research on the topic of hierarchical organization.[1] More details of the dataset are discussed in Sect. 8.4. The experimental results demonstrate the effectiveness of the proposed approach.

Moreover, the generated hierarchy is beneficial to a wide range of real-world applications. In this chapter, we investigate its usefulness in the application of opinion Question Answering (opinion-QA) on products. Opinion-QA on products seeks to uncover consumers' thinking and feeling about the products or aspects of products, such as "How do people think about the battery of Nokia N95?". It is different from traditional factual QA, where the questions ask for the fact, such as "Where is the capital of United States?" and the answer is "Washington, D.C."

For a product opinionated question, the answer should not be just a best answer. It should reflect the opinions of various users, and incorporate both positive and negative viewpoints. Hence the answer should be a summarization of public opinions and comments on the product or specific aspect asked in the question [22]. In addition, it should also include public opinions and comments on the sub-aspects. Such answers would help users to understand the inherent reasons of the opinions on the aspect asked. For example, the question "What do people think the camera of Nokia 5800?" asks for public positive and negative opinions on the aspect "camera" of product "Nokia 5800." The summarization of opinions on the sub-aspects such as "lens" and "resolution" would help users better understand that the public complaints on the aspect "camera" are due to the poor "lens" and/or low "resolution." Moreover, the answer should be presented following the

[1]http://www.comp.nus.edu.sg/~Jianxing/Products_Reviews.rar

general-to-specific logic, i.e., from general aspects to specific sub-aspects. This makes the answer easier to understand by the users [45].

Current Opinion-QA methods mainly include three components, including question analysis that identifies aspects and opinions asked in the questions, answer fragment retrieval, and answer generation which summarizes the retrieved fragments [35]. Although existing methods show encouraging performance, they are usually not able to generate satisfactory answers due to the following drawbacks. First, current methods often identify aspects as the noun phrases in the questions. However, noun phrases contain noise that are not aspects. This gives rise to imprecise aspect identification. For example, in the question "How can I persuade my wife that people prefer the battery of Nokia N95?" noun phrases "wife" and "people" are not aspects. Moreover, current methods relied on noun phrases are not able to reveal the implicit aspects, which are not explicitly asked in the questions. For example, the question "Is iPhone 4 expensive?" asks about the aspect "price", but the term "price" does not appear in the question. Second, current methods cannot discover sub-aspects of the aspect asked due to its ignorance of parent-child relations among aspects. Third, the answers generated by the existing methods do not follow the general-to-specific logic, leading to difficulty in understanding the answers.

To overcome these drawbacks in opinion-QA, we resort to the hierarchical organization of consumer reviews on products. In particular, the hierarchy organizes the product aspects, which can facilitate to identify aspects asked in the questions. While explicit aspects can be recognized naturally by referring to the hierarchy, implicit aspects can be inferred based on the associations between sentiment terms and aspects in the hierarchy [68]. The sentiment terms are discovered from the reviews on corresponding aspects. Moreover, by following the parent-child relations in the hierarchy, sub-aspects of the asked aspect can be directly acquired, and the answers can present aspects from general to specific. Evaluations are conducted the aforementioned product review dataset using 220 testing questions. Experimental results to demonstrate the effectiveness of our approach.

The main contributions of this chapter can be summarized as follows. First, we propose a framework to generate a hierarchical structure to organize the collaboratively constructed content, so as to facilitate users in understanding the knowledge embedded in the content. Second, we employ one example of the content (i.e. consumer reviews on products) as a case study, and develop a domain-assisted approach to generate the review hierarchical organization by exploiting domain knowledge and consumer reviews. Third, we apply the hierarchy to support the application of opinion-QA on products, and achieve satisfactory performance.

The rest of this chapter is organized as follows. Sect. 8.2 reviews related works and Sect. 8.3 elaborates the approach of generating the hierarchical organization. Sect. 8.4 presents the experimental results, while Sect. 8.5 introduces the application of opinion-QA by making use of the hierarchy. Sect. 8.6 concludes this chapter with future works.

## 8.2 Related Works

We first give an overview of the collaboratively constructed content, and review the work on the content. We then illustrate the work related to the topics of hierarchical organization of consumer reviews, and opinion-QA on products, respectively.

### *8.2.1 Overview of Collaboratively Constructed Content*

With the rapid development of Web 2.0, enormous collaboratively constructed content is emerging online. For example, Wikipedia hosts over 22 million articles in 285 languages [60], whereas the forum CNet.com involves more than seven million product reviews [8]. This content contains a large repository of collective knowledge, which can facilitate users to make informed decisions. Also, it provides a brilliant way for manufacturers to interact with their customers, so as to significantly promote their business.

In order to effectively leverage the knowledge in the constructed content, numerous works in the previous studies have been proposed. The works can be mainly summarized into two categories. The first one is the intrinsic direction, which aims to refine the content to make it more useful and reliable. Since the content is generated by regular users instead of the professional experts, it would be inaccurate and contains noise. Some researchers try to filter the low-quality content and detect the trustful content. For example, Agichtein et al. [2] proposed to classify the high-quality content in Question Answering forums. Multiple features were employed to train the classifier, including the lexical features such as N-grams, misspellings and typos etc, social media features like the interactions between content creators and other users, as well as the content usage statistics. Liu et al. [33] aimed to find the useful answers in the content of community Question Answering cites. They classified the useful answers by considering the askers' satisfactory, prior knowledge and corresponding experience. Accordingly, Adler et al. [1] estimated the trustfulness of Wikipedia articles by the information of edit history. In particular, the trust value of the newly inserted text is equal to the reputation of the original author of the text, as well as the reputation of all authors who edited text near the text. Also, the trust value of the text may increase if the reputation of its author gets higher. Respectively, Lu et al. [36] focused on predicting the helpfulness and quality of the consumer reviews. They proposed to exploit contextual information about authors' identities and social networks for improving the prediction performance. On the other hand, the content is unorganized, leading to difficulty in information accessibility. To tackle this problem, some research works focus on organizing the content. For example, Carenini et al. [6] organized the consumer reviews into a user-defined taxonomy. Such taxonomy is hand-crafted which is not scalable. Deng et al. [12] developed a knowledge ontology called *ImageNet* to organize numerous images by referring to the WordNet structure.

The second category is the extrinsic direction, which endeavors in utilizing the content to support the applications in the research filed of *AI*, *IR*, and *NLP*. In this category, the content is often viewed as a huge corpus to generate new dataset and obtain better term statistics. For example, Santamaria et al. [50] proposed to extend WordNet by leveraging the content of Web directories such as Open Directory Project (ODP). Given a word, they aimed to associate its sense to the Web directories by using the synonyms and hypernyms relations in WordNet. Subsequently, Davidov et al. [11] developed a system named *Accio* to construct the text categorization datasets based on ODP. Elsas and Dumais [15] utilized the dynamic characteristic of the content statistic to improve relevance ranking. Additionally, there are some works that develop applications based on the content of blogs, consumer reviews, etc. Zhang et al. [70] focused on the application of opinion retrieval. They retrieved documents from blogs that are relevant to the query topic, and simultaneously contain opinions about the query. Lu et al. [37] utilized the consumer reviews to support the application of summarization. They exploited the existing ontology online to identify the aspects in reviews, and generated a structured summary by selecting and ordering some opinionated sentences on aspects. In addition, some researchers view the content as a knowledge repository. They regard that every article in the content represents a concept, which can be utilized to support concept-based IR. Strube and Ponzetto [57] computed the semantic relatedness among concepts using the links in Wikipedia. Accordingly, the content provides an inventory of senses of ambiguous words/entity names, and a corpus of contexts containing ambiguous words. It can naturally be used to support the task of Word Sense Disambiguation, so as to resolve natural language polysemy.

In this chapter, we focus on automatically organizing the content of consumer reviews into a hierarchical structure. To generate a hierarchy from consumer reviews, there are mainly three basic tasks, including (a) identifying product aspects in the reviews; (b) classifying opinions on the aspects; and (c) determining the parent-child relations among the aspects. We next summarize the research work related to these three tasks.

### 8.2.2 Hierarchical Organization of Consumer Reviews

Consumer reviews usually convey various opinions on multiple product aspects. These aspects and corresponding opinions can help users better understand the products in details, and subsequently make informed decisions. To identify the aspects in reviews, existing techniques can be broadly classified into two major methods: supervised and unsupervised. Supervised methods usually learn an extraction model based on the pre-annotated training reviews. For example, Wong and Lam [62] proposed to learn an aspect extractor by employing *Hidden Markov Models* and *Conditional Random Fields*, respectively. They then utilized the extractor to identify aspects from auction text. However, it is time-consuming and tedious to obtain the training samples, and some unsupervised methods have emerged. Hu and Liu

[21] proposed to identify the product aspects by the technique of association rule mining. They assumed that product aspects are noun phrases. They extracted all frequent noun phrases as the aspect candidates, and employed the association rule mining algorithm to refine the candidates based on some compactness pruning rules and redundancy pruning rules. Subsequently, Popescu and Etzioni [48] proposed their system *OPINE*, which extracts the aspects based on the *KnowItAll* Web information extraction system [17]. Liu et al. [32] developed a supervised method based on language pattern mining to identify aspects in the reviews. Later, Mei et al. [39] utilized a probabilistic topic model to capture the mixture of aspects and sentiments simultaneously. Su et al. [58] designed a mutual reinforcement strategy to simultaneously cluster product aspects and opinion words by iteratively fusing both content information and sentiment link information. Afterwards, Wu et al. [63] utilized the dependency parser to extract the noun phrases and verb phrases from the reviews as aspect candidates. They then identified the aspects by a language model trained on the reviews.

To classify the opinion on the aspect, existing methods can also be categorized into supervised and unsupervised methods. Supervised methods often classify the opinions on the aspects by a sentiment classifier trained on corpus [46]. Respectively, unsupervised methods usually rely on a sentiment word dictionary, called sentiment lexicon. The lexicon typically contains a list of positive and negative words. The review is classified as positive opinion if it contains a majority of words in the positive word list. To generate the lexicon, the bootstrapping strategy is usually employed. For example, Hu and Liu [21] started with a set of adjective seed words for each opinion class (i.e. positive and negative). They then utilized synonym/antonym relations defined in *WordNet* to bootstrap the seed word set, and finally obtained the lexicon. Ding et al. [14] presented a holistic lexicon-based method to improve Hu and Liu's method [21] by addressing two issues: the opinions on the sentiment words would be content-sensitive, and may conflict in the review. They derived the lexicon by exploiting some constraints, such as *TOO*, *BUT*, *NEGATION*. For example, the opinions of two terms would be contrary if they are connected by the transitional term *BUT*. In addition, Wang et al. [59] proposed a generative Latent Rating Regression model (*LRR*) to infer opinion on the aspect based on the review and its associated overall rating.

To determine the parent-child relations among the aspects, we can refer to previous works in the field of ontology learning. Generally, there are two kinds of popular approaches, namely pattern-based approach and clustering-based approach.

Pattern-based method usually defines some lexical-syntactic patterns, and uses these patterns to discover instances of relations in text. As a pioneer, Hearst [20] proposed to identify the parent-child relations by defining six hand-crafted patterns, such as "$NP_y$ including $NP_x$," where $NP_y$ indicates the parent concept, and $NP_x$ represents the child concept. The paper then searched for the instances that were matched these patterns in the text corpus. Each matched instance contains a pair of noun phrases, filling the positions of $NP_x$ and $NP_y$. Once a matched instance of relation is identified, more patterns and instances can be found through a *bootstrapping* technique. In particular, the noun phrases in the newly identified instance were used to search frequent contexts in text, so as to yield new patterns indicating the parent-child relations. The new patterns were then used to discover more instances and continue the cycle to find new patterns. This approach can recognize the instances of relations with high accuracy when the patterns are carefully chosen. Also, the *bootstrapping* technique is effective and scalable to large datasets. It is a data-driven approach that helps to find more unknown patterns. However, such technique may be uncontrolled, and would generate undesired instances once a noisy pattern is brought into the bootstrap cycle [47]. Moreover, this approach identifies relations in concept pairs, which does not consider the global relations (i.e. ascendant and descendant) among the concepts. It may lead to the concept inconsistency problem. For example, it may infer "Apple" as the parent of a concept "iPhone", and "fruit" as the parent of concept "Apple". However, it is obvious that "fruit" should not be an ascendant of "iPhone" in this context.

Clustering-based approach usually organizes concepts into a hierarchy by the hierarchical clustering technique [5]. The technique first gathers the contexts of the concepts as features, and represents the concepts into feature vectors. Based on the vectors, it clusters the concepts into a hierarchy based on text similarities (e.g. Cosine similarity). The clustering can be performed by agglomerative [29], divisive [53], and incremental methods [55]. This approach determines the relations among concepts by similarity of their feature contexts. It thus can discover some new relations which the pre-defined patterns do not capture. In contrast with the pattern-based approach, this approach alleviates the concept inconsistency problem by a unified model that globally determines the relations among all concepts. However, the accuracy of the clustering-based approach is usually lower than the pattern-based approach. Also, it may fail to coherently produce clusters for the small corpus [47], and its performance is greatly influenced by the features used. Moreover, the new formed clusters do not have label, and naming clusters is a very challenging task.

Next, we summarize the work related to the application of opinion-QA.

### *8.2.3 Opinion QA*

Current methods on opinion-QA usually include three components, including question analysis, answer fragment retrieval, and answer generation.

Question analysis has to distinguish the opinion question from the factual one, and find the key points asked in the questions, such as the product aspect and product name. For example, Yu et al. [67] proposed to separate opinions from facts at both document and sentence level, and determine the polarity on the opinionated sentences in the answer documents. Similarly, Somasundaran et al. [56] utilized a SVM classifier to recognize opinionated sentences. The paper argued that the subjective types (i.e. sentiment and arguing) can improve the performance of opinion-QA. Later, Ku et al. [23] proposed a two-layered classifier for question analysis, and retrieved the answer-fragments by keyword matching. In particular, they first identified the opinion questions, and classified them into six predefined question types, including holder, target, attitude, reason, majority, and yes/no. These question types and corresponding polarity on the questions were used to filter non-relevant sentences in the answer fragments. $F_1$-measure was employed as the evaluation metric.

For the topic of answer generation in opinion-QA, Li et al. [28] formulated it as a sentence ranking task. They argued that the answers should be simultaneously relevant to topics and opinions asked in the questions. They thus designed the graph-based methods (i.e. PageRank and HITS) to select some high-ranked sentences to form answers. They first built a graph on the retrieved sentences, with each sentence as the node, and the similarity (i.e. Cosine similarity) between each sentences pair as the weight of the corresponding edge. Given a question, its similarity to each sentence in the graph was computed. Such similarity was viewed as the relevant score to the corresponding sentence. The sentences then were ranked based on three metric, i.e. relevant score to the query, similarity score obtained from the graph algorithm over sentences, and degree of opinion matching to the query. Respectively, Lloret et al. [35] proposed to form answers by re-ranking the retrieved sentences based on the metric of word frequency, non-redundancy and the number of noun phrases. Their method includes three components, including information retrieval, opinion mining and text summarization. Evaluations were conducted on the TAC 2008 Opinion Summarization track. Afterwards, Moghaddam et al. [41] developed a system called *AQA* to generate answers for opinion questions about products. It classifies the questions into five types, including target, attitude, reason, majority and yes/no. The *AQA* system includes five components, including question analysis, question expansion, high quality review retrieval, subjective sentence extraction, and answer grouping. The answers are generated by aggregating opinions in the retrieved fragments.

## 8.3 Hierarchical Organization Framework

As illustrated in Fig. 8.3, our approach mainly consists of four components, including (a) initial aspect hierarchy acquisition; (b) product aspect identification; (c) aspect hierarchy generation; and (d) aspect-level sentiment classification. We first define some notations and elaborate these components.

### *8.3.1 Preliminary and Notations*

*Preliminary: An aspect hierarchy is defined as a tree that consists of a set of unique product aspects* $\mathcal{A}$ *and a set of parent-child relations* $\mathcal{R}$ *among these aspects.*

Given the consumer reviews of a product, let $\mathcal{A} = \{a_1, \cdots, a_k\}$ denote the product aspects commented in the reviews. $\mathcal{H}^0(\mathcal{A}^0, \mathcal{R}^0)$ denotes the initial hierarchy acquired from domain knowledge. It contains a set of aspects $\mathcal{A}^0$ and relations $\mathcal{R}^0$. We aim to construct an aspect hierarchy $\mathcal{H}(\mathcal{A}, \mathcal{R})$, to cover all the aspects in $\mathcal{A}$ and their parent-child relations $\mathcal{R}$, so that all consumer reviews can be hierarchically organized. Note that $\mathcal{H}^0$ can be empty.

### 8.3.2 *Acquisition of Initial Aspect Hierarchy*

As aforementioned, product specifications in forum websites cover some product aspects and coarse-grained parent-child relations among these aspects. Such domain knowledge is useful to help organize aspects into a hierarchy. We here employ the approach proposed in [66] to automatically acquire an initial aspect hierarchy from the product specifications. The method first identifies the Web page region covering product descriptions, and removes the irrelevant content from the Web page. It then parses the region containing the product information to identify the aspects as well as their structure. Based on the aspects and their structure, it generates an initial aspect hierarchy.

### 8.3.3 *Product Aspect Identification*

As illustrated in Fig. 8.1, consumer reviews often consist of two common formats in most forum websites, including *Pros* and *Cons* reviews which reveals concise positive/negative opinions on the products, and the reviews in free text. For *Pros* and *Cons* reviews, we identify their aspects by extracting frequent noun phrases. Previous studies show that aspects are usually noun/noun phrases [31], and we can obtain highly accurate aspects by extracting frequent noun terms from *Pros* and *Cons* reviews [32]. For the free text reviews, we show the flowchart to identify the corresponding aspects in Fig. 8.4. We first split the reviews into sentences and parse each sentence by Stanford parser.[2] The frequent noun phrases (*NP*) are then extracted from the sentence parsing trees as the aspect candidates. While these candidates may contain noise (i.e. irrelevant terms), we propose to leverage *Pros* and *Cons* reviews to refine the candidates. In particular, we extract the frequent noun terms from *Pros* and *Cons* reviews as features, then train a one-class *SVM* [38] to identify the true aspects. As the identified aspects may contain synonym terms, such as "earphone" and "headphone", synonym clustering is further performed to obtain unique aspects. Technically, we collect synonym terms of the identified aspects as features, and represent each aspect into feature vector for clustering. The synonym terms are extracted from the synonym dictionary website.[3]

---

[2]http://nlp.stanford.edu/software/lex-parser.shtml

[3]http://thesaurus.com

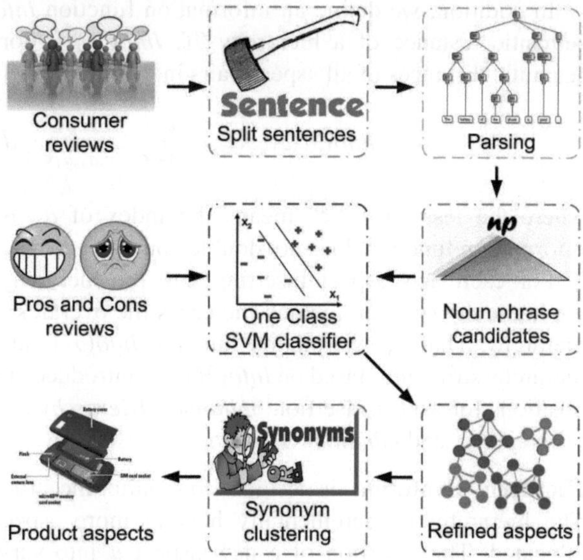


**Fig. 8.4** Procedure of product aspect identification on free text reviews

## 8.3.4 *Generation of Aspect Hierarchy*

To build the hierarchy, we propose a multi-criteria optimization approach to incrementally insert the newly identified aspects into the appropriate positions in the initial hierarchy. In the next subsections, we illustrate details of the approach.

### 8.3.4.1 Formulation

Given the aspects $\mathcal{A} = \{a_1, \cdots, a_k\}$ identified from the reviews and the initial hierarchy $\mathcal{H}^0(\mathcal{A}^0, \mathcal{R}^0)$ acquired from the domain knowledge, we here propose a multi-criteria optimization approach to generate an aspect hierarchy $\mathcal{H}^*$, which allocates all the aspects in $\mathcal{A}$, including those not in the initial hierarchy, i.e. $\mathcal{A} - \mathcal{A}^0$. The approach incrementally inserts the newly identified aspects into the appropriate positions in the initial hierarchy. The optimal positions are found by multiple criteria. The criteria have to guarantee that each aspect would most likely to be allocated under its parent aspect in the hierarchy.

Before introducing the criteria, we first define a metric, named *Semantic Distance*, $d(a_x, a_y)$, to quantify the parent-child relations between aspects $a_x$ and $a_y$. $d(a_x, a_y)$ is formulated as the weighted sum of some underlying features,

$$d(a_x, a_y) = \sum_j w_j f_j(a_x, a_y), \tag{8.1}$$

where $w_j$ is the weight for $j$-th feature function $f_j(\cdot)$. The estimation of the feature function $f(\cdot)$ will be described in Sect. 8.3.4.2, and learning of $d(a_x, a_y)$ (i.e. weight $w$) is introduced in Sect. 8.3.4.3.

In addition, we define an information function $Info(\mathcal{H})$ to measure the overall semantic distance of a hierarchy $\mathcal{H}$. $Info(\mathcal{H})$ is formulated as the sum of the semantic distances of all aspect-pairs in the hierarchy [65], as follows,

$$Info(\mathcal{H}(\mathcal{A}, \mathcal{R})) = \sum\nolimits_{x<y; a_x, a_y \in \mathcal{A}} d(a_x, a_y), \quad (8.2)$$

where the less sign "<" means the index of $a_x$ is less than that of $a_y$. The information function does not double count the distance of the aspect pairs.

For each new aspect inserting into the hierarchy, it introduces a change in the hierarchy structure, which increases the overall semantic distance of the entire hierarchy. That is, information function $Info(\mathcal{H})$ can be used to characterize the hierarchy structure. Based on $Info(\mathcal{H})$, we introduce three criteria to find the optimal positions for aspect insertion: *minimum Hierarchy Evolution*, *minimum Hierarchy Discrepancy* and *minimum Semantic Inconsistency*.

**Hierarchy Evolution** is designed to monitor the structure evolution of a hierarchy. The hierarchy is incrementally hosting more aspects until all the aspects are allocated. The insertion of a new aspect $a$ into various positions in the current hierarchy $\mathcal{H}^{(i)}$ leads to different new hierarchies. It gives rise to different increase of the overall semantic distance (i.e. $Info(\mathcal{H}^{(i)})$). When an aspect is placed into the optimal position in the hierarchy (i.e. as a child of its parent aspect), $Info(\mathcal{H}^{(i)})$ has the least increase. In other words, minimizing the change of $Info(\mathcal{H}^{(i)})$ is equivalent to searching for the best position to insert the aspect. Therefore among the new hierarchies, the optimal one $\hat{\mathcal{H}}^{(i+1)}$ should lead to least changes of overall semantic distance to $\mathcal{H}^{(i)}$, as follows,

$$\hat{\mathcal{H}}^{(i+1)} = \arg\min_{\mathcal{H}^{(i+1)}} \Delta Info(\mathcal{H}^{(i+1)} - \mathcal{H}^{(i)}). \quad (8.3)$$

The first criterion can be obtained by plugging the definition of $Info(\mathcal{H})$ in Eq. (8.2) and using *least square* as the loss function to measure the information changes,

$$obj_1 = \arg\min_{\mathcal{H}^{(i+1)}} (\sum\nolimits_{x<y; a_x, a_y \in \mathcal{A}_i \cup \{a\}} d(a_x, a_y) - \sum\nolimits_{x<y; a_x, a_y \in \mathcal{A}_i} d(a_x, a_y))^2, \quad (8.4)$$

**Hierarchy Discrepancy** is used to measure the global changes of the structure evolution. A good hierarchy should be the one that brings the least changes to the initial hierarchy in a macro-view, so as to avoid the algorithm falling into local minimum,

$$\hat{\mathcal{H}}^{(i+1)} = \arg\min_{\mathcal{H}^{(i+1)}} \Delta Info(\mathcal{H}^{(i+1)} - \mathcal{H}^{(0)})/(i+1). \quad (8.5)$$

By substituting the definition in Eq. (8.2), we then get the second criterion:

$$obj_2 = \arg\min_{\mathcal{H}^{(i+1)}} \frac{1}{i+1} (\sum\nolimits_{x<y; a_x, a_y \in \mathcal{A}_i \cup \{a\}} d(a_x, a_y) - \sum\nolimits_{x<y; a_x, a_y \in \mathcal{A}_0} d(a_x, a_y))^2. \quad (8.6)$$

**Semantic Inconsistency** is introduced to quantify the inconsistency between the semantic distance estimated via the hierarchy and that computed from the feature functions (i.e. Eq. (8.1)). We assume that a good hierarchy should precisely reflect the semantic distance among aspects. For two aspects, their semantic distance reflected by the hierarchy is computed as the sum of all adjacent interval distances along the shortest path between them,

$$d^{\mathcal{H}}(a_x, a_y) = \sum\nolimits_{p<q;(a_p,a_q)\in SP(a_x,a_y)} d(a_p, a_q), \tag{8.7}$$

where $SP(a_x, a_y)$ is the shortest path between the aspects $(a_x, a_y)$, $(a_p, a_q)$ are the adjacent nodes along the path.

The third criterion is then obtained to derive the optimal hierarchy,

$$obj_3 = \arg\min\nolimits_{\mathcal{H}^{(i+1)}} \sum\nolimits_{x<y;a_x,a_y\in \mathcal{A}_i \cup\{a\};} (d^{\mathcal{H}}(a_x, a_y) - d(a_x, a_y))^2, \tag{8.8}$$

where $d(a_x, a_y)$ is the distance computed by the feature function in Eq. (8.1).

**Multi-Criteria Optimization** Through integrating the above criteria, the multi-criteria optimization framework is formulated as,

$$\begin{aligned} obj &= \arg\min\nolimits_{\mathcal{H}^{(i+1)}} (\lambda_1 \cdot obj_1 + \lambda_2 \cdot obj_2 + \lambda_3 \cdot obj_3) \\ \lambda_1 &+ \lambda_2 + \lambda_3 = 1;\ \ 0 \le \lambda_1, \lambda_2, \lambda_3 \le 1. \end{aligned} \tag{8.9}$$

where $\lambda_1$, $\lambda_2$, $\lambda_3$ are the tradeoff parameters, which would be described in Sect. 8.4.1.

Given the semantic distance $d(a_x, a_y)$, we can find the optimal position for aspect insertion by Eq. (8.9). To summarize, our aspect hierarchy generation process starts from an initial hierarchy and inserts the aspects into it one-by-one until all the aspects are allocated. It is worth noting that the insertion order may influence the result. To avoid such influence, we select the aspect with the least objective value in Eq. (8.9) for insertion in each step. Based on resultant hierarchy, the consumer reviews are then organized to their corresponding aspect nodes in the hierarchy. We further prune out the nodes without reviews from the hierarchy.

In next subsections, we will introduce the estimation of the feature function $f(a_x, a_y)$ and semantic distance $d(a_x, a_y)$.

#### 8.3.4.2 Linguistic Features for Semantic Distance Estimation

Given two aspects $a_x$ and $a_y$, the feature is defined as a function $f(a_x, a_y)$ generating a numeric score or a vector of scores. By referring to the work of [65], we explore multiple features including *Contextual, Co-occurrence, Syntactic, Pattern* and *Lexical* features. These features are generated based on auxiliary documents collected from the Web. Specifically, we issue each aspect and aspect pair as queries to Google and Wikipedia respectively, and collect the top 100 returned documents for each query. Each document is split into sentences. Based on these documents and sentences, the features are generated as follows.

**Contextual features.** For each aspect, the hosted documents are collected and treated as context to build a unigram language model, with Dirichlet smoothing. Given two aspects $a_x$ and $a_y$, the KL-divergence [24] between their language models is computed as their *Global-Context* feature. Similarly, we collect the left 2 and right 2 words surrounding each aspect, and use them as context to build a unigram language model. The KL-divergence between the language models of two aspects $a_x$ and $a_y$ is defined as the *Local-Context* feature.

**Co-occurrence features.** We measure the co-occurrence of two aspects $a_x$ and $a_y$ by Pointwise Mutual Information (PMI):

$$PMI(a_x, a_y) = log(Count(a_x, a_y)/Count(a_x) \cdot Count(a_y)), \qquad (8.10)$$

where *Count*(·) stands for the number of documents or sentences containing the aspect(s), or the number of Google document hits for the aspect(s). Based on different definitions of *Count*(·), we define the features of *Document PMI*, *Sentence PMI*, and *Google PMI*, respectively.

**Syntactic features.** The sentences that contain both aspects $a_x$ and $a_y$ are collected, and parsed into the syntactic trees via the Stanford Parser. For each sentence, we compute the length of the shortest path between aspects $a_x$ and $a_y$ in the syntactic tree. The average length is took as *Syntactic-path* feature between $a_x$ and $a_y$. Accordingly, for each aspect, we parse its hosted sentences, and collect its modifier terms from the sentence parsing trees. The modifier terms are defined as the adjective and noun terms on the left side of the aspect. The modifier terms that share the same parent node with the aspect are selected. We then calculate the size of the overlaps between two modifiers sets for aspects $a_x$ and $a_y$ as the *Modifier Overlap* feature. In addition, we select the hosted sentences for each aspect, and perform semantic role labeling on the sentences by ASSERT parser.[4] The subject role terms are collected from the labeling sentences as the subject set. We then calculate overlaps between two subject sets for aspects $a_x$ and $a_y$ as the *Subject Overlap* feature. Similarly, for other semantic roles (i.e. objects and verbs), we define the features of *Object Overlap*, and *Verb Overlap* respectively.

**Relation pattern features.** Forty six relation patterns[5] are used in our work, including six patterns indicating the hypernym relations of two aspects in Hearst et al. [20], and 40 patterns measuring the part-of relations of two aspects in Girju et al. [18]. These pattern features are asymmetric, and they take into consideration

[4]http://cemantix.org/assert.html

[5]Available in http://www.aclweb.org/supplementals/D/D11/D11-1013.Attachment.zip

the parent-child relations among aspects. Based on these patterns, a 46-dimensional score vector is obtained for aspects $a_x$ and $a_y$. A score is 1 if two aspects match a pattern, and 0 otherwise.

**Lexical features.** The word length difference between aspects $a_x$ and $a_y$ is computed as *Length Difference* feature. In addition, we issue the query "define:aspect" to Google, and collect the definition of each aspect ($a_x/a_y$). We then count the word overlaps between the definitions of two aspects $a_x$ and $a_y$, as *Definition Overlap* feature.

#### 8.3.4.3 Estimation of Semantic Distance

As described in Sect. 8.3.4.1, the semantic distance $d(a_x, a_y)$ is formulated as $\sum_j w_j f_j(a_x, a_y)$, where $w$ denotes the weight, $f(a_x, a_y)$ is the feature function. To learn the weight $w$, we can employ the initial hierarchy as training data. We assume that the distance between two aspects $d^{\mathcal{G}}(a_x, a_y)$ reflected in the hierarchy is generated by summing up all the edge distances along the shortest path between $a_x$ and $a_y$, where the weight of every edge is viewed to be 1. The optimal weights are then estimated by solving the ridge regression optimization problem below,

$$\arg\min_{w_j|_{j=1}^m} \sum_{a_x, a_y \in \mathcal{A}^0; x<y} (d^{\mathcal{G}}(a_x, a_y) - \sum_{j=1}^m w_j f_j(a_x, a_y))^2 + \eta \cdot \sum_{j=1}^m w_j^2, \tag{8.11}$$

where $m$ represents the dimension of linguistic features, and $\eta$ is a tradeoff parameter.

Equation (8.11) can be re-written to matrix form, and the optimal solution is derived as,

$$\mathbf{w}_0^* = (\mathbf{f}^T \mathbf{f} + \eta \cdot \mathbf{I})^{-1} (\mathbf{f}^T \mathbf{d}) \tag{8.12}$$

where $\mathbf{w}_0^*$ is the optimal weight vector, $\mathbf{d}$ denotes the vector of the ground truth distance, $\mathbf{f}$ represents the feature function vector, and $\mathbf{I}$ is the identity metric.

The above learning algorithm can perform well when sufficient training data (i.e., aspect pair distance) are available. However, the initial hierarchy is usually too coarse and thus may not provide sufficient information. On the other hand, external linguistic resources (e.g. WordNet and Open Directory Project (ODP)) provide abundant hand-crafted hierarchies. We here propose to leverage these resources to assist semantic distance learning. A distance metric $\mathbf{w}_0$ is learned from the external linguistic resources by Eq. (8.12). Since $\mathbf{w}_0$ might be biased to the characteristics of the external linguistic resources, directly using $\mathbf{w}_0$ in our task may not perform well. Alternatively, we use $\mathbf{w}_0$ as prior knowledge to help learn the optimal distance metric $\mathbf{w}$ from the initial hierarchy. The learning problem is formulated as follows,

$$\arg\min_{\mathbf{w}} \left\| \mathbf{d} - \mathbf{f}^T \mathbf{w} \right\|^2 + \eta \cdot \|\mathbf{w}\|^2 + \gamma \cdot \|\mathbf{w} - \mathbf{w}_0\|^2, \tag{8.13}$$

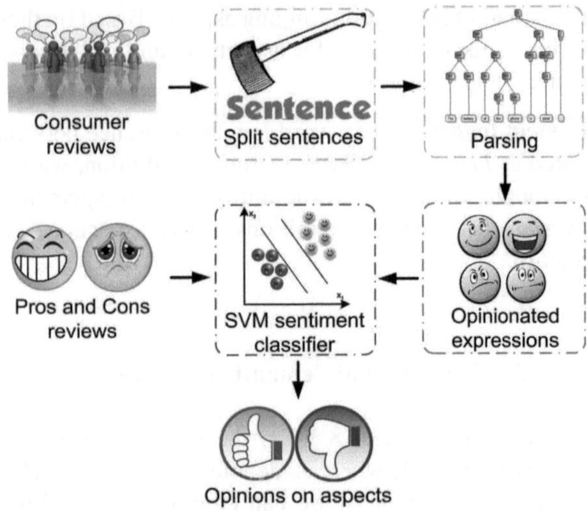


**Fig. 8.5** Procedure of aspect-level sentiment classification

where **d** denotes the ground truth distance in the initial hierarchy, $\eta$ and $\gamma$ are tradeoff parameters.

The optimal solution of **w** can be obtained as

$$\mathbf{w}^* = (\mathbf{f}^T\mathbf{f} + (\eta + \gamma) \cdot \mathbf{I})^{-1}(\mathbf{f}^T\mathbf{d} + \gamma \cdot \mathbf{w}_0). \tag{8.14}$$

As a result, we can compute the semantic distance $d(a_x, a_y)$ according to Eq. (8.1).

## 8.3.5 Aspect-Level Sentiment Classification

After generating a hierarchy to organize all the newly identified aspects and consumer reviews, we perform sentiment classification to determine opinions (i.e. positive and negative) on the corresponding aspects, and obtain the final hierarchical organization. The overview of our approach for sentiment classification is demonstrated in Fig. 8.5. We observe the *Pros* and *Cons* reviews (see Fig. 8.1) have explicitly categorized positive and negative opinions on the aspects. These reviews are valuable training samples to learn a sentiment classifier. We thus train a sentiment classifier based on *Pros* and *Cons* reviews, and employ the classifier to determine the opinions on aspects in the free text reviews.

Specifically, we first collect sentiment terms in *Pros* and *Cons* reviews based on the sentiment lexicon provided by MPQA project [61]. These terms are used as features and each review is represented into a feature vector. A sentiment classifier is then learned from the *Pros* reviews (i.e., positive samples) and *Cons* reviews (i.e., negative samples), and used to classify the opinions of free text reviews. The classifier used in previous studies [46] includes *SVM*, *Naïve Bayes* and *Maximum*

**Table 8.1** Statistics of the product review datasets

| Product name | Domain | Review# | Sentence# |
|---|---|---|---|
| Canon EOS 450D (Canon EOS) | Camera | 440 | 628 |
| Fujifilm Finepix AX245W (Fujifilm) | Camera | 541 | 839 |
| Panasonic Lumix DMC-TZ7 (Panasonic) | Camera | 650 | 1,546 |
| Apple MacBook Pro (MacBook) | Laptop | 552 | 4,221 |
| Samsung NC10 (Samsung) | Laptop | 2,712 | 4,946 |
| Apple iPod Touch 2nd (iPod Touch) | MP3 | 4,567 | 10,846 |
| Sony NWZ-S639 16GB (Sony NWZ) | MP3 | 341 | 773 |
| BlackBerry Bold 9700 (BlackBerry) | Phone | 4,070 | 11,008 |
| iPhone 3GS 16GB (iPhone 3GS) | Phone | 12,418 | 43,527 |
| Nokia 5800 XpressMusic (Nokia 5800) | Phone | 28,129 | 75,001 |
| Nokia N95 | Phone | 15,939 | 44,379 |

# Denotes the number of the reviews/sentences

*Entropy*, etc. Given a free text review that may cover multiple aspects, we first locate the opinionated expression that modifies the corresponding aspect, e.g. locating the expression "well" in the review "The battery of Nokia N95 works well." for the aspect "battery." Generally, an opinionated expression is associated with the aspect if it contains at least one sentiment term in the sentiment lexicon, and it is closest one to the aspect in the parsing tree within the context distance of 5. Finally, the learned sentiment classifier is leveraged to determine the opinion of the opinionated expression, i.e. the opinion on the aspect.

## 8.4 Evaluations

In this section, we evaluate the effectiveness of our approach on product aspect identification, aspect hierarchy generation, and aspect-level sentiment classification.

### *8.4.1 Data Set and Experimental Settings*

Table 8.1 shows the details of our product review dataset, which is publicly released.[6] This dataset contains consumer reviews on 11 popular products in four domains. These reviews were crawled from the prevalent forum websites, including cnet.com, viewpoints.com, reevoo.com, gsmarena.com and pricegrabber.com. All of the reviews were posted between June, 2009 and July 2011. Eight graduate students were invited to generate the ground truth on this dataset. They were asked to annotate the product aspects in each review, and also label consumer opinions expressed on the aspects. Each review was labeled by two annotators. The average inter-rater agreement in terms of Kappa statistics is 87 %. In addition,

[6] Available in http://www.comp.nus.edu.sg/~Jianxing/Product_Reviews.rar

**Table 8.2** Statistics of the external linguistic resources

| Statistic | WordNet | ODP |
|---|---|---|
| Total # hierarchies | 50 | 50 |
| Total # terms | 1,964 | 2,210 |
| Average # depth | 5.5 | 5.9 |
| Total # hierarchy topics | 12 | 16 |

three participants were asked to construct the gold standard hierarchy. For each product, they were provided the initial hierarchy and the aspects commented in the reviews. They were required to build a hierarchy to allocate all the aspects based on the initial hierarchy. In terms of Kappa statistics, the average inter-rater agreement of the parent-child relations among aspects is 73 %. The conflicts between participants were resolved through their discussions. For semantic distance learning, we collected 50 hierarchies from WordNet and ODP, respectively as external linguistic resources.[7] Specifically, we utilized the hypernym and meronym relations in WordNet to construct 50 hierarchies. Such relations indicate parent-child relations among concepts. We only used one word sense in WordNet to avoid word sense ambiguity. In addition, we parsed the topic lines in the ODP XML databases to obtain relations, and constructed another 50 hierarchies accordingly. Table 8.2 gives the details.

$F_1$-measure was employed as the evaluation metric for the experiments. It is the combination of *precision* and *recall*, as $F_1$*-measure* = *2*precision*recall/ (precision + recall)*. To evaluate the generated hierarchy, we defined *precision* as the percentage of correctly returned parent-child pairs out of the total number of returned pairs, and *recall* as the percentage of correctly returned parent-child pairs out of the total number of pairs in the gold standard. Throughout the experiments, we empirically set $\lambda_1 = 0.4$, $\lambda_2 = 0.3$, $\lambda_3 = 0.3$, $\eta = 0.4$ and $\gamma = 0.6$.

### *8.4.2 Evaluations on Product Aspect Identification*

We compared our aspect identification approach against two methods: (a) the method proposed by Hu et al. [21], which extracted the noun terms as aspect candidates, and refined the candidates by rules learned from association rule mining, and (b) the method proposed by Wu et al. [63], which extracted the noun phrases from a dependency parsing tree as aspect candidates, and refined the candidates by a language model built on the product reviews. From the results presented in Fig. 8.6, we can see that the proposed approach significantly outperforms Hu's method and Wu's method by over 8.84 %, 4.77 %, respectively in terms of average $F_1$-measure. This indicates the effectiveness of *Pros* and *Cons* reviews in assisting aspect identification on free text reviews.

[7] Available in http://www.aclweb.org/supplementals/D/D11/D11-1013.Attachment.zip

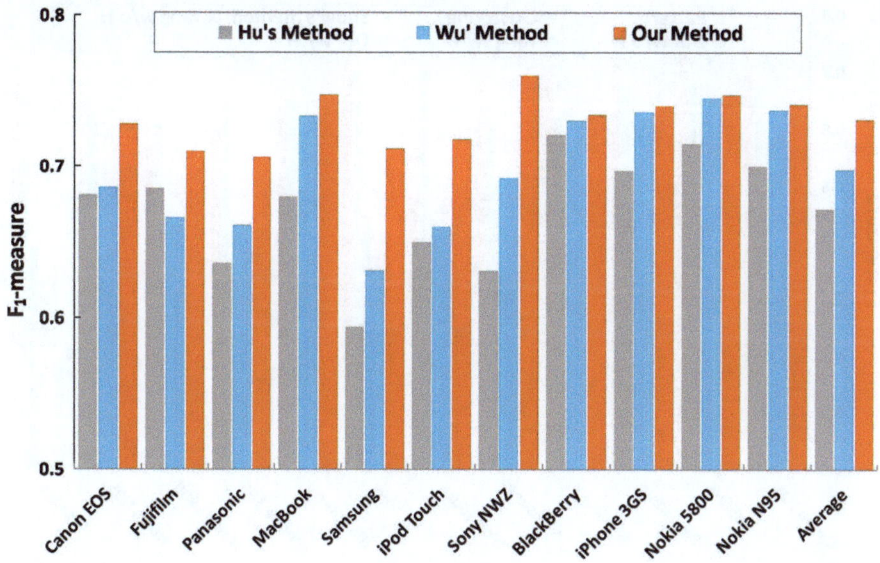


**Fig. 8.6** Performance of product aspect identification. The results are tested for statistical significance using T-Test, with p-values <0.05

## *8.4.3 Evaluations on Generation of Aspect Hierarchy*

We first compared the proposed approach against the state-of-the-arts methods, then evaluated the effectiveness of the components in our approach.

### 8.4.3.1 Comparisons to the State-of-the-Arts Methods

Four traditional methods in ontology learning for hierarchy generation are utilized for comparison.

- Pattern-based method [20] which explores the pre-defined patterns to identify parent-child relations and forms the hierarchy correspondingly.
- Clustering-based method [53] that builds the hierarchy by hierarchical clustering.
- The method proposed by Snow et al. [55] which generates the hierarchy based on a probabilistic model.
- The method proposed by Yang et al. [65], which defines multiple metric for the hierarchy generation.

Since our approach and Yang's method can utilize initial hierarchy to assist in hierarchy generation, we evaluated their performance with or without initial hierarchy, respectively. For the sake of fair comparison, Snow's, Yang's and our methods used the same linguistic features as described in Sect. 8.3.4.2.

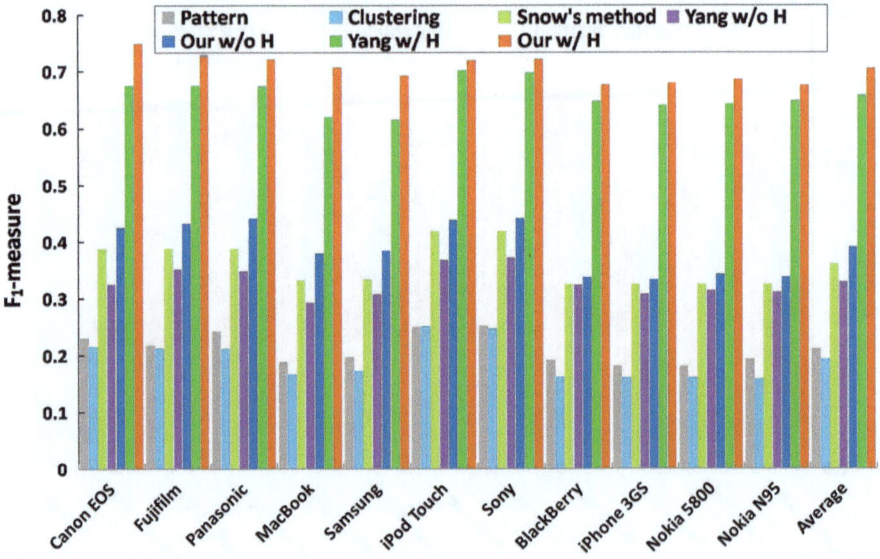


**Fig. 8.7** Performance of aspect hierarchy generation. T-Test, p-values <0.05. w/ H denotes the methods with initial hierarchy, accordingly, w/o H refers to the methods without initial hierarchy

As shown in Fig. 8.7, without the initial hierarchy, our approach outperforms the pattern-based, clustering-based, Snow's, and Yang's methods by the absolute gain of over 17.9 %, 19.8 %, 2.9 % and 6.1 % in terms of average $F_1$-measure, respectively. By exploiting initial hierarchy, our approach improves the performance significantly. As compared to the pattern-based, clustering-based and Snow's methods, our approach improves the average performance by the absolute gain of over 49.4 %, 51.2 % and 34.3 %, respectively. Compared to Yang's method with initial hierarchy, it achieves a significant absolute gain of 4.7 % in terms of average $F_1$-measure.

The results show that pattern-based and clustering-based methods perform poorly. Specifically, pattern-based method achieves higher precision but lower recall, while clustering-based method obtains both low precision and recall. A probable reason is that pattern-based method may suffer from the low coverage of patterns problem, especially when the patterns are pre-defined and may not include all patterns in the reviews. Respectively, the clustering-based method [53] is limitedly to the use of bisection clustering mechanism which only generates a binary-tree. In addition, we observe that the methods using heterogeneous features (i.e. Snow's, Yang's and Our) achieve high $F_1$-measure. We can speculate that the distinguishability of the parent-child relations among aspects would be enhanced by integrating multiple features. Also the results indicate that the methods with initial hierarchy (i.e. Yang's and Our) can significantly boost the performance. Such results further convince us that the initial hierarchy is valuable for hierarchy generation. Finally, the results show that our approach outperforms Yang's method when both utilize the initial hierarchy. A probable reason is that our approach is able to derive

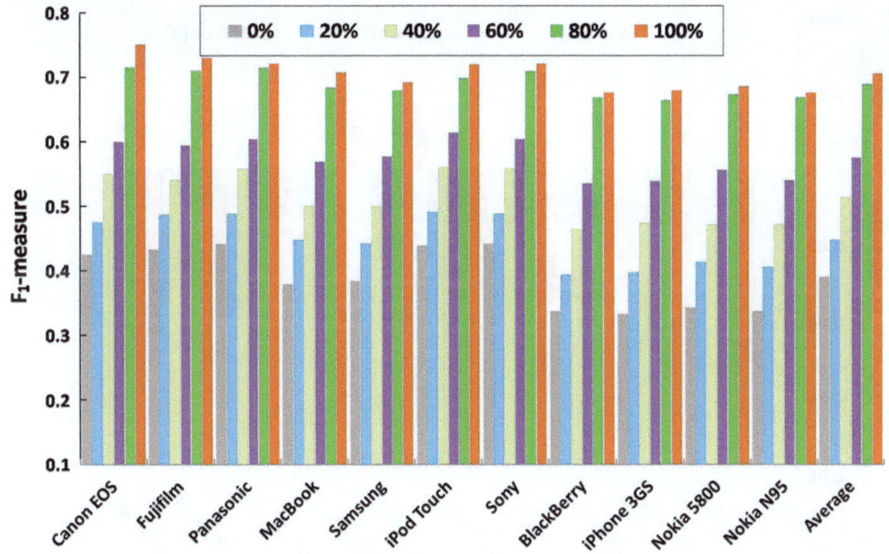


**Fig. 8.8** Evaluations on the impact of different proportion of initial hierarchy. T-Test, p-values $<0.05$

reliable semantic distance among aspects by exploiting external linguistic resources to assist distance learning, thereby improving the performance.

#### 8.4.3.2 Evaluations on the Effectiveness of Initial Hierarchy

We here show that by using different proportion of the initial hierarchy, our approach can still generate a satisfactory hierarchy. Different proportion of initial hierarchy were explored, including 0 %, 20 %, 40 %, 60 %, 80 %, and 100 % of the aspect pairs which were collected top-to-down, left-to-right from the initial hierarchy. As shown in Fig. 8.8, the performance increases when a larger proportion of the initial hierarchy is used. Thus, we can speculate that domain knowledge is valuable in aspect hierarchy generation.

#### 8.4.3.3 Evaluations on Multiple Optimization Criteria

A leave-one-out study is conducted to evaluate the effectiveness of each optimization criterion. In particular, we set one of the tradeoff parameters ($\lambda_1$, $\lambda_2$, $\lambda_3$) in Eq. (8.9) to zero, and distributed its weight to the rest of parameters proportionally. As illustrated in Fig. 8.9, we find that removing any optimization criterion would degrade the performance on most products. It is interesting to note that removing

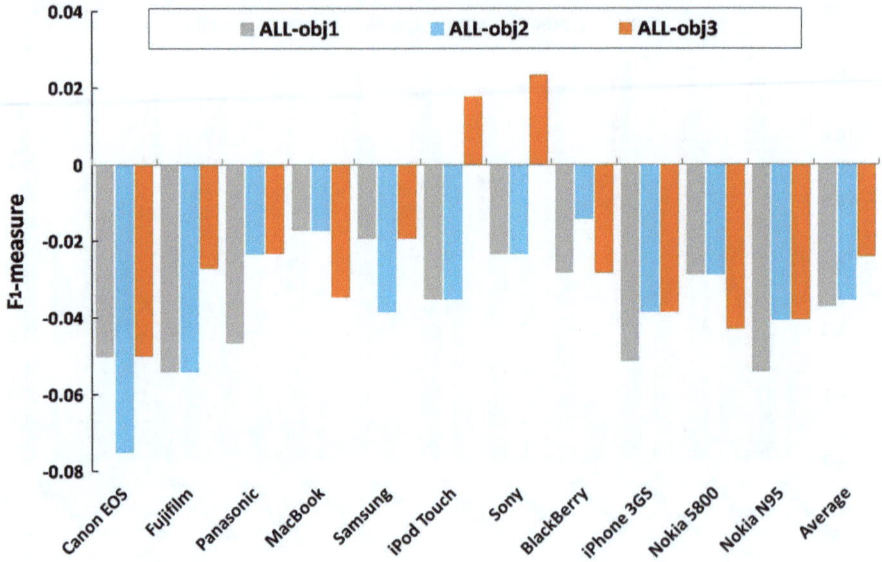


**Fig. 8.9** Evaluations of multiple optimization criteria. Changes in $F_1$-measure when a single criterion is removed. T-Test, p-values $<0.05$

the third optimization criterion, i.e., minimum semantic inconsistency, slightly increases the performance on two products (iPad touch and Sony MP3). The reason might be that the values of the three tradeoff parameters (empirically set in Sect. 8.4.1) are not suitable for these two products.

#### 8.4.3.4 Evaluations on Semantic Distance Learning

In this section, we evaluate the impact of the linguistic features and external linguistic resources for semantic distance learning. Five sets of features as described in Sect. 8.3.4.2 were investigated, including contextual, co-occurrence, syntactic, pattern and lexical features. As shown in Fig. 8.10, co-occurrence and pattern features outperform contextual and syntactic features. A probable reason is that co-occurrence and pattern features are effective to indicate parent-child relations among the aspects. Among these features, the lexical features perform the worst. We notice that the combination of all the features achieves the best performance. On average, the combined features outperform contextual features, co-occurrence features, syntactic features, pattern features, and lexical features by over 13.1 %, 10.0 %, 13.6 %, 9.7 %, and 24.3 %, respectively in terms of average $F_1$-measure. These results indicate that the heterogeneous features would be complementary and can assist to derive the semantic distance more accurately.

Next, we examine the effectiveness of using external linguistic resources (e.g. WordNet and ODP) on semantic distance learning. Our approach with or

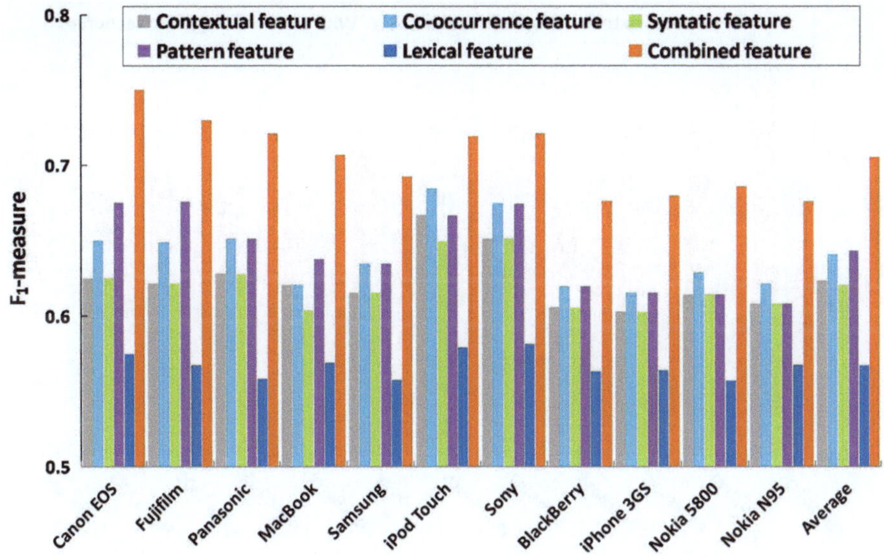


**Fig. 8.10** Evaluations the impact of linguistic features on semantic distance learning. T-Test, p-values <0.05

without external linguistic resources were examined. As illustrated in Fig. 8.11, by exploiting external linguistic resources, our approach significantly outperforms the method without external resources by over 4.2 % in terms of average $F_1$-measure. We can speculate that external linguistic resources help us obtain accurate semantic distance, which boosts the performance of hierarchy generation.

### *8.4.4 Evaluations on Aspect-Level Sentiment Classification*

In this experiment, we implemented the following sentiment classification methods:

- An unsupervised method. The opinion on each aspect is determined by referring to the sentiment lexicon *SentiWordNet* [44]. The lexicon contains a list of positive/negative words. The opinionated expression that is used to modify the aspect is classified as positive (or negative) if it contains a majority of words in the positive (or negative) list.
- Three supervised methods. We employed three supervised methods proposed in Pang et al. [46], including Naïve Bayes (*NB*), Maximum Entropy (*ME*), and Support Vector Machine (*SVM*). These classifiers were trained on *Pros* and *Cons* reviews as described in Sect. 8.2.3. *SVM* was implemented by using libSVM [7] with linear kernel, *NB* was implemented with Laplace smoothing, and *ME* was implemented with L-BFGS parameter estimation.

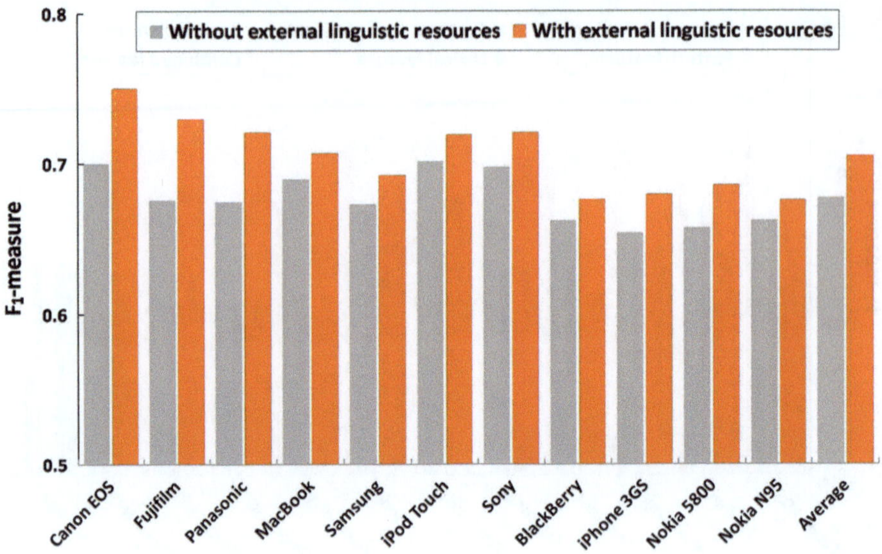


**Fig. 8.11** Evaluations the impact of external linguistic resources on semantic distance learning. T-Test, p-values <0.05

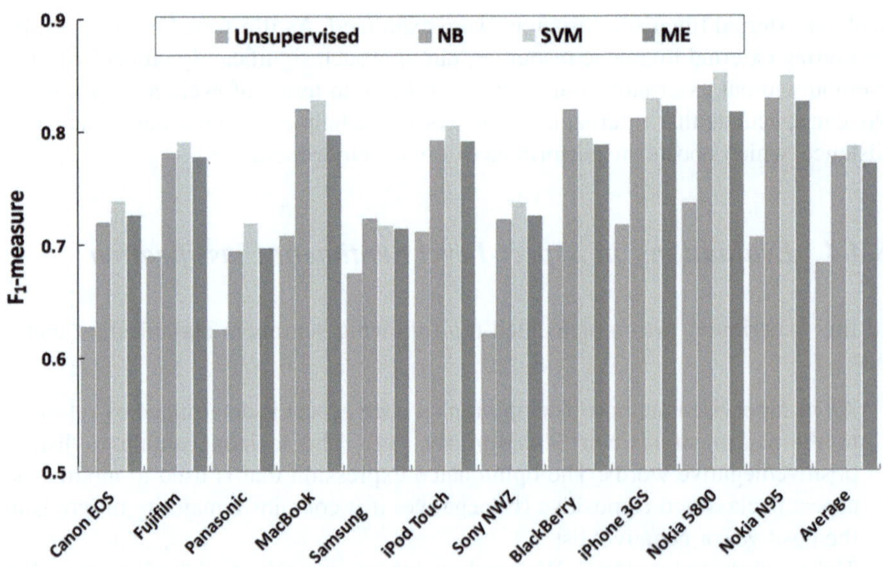


**Fig. 8.12** Performance of aspect-level sentiment classification. T-Test, p-values <0.05

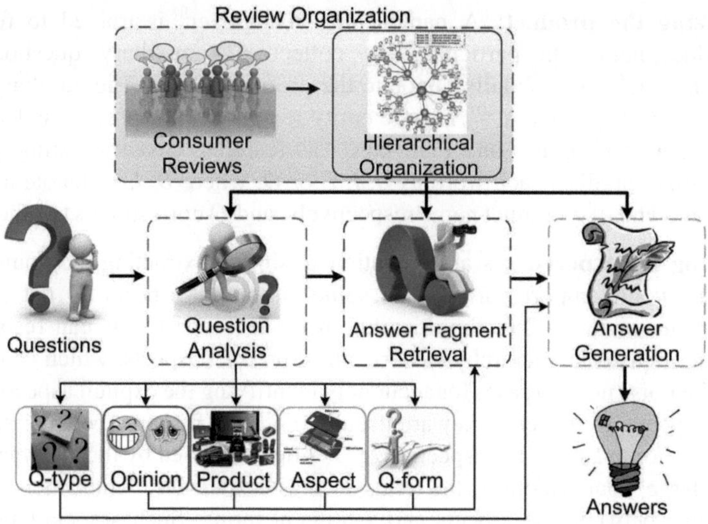


**Fig. 8.13** Flowchart of opinion-QA for products

Figure 8.12 shows the experimental results. We can see that the three supervised methods significantly outperform the unsupervised method. They achieve performance improvements on all the 11 products. In particular, *SVM* performs the best on nine products, *NB* obtains the best performance on two products. In terms of the average performance, *SVM* achieves slight improvements compared to *NB* and *ME*. These results are consistent with the previous research [46].

## 8.5 Application

In this section, we leverage the generated hierarchy to support the application of opinion-QA on products. [69] Figure 8.13 shows the framework of our approach. Generally, the framework includes three main components which are making use of the hierarchy, including question analysis, answer fragments retrieval, and answer generation. We elaborate each component as follows.

### *8.5.1 Question Analysis and Answer Fragment Retrieval*

Question analysis consists of five sub-tasks: recognizing product asked in the question; identifying aspects in the question; classifying opinions that the question asks for (the asked opinion could be positive, negative or both); identifying the question type (e.g. asking for public opinions, or the reason of the opinions, etc.); and identifying the question form (i.e. comparative question or single form question).

**Recognizing the product:** A name entity recognizer[8] is trained to recognize the product name. In particular, we collect 420 auxiliary questions from Yahoo!Answer,[9] and manually annotate the product names. The auxiliary corpus is available in Appendix A.[10] A name entity recognizer for product is learned on these data, with unigrams and POS tags[11] as features. Given a testing question, the recognizer predicts each word as *B, I, E* or *O*, where *B, I, E* denote the begin, internal, and end of a product name respectively, and *O* corresponds to other words.

**Identifying the aspects:** As aforementioned, simply extracting the noun phrases as aspects would import noise. Also, some "implicit" aspects do not explicitly appear in the reviews. One simple solution for these problems can resort to the review hierarchy. The hierarchy has organized product aspects, which can be used to filter the noise noun phrases for accurately identifying the explicit aspects. For the implicit aspects, we observe they are usually modified by some peculiar sentiment terms [58]. For example, the aspect "size" is often modified by the sentiment terms such as "large", but seldom by the terms such as "expensive." Thus, there are some associations between the aspects and sentiment terms. Such associations can be learned from the hierarchy and leveraged to infer the implicit aspects [68]. In order to simultaneously identify the (explicit/implicit) aspects, we adopt a hierarchical classification technique. The technique simultaneously learns to identify explicit aspects, and discovers the associations between aspects and sentiment terms by multiple classifiers. In particular, given a testing question, we identify its aspect by hierarchically classify [54] it into the appropriate aspect node of a particular product hierarchy. The classification greedily searches a path in the hierarchy from top to down. The search begins at the root node, and stops at the leaf node or a specific node where the relevance score is lower than a pre-defined threshold. The relevance score on each node is determined by a SVM classifier. Multiple SVM classifiers are trained on the hierarchy, one distinct classifier for a node. The reviews that are stored in the node and its child-nodes are used as training samples. We employ the features of noun terms, and sentiment terms in the sentiment lexicon provided by *MPQA* project [61].

**Classifying the opinions:** Given a set of testing questions, we first distinguish the opinion questions from the factual ones [69]. Since the opinion questions often contain one or more sentiment terms, we classify them by employing the sentiment terms in the sentiment lexicon provided from MPQA project [61]. Subsequently, we learn a SVM sentiment classifier to determine the opinion polarity of the opinion questions. In particular, the reviews and corresponding opinions stored in the hierarchy are used as training samples, which are represented by the unigram features.

---

[8]http://nlp.stanford.edu/software/CRF-NER.shtml

[9]http://answers.yahoo.com

[10]http://www.comp.nus.edu.sg/~Jianxing/auxiliary_material.zip

[11]Using stanford POS tagger, http://nlp.stanford.edu/software/tagger.shtml

**Identifying the question type:** Opinion questions are often categorized into four types [23],

- **Attitude** question, asking for public opinion on a product or product aspect, such as "What do people think about iPhone 3gs?"
- **Reason** question, asking for the reason of public opinion on a product or product aspect, such as "Why do people like iPhone 3gs?"
- **Target** question, asking for the object in the public opinion, such as "Which phone is better than Nokia N95?"
- **Yes/No** question, asking for whether a statement is correct, such as "Is Nokia N95 bad?"

We formulate the question type identification as a multi-class classification problem. A multi-class SVM classifier[12] is trained for the classification. We collect 420 auxiliary questions from Yahoo!Answer and manually annotate their types. The auxiliary corpus is available in Appendix B.[13] These questions are used for training, with POS tags and question words (i.e. why, what, how, do, is) as features.

**Identifying the question form:** Question form includes single and comparative. A question is viewed as comparative if it contains comparative adjectives and adverbs (e.g. cheaper, etc.), otherwise as the single form [41]. The POS tags are exploited to detect comparative adjectives (i.e. tag "JJR") and adverbs (i.e. tag "RBR").

After analyzing the question, we retrieve all review sentences on the asked aspect and all its sub-aspects from a certain product hierarchy, and choose the ones relevant to the opinion asked in the question. For the single form question, we view the retrieved sentences as the answer fragments. For the comparative questions, we select comparative sentences on the compared products from the retrieved sentences, and treat them as the answer fragments. Subsequently, question type is used to define the template for the answers. In particular, for the questions asking for reason and attitude, we generate the answers by summarizing corresponding answer fragments. For questions seeking for a target as the answer, we output the product names based on the majority voting of the opinions in the retrieved answer fragments. For the yes/no questions, we first generate the "yes/no" answer based on the consistency between the asked opinions and the major opinions in the answer fragments, and then summarize these fragments to form the answers.

### 8.5.2 Answer Generation

Answer generation aims to generate an appropriate answer for a given opinion question based on the retrieved answer fragments, i.e., review sentences. An answer

[12] http://svmlight.joachims.org/svm_multiclass.html

[13] http://www.comp.nus.edu.sg/~Jianxing/auxiliary_material.zip

is essentially a sequence of sentences. Hence, the task of answer generation is to select sentences from the retrieved answer fragments and order them appropriately. We formulate this task into a multi-criteria optimization problem. We incorporate multiple criteria in the answer generation process, including answer salience, coherence, and diversity. The parent-child relations between aspects are also incorporated to ensure the answer follow the general-to-specific logic. In the next subsections, we will introduce details of the proposed multi-criteria optimization approach.

### 8.5.2.1 Formulation

We first introduce the multiple criteria and then present the optimization problem.

**Salience** is used to measure the representativeness of the answer. A good answer should consist of salient review sentences. Let $\mathcal{S}$ denote the set of retrieved sentences. We define a binary variable $s_i \in \{0, 1\}$ to indicate the selection of sentence $i$ for the answer, i.e. $s_i = 1$ (or 0) indicates that $s_i$ is selected (or not). Let $\omega_i$ denote the salience of sentence $i$. The estimation of $\omega_i$ will be described in Sect. 8.5.2.2. The salience score of the answer (i.e., a set of sentences) is computed by summing up the scores of all its constituent sentences, as $\sum_{i \in \mathcal{S}} \omega_i s_i$.

**Coherence** is used to quantify the readability of an answer. To make the answer readable, the constituent sentences in the answer should be ordered properly. That is, the adjacent sentences should be coherent. We define $e_{i,j} \in \{0, 1\}$ to indicate whether the sentences $i$ and $j$ are adjacent in the answer; where $e_{i,j} = 1$ (or 0) means they are (or not) adjacent. The coherence between two adjacent sentences is measured by $c_{ij}$. The estimation of $c_{ij}$ will be described in Sect. 8.5.2.3. As aforementioned, the answer is expected to be presented in a general-to-specific manner, i.e. from general aspects to specific sub-aspects. We define $h_{i,j}$ in Eq. (8.15) to measure the general-to-specific coherence of sentences $i$ and $j$.

$$h_{i,j} = \begin{cases} e^{-\frac{1}{level_i - level_j}}; & \text{if } level_i \neq level_j; \\ 1; & \text{otherwise,} \end{cases} \tag{8.15}$$

where $level_i$ denotes level position of the aspect commented in sentence $i$ by referring to the hierarchy, with the root level being 0. The coherence score of the answer is computed by summing up the scores of all its adjacent sentences as, $\sum_{j \in \mathcal{S}} \sum_{i \in \mathcal{S}} h_{i,j} c_{i,j} e_{i,j}$.

**Diversity** A good answer should diversely cover all the important information. We introduce a matrix $\mathcal{M}$ in Eq. (8.16) to measure the pairwise diversities among sentences. $\mathcal{M}_{ij}$ corresponds to the diversity between sentences $i$ and $j$. When sentences $i$ and $j$ comment on the same aspects, $\mathcal{M}_{ij}$ will favor to select the pair of sentences that discusses on diverse content (i.e. low similarity). Otherwise, the pair of sentences commented on different aspects is viewed to be diverse, and $\mathcal{M}_{ij}$ is set as a constant bigger than one.

$$\mathcal{M}_{ij} = \begin{cases} 1 - similarity(i,j) \ \ if \ i,j \ commented \ on \ same \ aspect \\ \varphi \qquad\qquad\qquad\qquad otherwise, \end{cases} \tag{8.16}$$

where *similarity(i,j)* denotes the Cosine similarity between sentences $i$ and $j$, and $\varphi$ is a constant.[14]

**Multi-Criteria Optimization** We integrate the above criteria into the multi-criteria optimization formulation,

$$\begin{aligned} &\max\{\alpha_1 \cdot \textstyle\sum_{i\in\mathcal{S}} \omega_i s_i + \alpha_2 \cdot \sum_{j\in\mathcal{S}} \sum_{i\in\mathcal{S}} h_{i,j} c_{i,j} e_{i,j} + \alpha_3 \cdot \sum_{j\in\mathcal{S}} \sum_{i\in\mathcal{S}} s_i \mathcal{M}_{ij}; \\ &\begin{cases} s_i, e_{i,j} \in \{0,1\}, \forall i,j; \\ \alpha_1 + \alpha_2 + \alpha_3 = 1, \ 0 \le \alpha_1, \alpha_2, \alpha_3 \le 1, \end{cases} \end{aligned} \tag{8.17}$$

where $\alpha_1$, $\alpha_2$, $\alpha_3$ are the tradeoff parameters.

We further incorporate the following constrains into the optimization framework, so as to derive appropriate answers.

- The length of the answer is up to $K$,

$$\sum\nolimits_{i\in\mathcal{S}} l_i s_i \le K, \tag{8.18}$$

  where $l_i$ is the length of sentence $i$.
- When sentence $i$ is not selected (i.e. $s_i = 0$), the adjacency between any sentence to $i$ is set to zero (i.e. $\sum_{i\in\mathcal{S}} e_{i,j} = \sum_{i\in\mathcal{S}} e_{j,i} = 0$). When sentence $i$ is selected, there are two sentences adjacent to sentence $i$ in the answer, one before $i$ and another after $i$ (i.e. $\sum_{i\in\mathcal{S}} e_{i,j} = \sum_{i\in\mathcal{S}} e_{j,i} = 1$).

$$\sum\nolimits_{i\in\mathcal{S}} e_{i,j} = \sum\nolimits_{i\in\mathcal{S}} e_{j,i} = s_j, \quad \forall j. \tag{8.19}$$

- In order to avoid falling into a cycle in sentence selection, we employ the following constraints [13].

$$\begin{aligned} &\textstyle\sum_{i\in\mathcal{S}} f_{0,i} = n+1; \\ &\textstyle\sum_{i\in\mathcal{S}} f_{i,n+1} \ge 1; \\ &\textstyle\sum_{i\in\mathcal{S}} f_{i,j} - \sum_{i\in\mathcal{S}} f_{j,i} = s_j, \quad \forall j; \\ &0 \le f_{i,j} \le (n+1) \cdot e_{i,j}, \ \forall i,j, \end{aligned} \tag{8.20}$$

  where the variable $f_{i,j}$ is an integer to number the selected adjacent sentences from 1 to n + 1, and the first selected sentence is numbered $f_{0,i} = n+1$. If the last selected sentence obtains a number $f_{i,n+1}$ which is bigger then 1, then the selection has no cycle.

[14] Empirically set to 10 in the experiment.

**Solution** Given the salience weights $\omega_i|_{i=1}^{S}$, and coherence weights $c_{i,j}|_{i,j=1}^{S}$, the above multi-criteria optimization problem can be solved by *Integer Linear Programming* [51]. The optimal solutions $s_i|_{i=1}^{S}$ and $e_{i,j}|_{i,j=1}^{S}$ indicate the selected sentences and the order of them. In the next subsections, we will introduce the estimations of $\omega_i|_{i=1}^{S}$ and $c_{i,j}|_{i,j=1}^{S}$.

#### 8.5.2.2 Salience Weight Estimation

The salience weight of sentence $i$ is formulated as $\omega_i = \sum_{g=1}^{G} \varphi_g(i)/G$, where $\varphi(i)$ denotes the measurement for the importance of sentence $i$. We define seven measurements (i.e. $G = 7$) below.

**Helpfulness:** Many forum websites provide a helpfulness score, which is used to rate the quality of a review. The sentences that come from helpful reviews are often representative [40]. We compute $\varphi(i)$ of sentence $i$ by using helpfulness score from its host review.

**Timeliness:** The new coming sentence often contains more updated and useful information [34]. $\varphi(i)$ is the posting time of the review sentence $i$. We normalize it to [0, 1].

**Grammaticality:** The grammatical sentence is often more readable. We employ the method in Agichtein et al. [2] to calculate the grammar score. In particular, $\varphi(i)$ is calculated by the KL-divergence [24] between language models of sentence $i$ to Wikipedia articles.

**Position:** The first sentence in a review is usually informative [19]. $\varphi(i)$ is computed based on the position of the sentence in the review, i.e. $\varphi(i) = 1/position_i$.

**Aspect Frequency:** The sentence that contains the frequent aspects is often salient [43]. Hence, $\varphi(i)$ is computed as the sum of the frequency for aspects in sentence $i$.

**Centroid Distance:** Review sentences are stored in the corresponding aspect nodes in the hierarchy. The sentence that is close to the centroid of the reviews stored in an aspect node is more likely to be salient [16]. $\varphi(i)$ is computed as the Cosine similarity between sentence $i$ to the corresponding review cluster centroid based on the unigram features.

**Local Density:** The sentence would be informative when it is in the dense part of the aspect node in the feature space [52]. We employ *Multivariate Kernel Density Estimation* to estimate the density. We first represent all the sentences stored in each node into feature vectors, with unigram as features. The density of a sentence is then calculated as $\varphi(\boldsymbol{x}) = \sum_{i=1}^{n} K_H(\boldsymbol{x} - \boldsymbol{x}_i)/n$, where $\boldsymbol{x}$ denotes the feature vector of sentence $i$, $n$ is the size of sentences stored in the node, and $K_H(\boldsymbol{x}) = (2\pi)^{-1/2}\exp(-1/2(\boldsymbol{x}^T\boldsymbol{x}))$ represents the *Gaussian* kernel.

#### 8.5.2.3 Coherence Weight Estimation

The coherence $c_{i,j}$ between sentences $i$ and $j$ is formulated as $c_{i,j} = \boldsymbol{\mu} \cdot \boldsymbol{\psi}(i,j)$, where $\boldsymbol{\mu}$ is a weight vector, and $\boldsymbol{\psi}(i,j)$ denotes the feature function. $\boldsymbol{\psi}(i,j)$ takes two sentences $i$ and $j$ as input, and outputs a vector with each dimension indicating the present/absent of a feature. In order to capture the sequential relations among sentences, we utilize features as the *Cartesian* product over the terms of N-gram (N = 1,2) and POS tags generated from sentences $i$ and $j$ [25].

To learn the weight vector $\boldsymbol{\mu}$, we employ the *Passive-Aggressive* algorithm [10]. It is an online learning algorithm, so that we can update the weight when more consumer reviews are available. The algorithm takes up one training sample and outputs the solution that has the highest score under the current weight. If the output differs from training samples, the weight vector is updated according to Eq. (8.21). Since the consumer reviews often include multiple sentences, we can directly use the adjacency of these sentences as training samples. In particular, we treat the adjacent sentence pairs in the reviews as training samples (i.e. $c_{i,j} = 1$).

$$\min \left\| \boldsymbol{\mu}^{i+1} - \boldsymbol{\mu}^{i} \right\|$$
$$\begin{cases} \boldsymbol{\mu}^{i+1} \cdot \boldsymbol{\Psi}(\mathbf{p}, \mathbf{q}^*) - \boldsymbol{\mu}^{i+1} \cdot \boldsymbol{\Psi}(\mathbf{p}, \hat{\mathbf{q}}) \geq \tau(\hat{\mathbf{q}}, \mathbf{q}^*); \\ \tau(\hat{\mathbf{q}}, \mathbf{q}^*) = \frac{2 \cdot T(\hat{\mathbf{q}}, \mathbf{q}^*)}{m(m-1)/2}, \end{cases} \tag{8.21}$$

where $\boldsymbol{\mu}^i$ is the current weight vector and $\boldsymbol{\mu}^{i+1}$ is the updated vector, $\mathbf{q}^*$ and $\hat{\mathbf{q}}$ are the gold standard and predicted sequence of sentences, respectively, $\mathbf{p}$ denotes a set of sentences, $\boldsymbol{\Psi}(\cdot)$ is the feature function on the whole feature space (i.e. $\sum \boldsymbol{\psi}(\cdot)$), $\tau(\cdot,\cdot)$ is a *Kendall's tau* lost function [26], $T(\cdot,\cdot)$ represents the number of inversion operations that needs to bring $\hat{\mathbf{q}}$ to $\mathbf{q}^*$, and $m$ denotes the number of sentences.

### *8.5.3 Evaluations on Question Analysis*

We employed the product review dataset as described in Sect. 8.4.1 as corpus. In addition, we created 220 questions for these products by referring to real questions in Yahoo!Anwser service. We corrected the typos and grammar errors for these real questions. Each product contains 15 opinion questions and 5 factual questions, respectively. All questions were shown in Appendix C.[15] Three annotators were invited to generate the gold standard. Each question was labeled by two annotators. The labels include product name, product aspect, opinion, question type and question form. The average inter-rater agreement in terms of Kappa statistics is 89 %. These annotators were then invited to read the reviews, and create the ground truth answers by selecting and ordering some review sentences. Such process is

[15]http://www.comp.nus.edu.sg/~Jianxing/auxiliary_material.zip

**Table 8.3** Performance of question analysis

| Evaluated topics | P | R | $F_1$ |
|---|---|---|---|
| Product recognition | 0.755 | 0.618 | 0.680 |
| Opinionated/factual | 0.897 | 0.895 | 0.893 |
| Opinion classification | 0.755 | 0.745 | 0.748 |
| Question type | 0.800 | 0.775 | 0.783 |
| Question form | 0.910 | 0.903 | 0.905 |

time consuming and labor-intensive. We speed up the annotation process as follows. We first collected all the review sentences in the answers generated by three evaluated methods to be discussed in Sect. 8.5.4.1. In addition, we sampled the top-N (N = 20) sentences on each asked aspect and its sub-aspects respectively, where the sentences were ranked based on their salient weights in Sect. 8.5.2.2. We then provided such subset of review sentences to the three annotators, and let them individually create an answer of up to 100 words (i.e. K = 100) for each question.

We employed *precision* (P), *recall* (R) and $F_1$-measure ($F_1$) as the evaluation metric for question analysis, and utilized *ROUGE* [30] as the metric to evaluate the quality of answer generation. *ROUGE* is a widely accepted standard for summarization, which measures the quality of the summarized answers by counting the overlapping N-grams between the answers generated by machine and human, respectively. In the experiment, we reported the $F_1$-measure of *ROUGE-1*, *ROUGE-2* and *ROUGE-SU4*, which count the overlapping unigrams, bigrams and skip-4 bigrams[16] respectively. *ROUGE-1* can measure informativeness of the answers, while higher order *ROUGE-N* (*N* = *2,4*) captures the matching of subsequences, which can measure the fluency and readability of the answers. For the trade-off parameters, we empirically set $\lambda_1 = 0.4$, $\lambda_2 = 0.3$ and $\lambda_3 = 0.3$.

We first evaluated the performance of product recognition, opinionated/factual classification, opinion classification, question type and question form identification. The experimental results are shown in Table 8.3. The results show that traditional methods achieve encouraging performance on the aforementioned tasks.

We next examined the performance of our approach on aspect identification. The method proposed by Balahur et al. [3] was reimplemented as the baseline, which identifies aspects based on noun phrase extraction. This method achieved good performance on the opinion QA task in TAC 2008 and was employed in subsequent works. As demonstrated in Table 8.4, our approach significantly outperforms Balahur's method by over 49.4 % in terms of average $F_1$-measure. A probable reason is that Balahur's method relies on noun phrases, which may mis-identify some noise noun phrases as aspects, while our approach performs hierarchical classification based on the hierarchy, which can leverage the prior knowledge encoded in the hierarchy to filter out the noise and obtain accurate aspects.

[16]It represents any pair of words in their sentence order, allowing at most two gaps in between.

**Table 8.4** Performance of aspect identification for question analysis

| Methods | P | R | $F_1$ |
|---|---|---|---|
| Our method | **0.851*** | **0.763*** | **0.805*** |
| Balahur's method | 0.825 | 0.400 | 0.538 |

*denotes the results are tested for statistical significance using T-Test, p-values <0.05

**Table 8.5** Performance of implicit aspect identification for question analysis

| Methods | P | R | $F_1$ |
|---|---|---|---|
| Our method | **0.726*** | **0.643*** | **0.682*** |
| Su's method | 0.689 | 0.571 | 0.625 |

*T-Test, p-values <0.05

Moreover, we evaluated the effectiveness of our approach on implicit aspect identification. The 70 implicit aspect questions in our question corpus were used here. The method proposed by Su et al. [58] was reimplemented as the baseline. It identifies implicit aspects by mutual clustering, and it was evaluated in [68]. As shown in Table 8.5, our approach significantly outperforms Su's method by over 9.1 % in terms of average $F_1$-measure. The results show that the hierarchy can help to identify implicit aspects by exploiting the underlying associations among sentiment terms and aspects.

## 8.5.4 *Evaluations on Answer Generation*

### 8.5.4.1 Comparisons to the State-of-the-Arts Methods

We compared our multi-criteria optimization approach against two state-of-the-arts methods: (a) the method presented in Li et al. [27], which selects some retrieved sentences to generate the answers based on a graph-based algorithm; (b) the method proposed by Lloret et al. [35] that forms the answers by re-ranking the retrieved sentences.

As shown in Table 8.6, our approach outperforms Li's method and Lloret's method by the significant absolute gains of over 23.7 % and 21.5 % respectively, in terms of average *ROUGE-1*. It improves the performance over these two methods in terms of average *ROUGE-2* by the absolute gains of over 9.41 % and 7.87 %, respectively; and in terms of *ROUGE-SU4* by the absolute gains of over 8.86 % and 7.31 %, respectively. By analyzing the results, we find that the improvements come from the use of the hierarchical organization and the answer generation algorithm which exploits multiple criteria, especially the parent-child relation among aspects. In addition, our approach can generate the answers by following the general-to-specific logic, while Li's and Lloret's methods fail to do so due to their ignorance of parent-child relations among aspects.

**Table 8.6** Performance of answer generation

| Methods | ROUGE1 | ROUGE2 | ROUGE-SU4 |
|---|---|---|---|
| Our method | **0.364*** | **0.137*** | **0.138*** |
| Li's method | 0.127 | 0.043 | 0.049 |
| Lloret's method | 0.149 | 0.058 | 0.065 |

* T-Test, p-values <0.05

**Table 8.7** Sample answers of our approach on opinion-QA for products

| |
|---|
| *Question 1: What reasons do people give for preferring iPhone 3gs?* |
| There are 9,928 opinionated reviews about product "iphone 3gs", with 5,717 positive and 4,221 negative reviews |
| This phone is amazing and I would recommend it to anyone. It looks funky and cool. It is worth the money. It's great organiser, simple easy to use software. It is super fast, excellent connection via wifi or 3G. It is able to instantly access email. It's amazing and has so many free apps. The design is so simple and global. The hardware is good and reliable. The camera is a good and colors are vibrant. The touch screen is user friendly and the aesthetics are top notch. Battery is charged quickly, and power save right after stop using |
| *Question 2: Does anyone think it is expensive to get a iPhone 3GS?* |
| Yes |
| There are 2,645 opinionated reviews on aspect "price" about product "iphone 3gs", with 889 positive and 1,756 negative reviews |
| Throw the costly phone, apple only knows to sell stupid stuff expensively. Don't fool yourself with iPhone 3gs, believing that it costs much by Apple luxurious advertising. Apple is so greedy and it just wants to earn easy & fast money by selling its techless product expensively. The phone will charge once you insert any sim card. iPhone 3gs is high-priced due to the capacitive and Apple license. You need to pay every application at the end it costs too much. The network provider will make up some of the cost of the phone on your call charges |

#### 8.5.4.2 Evaluations on the Effectiveness of Multiple Criteria

We further evaluated the effectiveness of each optimization criterion by tuning the trade-off parameters (i.e. $\lambda_1$, $\lambda_2$, and $\lambda_3$). We fixed $\lambda_1$ as a constant in [0, 1] with 0.1 as an interval, and updated $\lambda_2$ from 0 to $1 - \lambda_1$, $\lambda_3 = 1 - \lambda_1 - \lambda_2$, correspondingly. The performance change is shown in Fig. 8.14 in terms of *ROUGE-1*, *ROUGE-2*, and *ROUGE-SU4*, respectively. The best performance is achieved at $\lambda_1 = 0.4, \lambda_2 = 0.3, \lambda_3 = 0.3$. We observe the performance drops dramatically when any parameter (i.e. $\lambda_1, \lambda_2, \lambda_3$) is close to 0 (i.e. remove any of the corresponding criterion). Thus, we can conclude that all the criteria are useful in answer generation. We also find that the performance change is sharp when $\lambda_1$ changes. This indicates that the salience criterion is crucial for answer generation.

Table 8.7 shows the exemplar answers generated by our approach. Each answer first gives the statistic of positive and negative reviews. This helps user to quickly get an overview of public opinions. The summary of relevant review sentences is then presented in the answer. The answer diversely comments the asked aspect and all its available sub-aspects following the general-to-specific logic. Moreover, we feel that the answers are informative and readable.

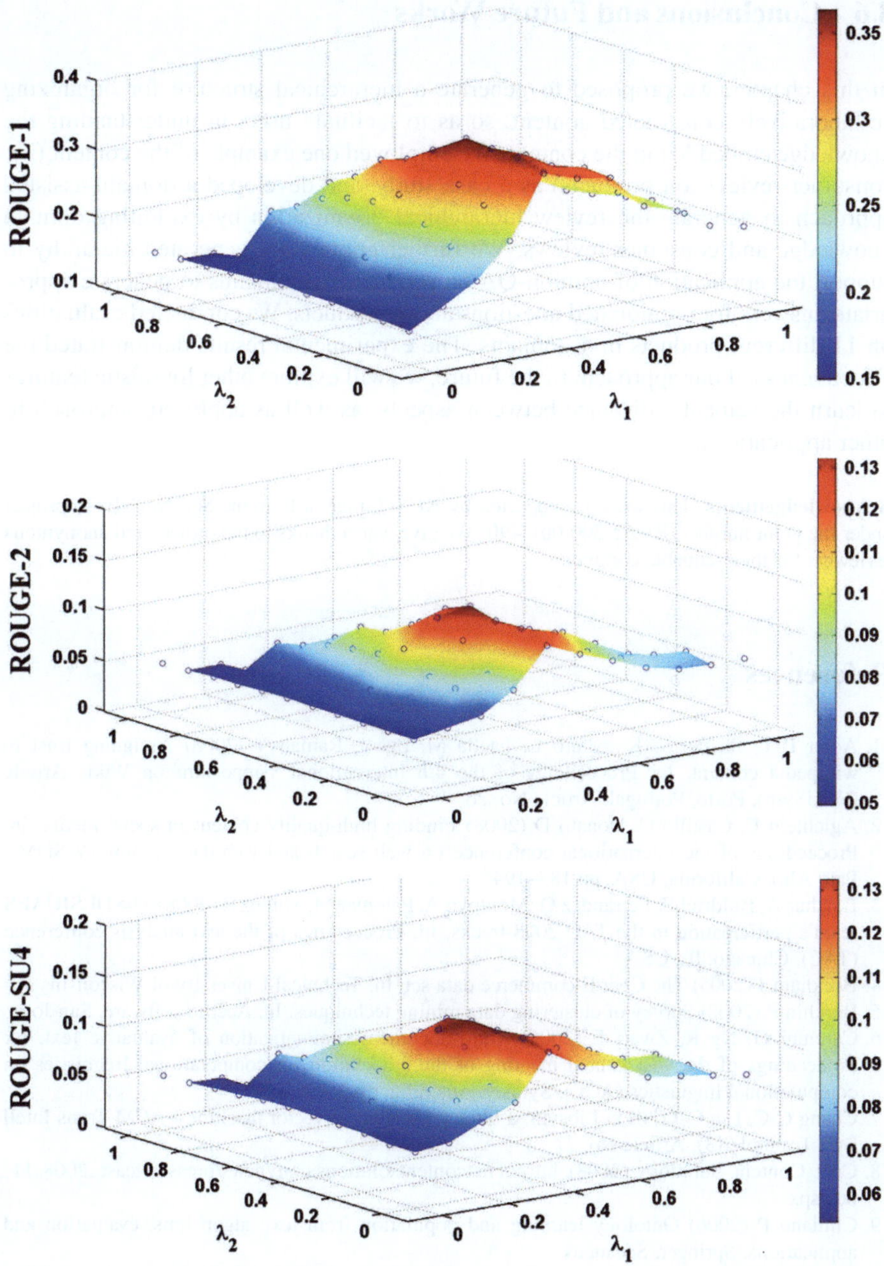


**Fig. 8.14** Evaluations on multiple optimization criteria for answer generation in terms of *ROUGE-1*, *ROUGE-2*, and *ROUGE-SU4*, respectively

## 8.6 Conclusions and Future Works

In this chapter, we proposed to generate a hierarchical structure for organizing collaboratively constructed content, so as to facilitate users in understanding the knowledge embedded in the content. We employed one example of the content (i.e. consumer reviews on products) as a case study, and developed a domain-assisted approach to generate the review hierarchical organization by exploiting domain knowledge and consumer reviews. We further applied the generated hierarchy to support the application of opinion-QA on products, which aims to generate appropriate answers for opinionated questions about products. We conducted evaluations on 11 different products in 4 domains. The experimental results demonstrated the effectiveness of our approach. In the future, we will explore other linguistic features to learn the semantic distance between aspects, as well as apply our approach to other applications.

**Acknowledgements** This work is supported by NUS-Tsinghua Extreme Search (NExT) project under the grant number: R-252-300-001-490. We give warm thanks to the project and anonymous reviewers for their valuable comments.

## References

1. Adler B-T, Chatterjee K, Alfaro L, Faella M, Pye I, Raman V (2008) Assigning trust to wikipedia content. In: Proceedings of the 4th international symposium on Wikis Article (WikiSym), Porto, Portugal. Article No. 26
2. Agichtein E, Castillo C, Donato D (2008) Finding high-quality content in social media. In: Proceedings of the international conference on web search and web data mining (WSDM), Palo Alto, California, USA, pp 183–194
3. Balahur A, Boldrini E, Ferrandez O, Montoyo A, Palomar M, Munoz R (2008) The DLSIUAES team's participation in the TAC 2008 tracks. In: Proceedings of the text analysis conference (TAC), Chicago, IL, USA
4. Beckham J (2005) The Cnet E-commerce data set. In: Technical University of Wisconsin
5. Berkhin P (2002) Survey of clustering data mining techniques. In: Accrue software, San Jose
6. Carenini G, Ng R, Zwart E (2006) Multi-document summarization of evaluative text. In: Proceedings of the 44st annual meeting of the association for computational linguistics on computational linguistics (ACL), Sydney, Australia, pp 3–7
7. Chang C-C, Lin C-J (2011) Libsvm: a library for support vector machines. ACM Trans Intell Syst Technol 2(3), Article No. 27
8. Cnet Content Solutions (2008) http://cnetcontentsolutions.com/news/press_release_2008_11_06.aspx
9. Cimiano P (2006) Ontology learning and population from text: algorithms, evaluation and applications. Springer, Secaucus
10. Crammer K, Dekel O, Keshet J, Shwartz S-S, Singer Y (2006) Online passive aggressive algorithms. J Mach Learn Res 7:551–585
11. Davidov D, Gabrilovich E, Markovitch S (2004) Parameterized generation of labeled datasets for text categorization based on a hierarchical directory. In: Proceedings of 27th annual international ACM SIGIR conference on research and development in information retrieval (SIGIR), Sheffield, UK, pp 250–257

12. Deng J, Dong W, Socher R, Li L-J, Li K, Li F-F (2009) ImageNet: a large-scale hierarchical image database. In: Proceedings of IEEE conference on computer vision and pattern recognition (CVPR), Miami, Florida, USA, pp 248–255
13. Deshpande P, Barzilay R, Karger D-R (2007) Randomized decoding for selection-and-ordering problems. In: Proceedings of the conference of the North American chapter of the association for computational linguistics (NAACL), Rochester, New York, USA, pp 444–451
14. Ding X, Liu B, Yu P-S (2008) A holistic lexicon-based approach to opinion mining. In: Proceedings of first ACM international conference on web search and data mining (WSDM), Palo Alto, California, USA, pp 231–240
15. Elsas J, Dumais S-T (2010) Leveraging temporal dynamics of document content in relevance ranking. In: Proceedings of the 3rd ACM international conference on web search and data mining (WSDM), New York, NY, USA, pp 1–10
16. Erkan G, Radev DR (2004) LexRank: graph-based lexical centrality as salience in text summarization. J Artif Intell Res 22(1):457–479
17. Etzioni O, Cafarella M, Downey D, Popescu A, Shaked T, Soderland S, Weld D, Yates A (2005) Unsupervised named-entity extraction from the web: an experimental study. J Artif Intell 165(1):91–134
18. Girju R, Badulescu A (2006) Automatic discovery of part-whole relations. J Comput Linguist 32(1):83–135
19. He J, Dai D (2011) Summarization of yes/no questions using a feature function model. J Mach Learn Res 20:351–366
20. Hearst M-A (1992) Automatic acquisition of hyponyms from large text Corpora. In: Proceedings of the 14th international conference on computational linguistics (COLING), Nantes, France, pp 539–545
21. Hu M, Liu B (2004) Mining and summarizing customer reviews. In: Proceedings of the 10th ACM SIGKDD international conference on knowledge discovery and data mining, Seattle, WA, USA, pp 168–177
22. Jiang P, Fu H, Zhang C, Niu Z (2010) A framework for opinion question answering. In: Advanced information management and service (IMS), Seoul, Korea, pp 424–427
23. Ku L-W, Liang Y-T, Chen H-H (2008) Question analysis and answer passage retrieval for opinion question answering systems. Int J Comput Linguist Chin Lang Process 13:307–326
24. Kullback S (1951) On information and sufficiency. Ann Math Stat 22(1):79–6
25. Lapata M (2003) Probabilistic text structuring: experiments with sentence ordering. In: Proceedings of the 41st annual meeting of the association for computational linguistics on computational linguistics (ACL), Sapporo, Japan, pp 545–552
26. Lapata M (2006) Automatic evaluation of information ordering: Kendallś Tau. J Comput Linguist 32(4):471–484
27. Li F, Tang Y, Huang M, Zhu X (2009) Answering opinion questions with random walks on graphs. In: Proceedings of the joint conference of the 47th annual meeting of the ACL and the 4th international joint conference on natural language processing of the AFNLP (ACL/AFNLP), Singapore, pp 737–745
28. Li T, Zhang Y, Sindhwani V (2009) A non-negative matrix tri-factorization approach to sentiment classification with lexical prior knowledge. In: Proceedings of the 47th annual meeting of the association for computational linguistics on computational linguistics (ACL), Singapore, pp 244–252
29. Lin D (1998) Automatic retrieval and clustering of similar words. In: Proceedings of the 17th international conference on computational linguistics (COLING), Montreal, Quebec, Canada, pp 768–774
30. Lin C-Y, Hovy E (2003) Automatic evaluation of summaries using N-gram co-occurrence statistics. In: Proceedings of the 2003 conference of the North American chapter of the association for computational linguistics on human language (HLT-NAACL), Edmonton, Canada, pp 71–78

31. Liu B (2009) Sentiment analysis and subjectivity. In: Handbook of natural language processing. Marcel Dekker, New York
32. Liu B, Hu M, Cheng J (2005) Opinion observer: analyzing and comparing opinions on the web. In: Proceedings of the 14th international conference on world wide web (WWW), Chiba, Japan, pp 342–351
33. Liu Y, Bian J, Agichtein E (2008) Predicting information seeker satisfaction in community question answering. In: Proceedings of the 31st annual international ACM SIGIR conference on research and development in information retrieval (SIGIR), Singapore, pp 483–490
34. Liu Y, Huang X, An A, Yu X (2008) Modeling and predicting the helpfulness of online reviews. In: Proceedings of the 18th IEEE international conference on data mining (ICDM), Pisa, Italy, pp 443–452
35. Lloret E, Balahur A, Palomar M, Montoyo A (2011) Towards a unified approach for opinion question answering and summarization. In: Proceedings of the 49th annual meeting of the association for computational linguistics on computational linguistics (ACL), Portland, Oregon, USA, pp 168–174
36. Lu Y, Tsaparas P, Ntoulas A, Polanyi L (2010) Exploiting social context for review quality prediction. In: Proceedings of the 19th international world wide web conference (WWW), Raleigh, North Carolina, USA, pp 691–700
37. Lu Y, Duan H, Wang H, Zhai C-X (2010) Exploiting structured ontology to organize scattered online opinions. In: Proceedings of the 14th international conference on computational linguistics (COLING), Beijing, China, pp 734–742
38. Manevitz L-M, Yousef M (2002) One-class SVMs for document classification. J Mach Learn Res 2:139–154
39. Mei Q, Ling X, Wondra M, Su H, Zhai C-X (2007) Topic sentiment mixture: modeling facets and opinions in weblogs. In: Proceedings of the 16th international conference on world wide web (WWW), Banff, Alberta, Canada, pp 171–180
40. Mizil C-D, Kossinets G, Kleinberg J, Lee L (2009) How opinions are received by online communities: a case study on Amazon.com helpfulness votes. In: Proceedings of the 18th international conference on world wide web (WWW), Madrid, Spain, pp 141–150
41. Moghaddam S, Ester M (2011) AQA: aspect-based opinion question answering. In: IEEE international conference on data mining, Vancouver, BC, Canada, pp 89–96
42. Murthy K, Faruquie T-A, Subramaniam LV, Prasad KH, Mohania M (2010) Automatically generating term-frequency-induced taxonomies. In: Proceedings of the 48th annual meeting of the association for computational linguistics on computational linguistics (ACL), Uppsala, Sweden, pp 126–131
43. Nishikawa H, Hasegawa T, Matsuo Y, Kikui G (2010) Optimizing informativeness and readability for sentiment summarization. In: Proceedings of the 48th annual meeting of the association for computational linguistics on computational linguistics (ACL), Uppsala, Sweden, pp 325–330
44. Ohana B, Tierney B (2009) Sentiment classification of reviews using SentiWordNet. In: Proceedings of the 9th IT&T conference, Dublin, Ireland
45. Ouyang Y, Li W, Lu Q (2009) An integrated multi-document summarization approach based on word hierarchical representation. In: Proceedings of the ACL-IJCNLP 2009 conference, Singapore, pp 113–116
46. Pang B, Lee L, Vaithyanathan S (2002) Thumbs up? Sentiment classification using machine learning techniques. In: Proceedings of the conference on empirical methods on natural language processing (EMNLP), Philadelphia, USA, pp 79–86
47. Pantel P, Pennacchiotti M (2006) Espresso: leveraging generic patterns for automatically harvesting semantic relations. In: Proceedings of the 44th annual meeting of the association for computational linguistics on computational linguistics (ACL), Sydney, Australia, pp 113–120

48. Popescu A-M, Etzioni O (2005) Extracting product features and opinions from reviews. In: Proceedings of the conference on human language technology and empirical methods in natural language processing (HLT/EMNLP), Vancouver, BC, Canada, pp 339–346
49. Ramakrishnan R, Tomkins A (2007) Toward a PeopleWeb. Computer 40(8):63–72
50. Santamaria C, Gonzalo J, Verdejo F (2003) Automatic association of Web directories with word senses. J Comput Linguist 29(3):485–502
51. Schrijver A (1998) Theory of linear and integer programming. Wiley, Chichester/New York
52. Scott D-W (1992) Multivariate density estimation: theory, practice, and visualization. Wiley, New York
53. Shi B, Chang K (2008) Generating a concept hierarchy for sentiment analysis. In: IEEE international conference on systems man and cybernetics, Singapore, pp 312–317
54. Silla C, Freitas A (2011) A survey of hierarchical classification across different application domains. J Data Min Knowl Disc 22(1–2):31–72
55. Snow R, Jurafsky D (2006) Semantic taxonomy induction from heterogenous evidence. In: Proceedings of the 44th annual meeting of the association for computational linguistics on computational linguistics (ACL), Sydney, Australia, pp 801–808
56. Somasundaran S, Wilson T, Wiebe J, Stoyanov V (2007) QA with attitude: exploiting opinion type analysis for improving question answering in online discussions and the News. In: Proceedings of the conference on weblogs and social (ICWSM), Boulder, Colorado, USA
57. Strube M, Ponzetto S-P (2006) WikiRelate! computing semantic relatedness using Wikipedia. In: Proceedings of the 21st national conference on artificial intelligence (AAAI), Boston, Massachusetts, USA, pp 1419–1424
58. Su Q, Xu X, Guo H, Wu X, Zhang X, Swen B, Su Z (2008) Hidden sentiment association in Chinese Web opinion mining. In: Proceedings of the 17th international conference on world wide web (WWW), Beijing, China, pp 959–968
59. Wang H, Lu Y, Zhai C-X (2010) Latent aspect rating analysis on review text data: a rating regression approach. In: Proceedings of the 16th ACM SIGKDD international conference on knowledge discovery and data mining, Washington, DC, USA, pp 783–792
60. Wikipedia (2012) http://en.wikipedia.org/wiki/Wikipedia
61. Wilson T, Wiebe J, Hoffmann P (2005) Recognizing contextual polarity in phrase-level sentiment analysis. In: Proceedings of the conference on human language technology and empirical methods in natural language processing (HLT/EMNLP), Vancouver, BC, Canada, pp 347–354
62. Wong T-L, Lam W (2005) Hot item mining and summarization from multiple auction Web sites. In: Proceedings of the 2005 eighth IEEE international conference on data mining (ICDM), Washington, DC, USA, pp 797–800
63. Wu Y, Zhang Q, Huang X, Wu L (2009) Phrase dependency parsing for opinion mining. In: Proceedings of the 47th annual meeting of the association for computational linguistics on computational linguistics (ACL), Singapore, pp 1533–1541
64. Yang H (2011) Personalized concept hierarchy construction. Ph.D. thesis, Carnegie Mellon University
65. Yang H, Callan J (2009) A metric-based framework for automatic taxonomy induction. In: Proceedings of the 47th annual meeting of the association for computational linguistics on computational linguistics (ACL), Singapore, pp 271–279
66. Ye S, Chua T-S (2006) Learning object models from semi-structured web documents. IEEE Trans Knowl Data Eng 18(3):334–349
67. Yu H, Hatzivassiloglou V (2003) Towards answering opinion questions: separating facts from opinions and identifying the polarity of opinion sentences. In: Proceedings of the conference on empirical methods on natural language processing (EMNLP), Sapporo, Japan, pp 129–136

68. Yu J, Zha Z-J, Wang M, Wang K, Chua T-S (2011) Domain-assisted product aspect hierarchy generation: towards hierarchical organization of unstructured consumer reviews. In: Proceedings of the conference on empirical methods on natural language processing (EMNLP), Edinburgh, UK, pp 140–150
69. Yu J, Zha Z-J, Wang M, Chua T-S (2012) Answering opinion questions on products by exploiting hierarchical organization of consumer reviews. In: Proceedings of the conference on empirical methods on natural language processing (EMNLP), Jeju, Korea, pp 391–401
70. Zhang W, Yu C, Meng W (2007) Opinion retrieval from blogs. In: Proceedings of the 18th ACM international conference on information and knowledge management (CIKM), Lisboa, Portugal, pp 831–840

# Chapter 9
# Word Sense Disambiguation Using Wikipedia

**Bharath Dandala, Rada Mihalcea, and Razvan Bunescu**

**Abstract** This paper describes explorations in word sense disambiguation using Wikipedia as a source of sense annotations. Through experiments on four different languages, we show that the Wikipedia-based sense annotations are reliable and can be used to construct accurate sense classifiers.

## 9.1 Introduction

Ambiguity is inherent to human language. In particular, word sense ambiguity is prevalent in all natural languages, with a large number of the words in any given language carrying more than one meaning. For instance, the English noun *plant* can mean *green plant* or *factory*; similarly the French word *feuille* can mean *leaf* or *paper*. The correct sense of an ambiguous word can be selected based on the context where it occurs, and correspondingly the problem of *word sense disambiguation* is defined as the task of automatically assigning the most appropriate meaning to a polysemous word within a given context.

Two well studied categories of approaches to WSD are represented by knowledge-based [16, 26, 40] and data-driven [41, 44, 54] methods. Knowledge-based methods rely on information drawn from wide-coverage lexical resources such as WordNet [35]. Their performance has been generally constrained by the limited amount of lexical and semantic information present in these resources. In a recent effort to alleviate the semantic information bottleneck, Ponzetto and Navigli [46] created WordNet++, an extended version of WordNet that was

B. Dandala · R. Mihalcea (✉)
Department of Computer Science, University of North Texas, Denton, TX, USA
e-mail: BharathDandala@my.unt.edu; rada@cs.unt.edu

R. Bunescu
School of EECS, Ohio University, Athens, OH, USA
e-mail: bunescu@ohio.edu

I. Gurevych and J. Kim (eds.), *The People's Web Meets NLP*, Theory and Applications of Natural Language Processing, DOI 10.1007/978-3-642-35085-6_9, 

augmented with unlabeled relations extracted from Wikipedia. The resulting knowledge-based system was shown to be competitive with state-of-the-art supervised approaches in a coarse grained all-words setting and on domain-specific datasets. Knowledge-based methods have also been observed to be robust when tested on data with different sense distributions [3], a setting where supervised methods would normally need to use domain adaptation [1].

Among the various data-driven word sense disambiguation methods proposed to date, supervised systems have been observed to lead to highest performance. In these systems, the sense disambiguation problem is formulated as a supervised learning task, where each sense-tagged occurrence of a particular word is transformed into a feature vector which is then used in an automatic learning process. Despite their high performance, these supervised systems have an important drawback: their applicability is limited to those few words for which sense tagged data is available, and their accuracy is strongly connected to the amount of labeled data available at hand. To address the sense-tagged data bottleneck problem, different methods have been proposed in the past, with various degrees of success. This includes the automatic generation of sense-tagged data using monosemous relatives [2, 24, 34], automatically bootstrapped disambiguation patterns [31, 54], parallel texts as a way to point out word senses bearing different translations in a second language [11, 12, 42], and the use of volunteer contributions over the Web [8].

In this paper, we present experiments with a method for building sense tagged corpora using Wikipedia as a source of sense annotations. Starting with the hyperlinks available in Wikipedia, we generate sense annotated corpora that can be used for building accurate and robust sense classifiers. Through word sense disambiguation experiments performed on the Wikipedia-based sense tagged corpus generated for four different languages, we show that the Wikipedia annotations are reliable, and the quality of a sense tagging classifier built on this data set exceeds by a large margin the accuracy of an informed baseline that selects the most frequent word sense by default. Note that we are performing the traditional word sense disambiguation task, as typically done under the SENSEVAL/SEMEVAL evaluations, where we attempt to define a sense inventory based on all the sense occurrences in Wikipedia. This is related, but somehow different than the Wikification task [33, 37], where all the articles in Wikipedia are considered as potential senses for a word. Also, unlike Wikification, we are not performing a keyphrase extraction step prior to the disambiguation.

This work follows a growing line of research from recent years, where Wikipedia has been used as a resource of world knowledge in many natural language processing applications [28]. The vast amount of knowledge available in Wikipedia has been shown to benefit a diverse set of tasks including text categorization [14], information extraction [52, 53], coreference resolution [6, 18, 48, 50], information retrieval [9, 27, 36, 47], question answering [4, 13, 22], semantic relatedness [15], and named entity recognition [7, 10]. The proposed approaches use the semi-structured information available in Wikipedia either directly or indirectly by mapping automatically to resources such as DBPedia [5] or YAGO [51] that distill information from Wikipedia and other knowledge repositories.

The paper is organized as follows. We first provide a brief overview of Wikipedia, and describe the view of Wikipedia as a sense tagged corpus. We then show how the hyperlinks defined in this resource can be used to derive sense annotated corpora, and we show how a word sense disambiguation system can be built on this dataset. We present the results obtained in the word sense disambiguation experiments in four languages, and conclude with a discussion of the results.

## 9.2 Wikipedia

Wikipedia is a free online encyclopedia, representing the outcome of a continuous collaborative effort of a large number of volunteer contributors. Virtually any Internet user can create or edit a Wikipedia webpage, and this "freedom of contribution" has a positive impact on both the quantity (fast-growing number of articles) and the quality (potential mistakes are quickly corrected within the collaborative environment) of this online resource.

The basic entry in Wikipedia is an *article* (or *page*), which defines and describes a concept, an entity, or an event, and consists of a hypertext document with hyperlinks to other pages within or outside Wikipedia. The role of the hyperlinks is to guide the reader to pages that provide additional information about the entities or events mentioned in an article. Articles are organized into *categories*, which in turn are organized into category hierarchies. For instance, the article on ALAN TURING shown partially in Fig. 9.1 is included in the category BRITISH CRYPTOGRAPHERS, which in turn has a parent category named BRITISH SCIENTISTS, and so forth. The left pane of the page in this figure contains a set of links, of which more important for this work are the *interlingua links* that map to equivalent articles in other languages. The right pane contains the first two paragraphs of the actual article, with the hyperlinks shown in gray, whereas the bottom part of the figure shows the first ten categories associated with the article.

Each article in Wikipedia is uniquely referenced by an identifier, consisting of one or more words separated by spaces or underscores and occasionally a parenthetical explanation. For example, the article for the entity Turing that refers to the "English computer scientist" has the unique identifier ALAN TURING, whereas the article on Turing with the "stream cipher" meaning has the unique identifier TURING (CIPHER).

The hyperlinks within Wikipedia are created using these unique identifiers, together with an *anchor text* that represents the surface form of the hyperlink. For instance, "Alan Mathison Turing, [[Order of the British Empire|OBE]], [[Fellow of the Royal Society|FRS]] was an English [[mathematician]]" is the wiki source for the first sentence in the example page on ALAN TURING, containing links to the articles ORDER OF THE BRITISH EMPIRE, FELLOW OF THE ROYAL SOCIETY, and MATHEMATICIAN. If the surface form and the unique identifier of an article coincide, then the surface form can be turned directly into a hyperlink in the HTML version by placing double brackets around it (e.g. *[[mathematician]]*).

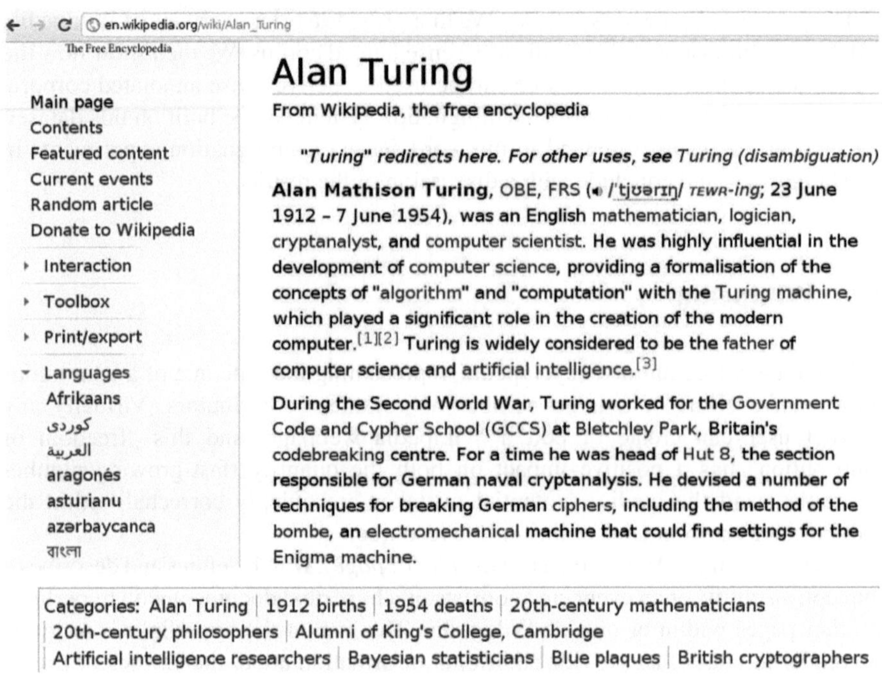


**Fig. 9.1** Snapshot of a fragment from the Wikipedia article on Alan Turing

Alternatively, if the surface form should be hyperlinked to an article with a different unique identifier, e.g. link the acronym *FRS* to the article on FELLOW OF THE ROYAL SOCIETY, then a piped link is used instead, as in *[[Fellow of the Royal Society|FRS]]*.

One of the implications of the large number of contributors editing the Wikipedia articles is the occasional lack of consistency with respect to the unique identifier used for a certain entity. For instance, *Alan Turing* is also referred to using the last name *Turing*, or the full name *Alan Mathison Turing*. This has led to the so-called *redirect pages*, which consist of a redirection hyperlink from an alternative name (e.g. *Turing*) to the article actually containing the description of the entity (e.g. *Alan Turing*), as shown in Fig. 9.2.

Another structure that is particularly relevant to the work described in this paper is the *disambiguation page*. Disambiguation pages are specifically created for ambiguous entities, and consist of links to articles defining the different meanings of the entity. The unique identifier for a disambiguation page typically consists of the parenthetical explanation *(disambiguation)* attached to the name of the ambiguous entity, as in e.g. SENSE_(DISAMBIGUATION), which is the unique identifier for the disambiguation page of the noun Sense, as shown in Fig. 9.3.

Wikipedia editions are available for more than 280 languages, with a number of entries varying from a few pages to three millions articles or more per language.

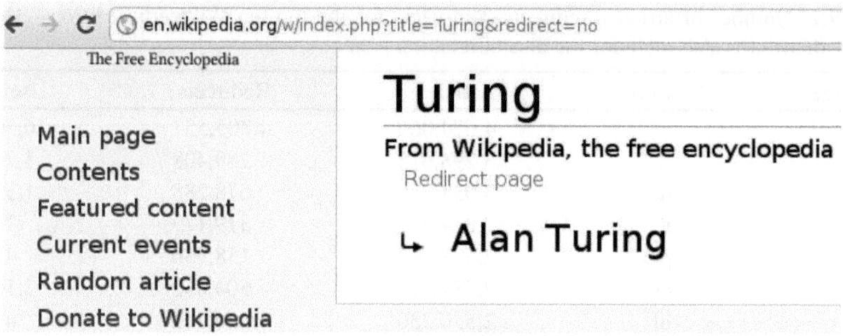


**Fig. 9.2** Example redirect page, from Turing to Alan Turing

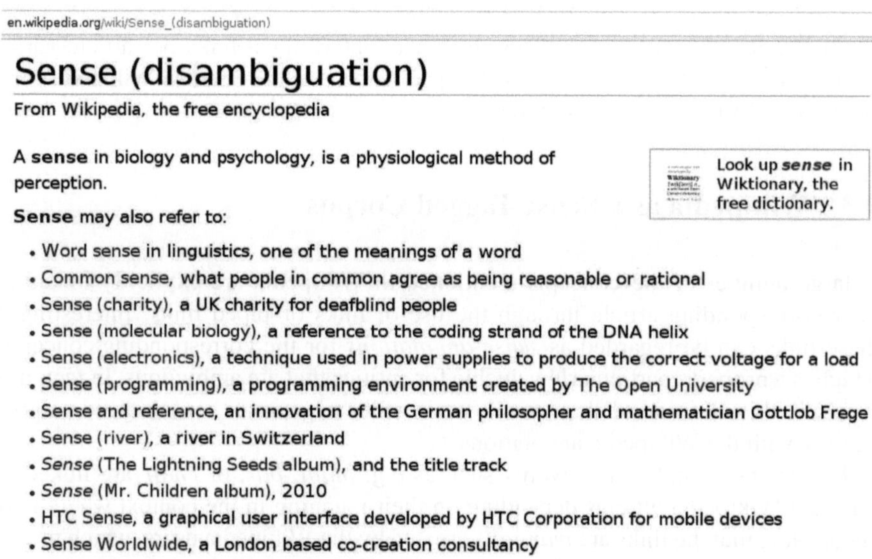


**Fig. 9.3** Disambiguation page for the noun Sense

Table 9.1 shows the ten largest Wikipedias (as of March 2012), along with the number of articles and approximate number of contributors.[1]

Finally, also relevant for the work described in this paper are the *interlingual links*, which explicitly connect articles in different languages. For instance, the English article for the noun SENSE is connected, among others, to the Spanish article SENTIDO (PERCEPCIÓN) and the Latin article SENSUS (BIOLOGIA). On average, about half of the articles in a Wikipedia version include interlingual links

[1] http://meta.wikimedia.org/wiki/List_of_Wikipedias

**Table 9.1** Number of articles, redirects, and users for the top ten Wikipedia editions. The total number of articles also includes the disambiguation pages

| Language | Code | Articles | Redirects | Users |
|---|---|---|---|---|
| English | en | 4,674,066 | 4,805,557 | 16,503,562 |
| French | fr | 3,298,615 | 789,408 | 1,250,266 |
| German | de | 3,034,238 | 678,288 | 1,398,424 |
| Italian | it | 2,874,747 | 319,179 | 731,750 |
| Polish | pl | 2,598,797 | 158,956 | 481,079 |
| Spanish | es | 2,587,613 | 504,062 | 2,162,925 |
| Dutch | nl | 2,530,250 | 226,201 | 446,458 |
| Russian | ru | 2,300,769 | 682,402 | 819,812 |
| Japanese | jp | 1,737,565 | 372,909 | 607,152 |
| Portuguese | pt | 719,944 | 100,000 | 919,782 |

to articles in other languages. The number of interlingual links per article varies from an average of 5 in the English Wikipedia, to 10 in the Spanish Wikipedia, and as many as 23 in the Arabic Wikipedia.

## 9.3 Wikipedia as a Sense Tagged Corpus

A large number of the concepts mentioned in Wikipedia are explicitly linked to their corresponding article through the use of links or piped links. Interestingly, these links can be regarded as *sense annotations* for the corresponding concepts, which is a property particularly valuable for entities that are ambiguous. In fact, it is precisely this observation that we rely on in order to generate sense tagged corpora starting with the Wikipedia annotations.

For example, ambiguous words such as e.g. *plant*, *bar*, or *chair* are linked to different Wikipedia articles depending on their meaning in the context where they occur. Note that the links are *manually* created by the Wikipedia users, which means that they are most of the time accurate and referencing the correct article. The following represent five example sentences for the ambiguous word *bar*, with their corresponding Wikipedia annotations (links):

(a) In 1834, Sumner was admitted to the **[[bar (law)|bar]]** at the age of 23, and entered private practice in Boston.
(b) It is danced in 3/4 time (like most waltzes), with the couple turning approx. 180° every **[[bar (music)|bar]]**.
(c) Vehicles of this type may contain expensive audio players, televisions, video players, and **[[bar (counter)|bar]]**s, often with refrigerators.
(d) Jenga is a popular beer in the **[[bar (establishment)|bar]]**s of Thailand.
(e) This is a disturbance on the water surface of a river or estuary, often caused by the presence of a **[[bar (landform)|bar]]** or dune on the riverbed.

To derive sense annotations for a given ambiguous word, we use the links extracted for all the hyperlinked Wikipedia occurrences of the given word, and map these annotations to word senses. For instance, for the *bar* example above, we extract five possible annotations: *bar (counter), bar (establishment), bar (landform), bar (law)*, and *bar (music)*.

Although Wikipedia provides the so-called disambiguation pages that list the possible meanings of a given word, we decided to use instead the annotations collected directly from the Wikipedia links. This decision is motivated by two main reasons. First, a large number of the occurrences of ambiguous words are not linked to the articles mentioned by the disambiguation page, but to related concepts. This can happen when the annotation is performed using a concept that is similar, but not identical to the concept defined. For instance, the annotation for the word *bar* in the sentence "The blues uses a rhythmic scheme of twelve 4/4 [[measure (music)|bars]]" is *measure (music)*, which, although correct and directly related to the meaning of *bar (music)*, is not listed in the disambiguation page for *bar*.

Second, there are several inconsistencies that make it difficult to use the disambiguation pages in an automatic system. For example, for the word *bar*, the Wikipedia page with the identifier *bar* is a disambiguation page, whereas for the word *paper*, the page with the identifier *paper* contains a description of the meaning of paper as "material made of cellulose," and a different page *paper_(disambiguation)* is defined as a disambiguation page. Moreover, in other cases such as e.g. the entries for the word *organization*, no disambiguation page is defined; instead, the articles corresponding to different meanings of this word are connected by links labeled as "alternative meanings."

Therefore, rather than using the senses listed in a disambiguation page as the sense inventory for a given ambiguous word, we chose instead to collect all the annotations available for that word in the Wikipedia pages, and then cluster these together to form the sense inventory.

### 9.3.1 Wikipedia and WordNet

It is interesting to note that Wikipedia has a different sense coverage and distribution compared to more "traditional" lexical resources such as WordNet [35]. For instance, the meaning of *ambiance* for the ambiguous word *atmosphere* does not appear at all in the Wikipedia corpus, although it has the highest frequency in other annotated data such as SENSEVAL. This is partly due to the coarser sense distinctions made in Wikipedia; for instance, Wikipedia does not make the distinction between the act of grasping and the actual hold for the noun *grip*, and occurrences of both of these meanings are annotated with the label *grip_(handle)*.

There are also cases when Wikipedia makes different or finer sense distinctions than WordNet. For instance, there are several Wikipedia annotations for *image* as *copy*, but this meaning is not even defined in WordNet. Similarly, Wikipedia makes the distinction between *dance performance* and *theatre performance*, but both these meanings are listed under one single entry in WordNet (*performance* as *public presentation*).

### 9.3.2 Building Sense Tagged Corpora

Starting with a given ambiguous word, we derive a sense-tagged corpus following three main steps:

First, we extract all the paragraphs in Wikipedia that contain an occurrence of the ambiguous word as part of a link or a piped link. We select paragraphs based on the Wikipedia paragraph segmentation, which typically lists one paragraph per line.[2] To focus on the problem of word sense disambiguation, rather than named entity recognition, we explicitly avoid named entities by considering only those word occurrences that are spelled with a lower case. Although this simple heuristic will also eliminate examples where the word occurs at the beginning of a sentence (and therefore are spelled with an upper case), we decided nonetheless to not consider these examples so as to avoid any possible errors.

Next, we collect all the possible labels for the given ambiguous word by extracting the leftmost component of the links. For instance, in the piped link *[[musical_notation|bar]]*, the label *musical_notation* is extracted. In the case of simple links (e.g. *[[bar]]*), the word itself can also play the role of a valid label if the page it links to is not determined as a disambiguation page.

Finally, the labels are clustered into word senses, by linking those labels that refer to the same word meaning. This step is mainly motivated by the fact that words have often a large number of labels (e.g., more than 100 labels for the word "bar" in the 2012 Wikipedia), which cannot be directly used as senses. These labels are often redundant (e.g., both *musical notation* and *Bar (music)* are used as labels for the word "bar"), or refer to senses that may be too fine grained. Thus, using the labels as senses without the clustering step is likely to result in significant noise in the word sense classifier.

For the purpose of the experiments reported in this paper, the clustering has been done primarily manually, by using several heuristics. One heuristic, for instance, creates a cluster from labels associated with entities that are instances of a more general concept. Based on this heuristic, the labels *atmosphere of Earth* and *atmosphere of Mars* are clustered together with the more general label *atmosphere*. We are currently exploring automatic techniques to identify clusters of labels that refer to the same word sense. From the resulting clusters of labels we keep only the clusters that have enough support in Wikipedia i.e., clusters for which the total number of disambiguated instances exceeds a predefined threshold. In this way, we ensure that each sense contributes a minimal number of training examples in the machine learning approach to WSD described in Sect. 9.4.

[2]The average length of a paragraph is 80 words.

**Table 9.2** Word senses for the word *bar*, based on annotation labels used in Wikipedia

| Word sense | Labels in Wikipedia | Wikipedia definition |
|---|---|---|
| Bar (establishment) | Bar_(establishment), nightclub gay_club, pub | A retail establishment which serves alcoholic beverages |
| Bar (counter) | Bar_(counter) | The counter from which drinks are dispensed |
| Bar (unit) | Bar_(unit) | A scientific unit of pressure |
| Bar (music) | Bar_(music), measure_music musical_notation | A period of music |
| Bar (law) | Bar_association, bar_law law_society_of_upper_canada state_bar_of_california | The community of persons engaged in the practice of law |
| Bar (landform) | Bar_(landform) | A type of beach behind which lies a lagoon |
| Bar (metal) | Bar_metal, pole_(object) | |
| Bar (sports) | Gymnastics_uneven_bars, handle_bar | |
| Bar (solid) | Candy_bar, chocolate_bar | |

### *9.3.3 An Example*

As an example, consider the ambiguous word *bar*, with 3,784 examples extracted from Wikipedia where *bar* appeared as the rightmost component of a piped link or as a word in a simple link. Since the page with the identifier *bar* is a disambiguation page, all the examples containing the single link *[[bar]]* are removed, as the link does not remove the ambiguity. From the remaining examples, we extract their labels and cluster them into 23 different senses. Table 9.2 shows a subset of these labels, to illustrate the senses that can be extracted from the Wikipedia annotations.

## 9.4 Word Sense Disambiguation

Provided a set of sense-annotated examples for a given ambiguous word, the task of a word sense disambiguation system is to automatically learn a disambiguation model that can predict the correct sense for a new, previously unseen occurrence of the word. Assuming that such a system can be reliably constructed, the implications are two-fold. First, accurate disambiguation models suggest that the data is reliable and consists of correct sense annotations. Second, and perhaps more importantly, the availability of a system able to correctly predict the sense of a word can have important implications for applications that require such information, including machine translation and automatic reasoning.

We use a word sense disambiguation system that integrates local and topical features within a machine learning framework, similar to several of the top-performing supervised word sense disambiguation systems participating in the recent SENSEVAL evaluations.[3]

The disambiguation algorithm starts with a preprocessing step, where the text is tokenized and annotated with part-of-speech tags. Collocations are identified using a sliding window approach, where a collocation is defined as a sequence of words that forms a compound concept defined in Wikipedia.

Next, local and topical features are extracted from the context of the ambiguous word. Specifically, we use the current word and its part-of-speech, a local context of three words to the left and right of the ambiguous word, the parts-of-speech of the surrounding words, the verb and noun before and after the ambiguous words, and a global context implemented through sense-specific keywords determined as a list of at most five words occurring at least three times in the contexts defining a certain word sense.

This feature set is similar to the one used by Ng and Lee [41] and Mihalcea [32], as well as by a number of state-of-the-art word sense disambiguation systems participating in the SENSEVAL-2 and SENSEVAL-3 evaluations. The features are integrated in a Naive Bayes classifier, which was selected mainly for its performance in previous work showing that it can lead to a state-of-the-art disambiguation system given the features we consider [25].

## 9.5 Experiments and Results

To evaluate the quality of the sense annotations generated using Wikipedia, we performed a word sense disambiguation experiment on a subset of 30 ambiguous words used during the SENSEVAL-2 and SENSEVAL-3 evaluations. Since the Wikipedia annotations are focused on nouns (associated with the entities typically defined by Wikipedia), the sense annotations we generate and the word sense disambiguation experiments are also focused on nouns. The 30 words that have been determined to be interesting for the task of word sense disambiguation (and thus their listing in the SENSEVAL tasks) have also been found to have enough annotated data in previous experiments based on Wikipedia [32].

We generate sense tagged datasets for the 30 ambiguous words following the approach described in Sect. 9.3.2, and use these datasets for two main disambiguation experiments.

[3]http://www.senseval.org

### 9.5.1 Word Sense Disambiguation on Two English Wikipedias

First, while focusing on English, we run the disambiguation algorithm on the 30 ambiguous words, using data collected from a 2007 version of Wikipedia as well as a more recent version from 2012.

Tables 9.3 and 9.4 show the disambiguation results using the word sense disambiguation system described in Sect. 9.4, in a ten-fold cross-validation evaluation applied on the 2007 and 2012 Wikipedia data. For each word, the table also shows the number of senses, the total number of examples, and a simple baseline that selects the most frequent sense by default.[4]

Overall, the Wikipedia-based sense annotations were found reliable, leading to accurate sense classifiers with an average relative error rate reduction of 44 % compared to the most frequent sense baseline in the Wikipedia 2007 dataset. There were a few exceptions to this general trend. For instance, considering the initial evaluations run on the 2007 Wikipedia, for some of the words for which only a small number of examples could be collected from Wikipedia, e.g. *restraint* or *shelter*, no accuracy improvement was observed compared to the most frequent sense baseline. Similarly, several words in that data set had highly skewed sense distributions, such as e.g. *bank*, which has a total number of 1,074 examples out of which 1,044 examples pertain to the meaning of *financial institution*, or the word *material* with 213 out of 223 examples annotated with the meaning of *substance*.

In the Wikipedia 2012 experiments, the average error reduction with respect to the most frequent sense baseline was 19.5 %. Both the baseline and the Naive Bayes classifier had lower disambiguation accuracies in the 2012 experiments, a consequence of a significantly higher average polysemy. As a result of a richer repository of senses in Wikipedia 2012 and a more fine grained clustering of labels, the average number of senses per word increased from 3.31 in the 2007 experiments to 9.63 in the 2012 experiments. Furthermore, while the number of examples for every word increased in the 2012 datasets, the average number of examples per word sense decreased from 95.4 in 2007 to 87.4 in 2012. A higher number of examples per word sense is generally expected to lead to better disambiguation accuracy, where the expected rate of improvement can be estimated from the slope of the learning curve of the WSD system.

One aspect that is particularly relevant for any supervised system is the learning rate with respect to the amount of available data. To determine the learning curve, we measured the disambiguation accuracy under the assumption that only a fraction of the data were available. We ran ten fold cross-validation experiments using 10 %, 20 %, ..., 100 % of the data, and averaged the results over all the words in the English data set. The resulting learning curve is plotted in Fig. 9.4. Overall, in particular for the 2012 data, the curve indicates a continuously growing accuracy

[4]Note that this baseline assumes the availability of a sense tagged corpus in order to determine the most frequent sense of a word. The baseline is therefore "informed," as compared to a random, "uninformed" sense selection.

**Table 9.3** Word sense disambiguation results, including one baseline (MFS = most frequent sense) and the word sense disambiguation system based on Wikipedia 2007 data. Number of senses (#s) and number of examples (#ex) are also indicated

| Word | #s | #ex | Baseline MFS(%) | Word sense Disambig.(%) |
|---|---|---|---|---|
| Argument | 2 | 114 | 70.17 | **89.47** |
| Arm | 3 | 291 | 61.85 | **84.87** |
| Atmosphere | 3 | 773 | 54.33 | **71.66** |
| Bank | 3 | 1,074 | **97.20** | **97.20** |
| Bar | 10 | 1,108 | 47.38 | **83.12** |
| Chair | 3 | 194 | 67.57 | **80.92** |
| Channel | 5 | 366 | 51.09 | **71.85** |
| Circuit | 4 | 327 | 85.32 | **87.15** |
| Degree | 7 | 849 | 58.77 | **85.98** |
| Difference | 2 | 24 | **75.00** | **75.00** |
| Disc | 3 | 73 | 52.05 | **71.23** |
| Dyke | 2 | 76 | 77.63 | **89.47** |
| Fatigue | 3 | 123 | 66.66 | **93.22** |
| Grip | 3 | 34 | 44.11 | **70.58** |
| Image | 2 | 84 | 69.04 | **80.28** |
| Material | 3 | 223 | **95.51** | **95.51** |
| Mouth | 2 | 409 | 94.00 | **95.35** |
| Nature | 2 | 392 | **98.72** | 98.21 |
| Paper | 5 | 895 | **96.98** | **96.98** |
| Party | 3 | 764 | 68.06 | **75.91** |
| Performance | 2 | 271 | **95.20** | **95.20** |
| Plan | 3 | 83 | 77.10 | **81.92** |
| Post | 5 | 33 | **54.54** | 51.51 |
| Restraint | 2 | 9 | **77.77** | **77.77** |
| Sense | 2 | 183 | **95.10** | **95.10** |
| Shelter | 2 | 17 | **94.11** | **94.11** |
| Sort | 2 | 11 | 81.81 | **90.90** |
| Source | 3 | 78 | 55.12 | **92.30** |
| Spade | 3 | 46 | 60.86 | **80.43** |
| Stress | 3 | 565 | 53.27 | **86.37** |
| AVERAGE | 3.31 | 316 | 72.58 | **84.65** |

with increasingly larger amounts of data, which suggests that more data is likely to lead to increased accuracy. Note that this is true for a given number of senses (e.g., the senses from the 2007 Wikipedia, or the senses from the 2012 Wikipedia), which benefit from larger amounts of data. The overall trend, however, as noticed by comparing the results in Tables 9.3 and 9.4, is a drop in performance due to a growing number of senses.

**Table 9.4** Word sense disambiguation results, including one baseline (MFS = most frequent sense) and the word sense disambiguation system based on Wikipedia 2012 data. Number of senses (#s) and number of examples (#ex) are also indicated

| | | | Baseline | Word sense |
|---|---|---|---|---|
| Word | #s | #ex | MFS(%) | Disambig.(%) |
| Argument | 7 | 458 | 66.81 | **76.2** |
| Arm | 5 | 346 | 66.76 | **70.8** |
| Atmosphere | 7 | 1,673 | 62.7 | **66.64** |
| Bank | 8 | 2,440 | 91.02 | **92.66** |
| Bar | 23 | 3,784 | 24.37 | **38.21** |
| Chair | 7 | 624 | 47.28 | **63.62** |
| Channel | 19 | 1,127 | 27.95 | **41.96** |
| Circuit | 18 | 561 | 25.85 | **32.08** |
| Degree | 17 | 2,004 | 55.49 | **75.89** |
| Difference | 7 | 65 | 35.38 | **49.23** |
| Disc | 13 | 316 | 23.73 | **36.07** |
| Dyke | 4 | 305 | 48.52 | **57.7** |
| Fatigue | 5 | 691 | 68.45 | **73.37** |
| Grip | 9 | 106 | 51.89 | **53.77** |
| Image | 16 | 908 | 60.02 | **66.07** |
| Material | 8 | 555 | 74.59 | **75.13** |
| Mouth | 4 | 678 | 80.53 | **85.69** |
| Nature | 8 | 2,454 | 50.45 | **67.35** |
| Paper | 10 | 1,551 | 90.33 | **91.29** |
| Party | 13 | 909 | 33 | **38.72** |
| Performance | 14 | 838 | 70.29 | **71.12** |
| Plan | 10 | 186 | 62.37 | **63.97** |
| Post | 10 | 216 | 21.76 | **34.72** |
| Restraint | 4 | 31 | 48.39 | **54.83** |
| Sense | 7 | 345 | 79.42 | **84.92** |
| Shelter | 10 | 231 | 26.41 | **30.73** |
| Sort | 2 | 22 | 63.64 | **68.18** |
| Source | 12 | 327 | **58.1** | **58.1** |
| Spade | 2 | 105 | 75.23 | **86.66** |
| Stress | 10 | 1,428 | 48.67 | **56.23** |
| AVERAGE | 9.63 | 842 | 54.65 | **62.07** |

### *9.5.2 Word Sense Disambiguation on Multiple Languages*

Second, we test the applicability of the disambiguation system in several languages by applying it on data gathered from Wikipedias in four different languages. Starting with the set of 30 ambiguous English words, we generate corresponding sets in Italian, Spanish, and German, by translating the words using Google translate.[5]

[5]http://translate.google.com

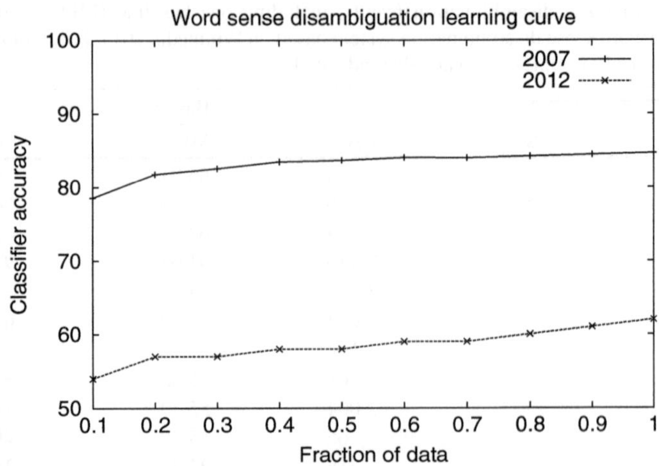


**Fig. 9.4** Learning curve on the Wikipedia 2007 and 2012 data sets

In this way, we generate words that share a similar semantic space, but which have their own ambiguities. We use Google translate mainly as a substitute for a bilingual dictionary, which, unlike the use of dictionaries from various sources, also offers consistency across the three languages used in our experiments. As with any bilingual dictionary, the various translations for a word are ordered in reversed order of their frequency, and thus the top translation that we use is often ambiguous.

Tables 9.5–9.7 show the words considered in each language, along with their number of senses, number of examples gathered from Wikipedia, and disambiguation results. With three exceptions, all the words are ambiguous, which further supports our choice of the Google translate resource as a way to generate ambiguous words in the three target languages. The average number of senses per word is similar among the three languages: from 4.2 for Spanish and German, to 4.4 for Italian. This is significantly smaller than the 9.6 senses per word for English and goes to explain the relatively higher accuracy of word sense disambiguation in these languages in comparison with English. The Naive Bayes classifier continues to obtain higher accuracy than the most frequent sense baseline in each of the three languages, albeit the error reduction is not as substantial as for English. One explanation for the smaller error reduction may be given by the relatively smaller accuracy of the text processing tools in the three languages compared to English. Tokenization and part-of-speech tagging are used to preprocess the text and generate informative features for the Naive Bayes classifier. Errors in these text processing steps are likely to compound and thus have a negative impact on the final WSD performance.

**Table 9.5** Word sense disambiguation results on Spanish, using Wikipedia 2012 data. In addition to a baseline (MFS = most frequent sense) and the word sense disambiguation system results, the number of senses (#s) and number of examples (#ex) are also indicated

| | | | Baseline | Word sense |
|---|---|---|---|---|
| Word | #s | #ex | MFS(%) | Disambig.(%) |
| Abrigo | 2 | 38 | 84.21 | **86.84** |
| Ambiente | 2 | 10 | **90** | **90** |
| Argumento | 4 | 167 | **85.03** | 83.23 |
| Banco | 4 | 565 | 91.5 | **91.68** |
| Bar | 3 | 440 | 83.18 | **87.27** |
| Boca | 2 | 404 | 77.23 | **82.67** |
| Brazo | 2 | 235 | **99.57** | **99.57** |
| Canal | 9 | 457 | **43.54** | **43.54** |
| Circuito | 5 | 162 | 64.2 | **70.98** |
| Control | 4 | 59 | **77.97** | 74.57 |
| Diferencia | 2 | 31 | 77.42 | **87.09** |
| Dique | 2 | 316 | 87.34 | **87.65** |
| Disco | 15 | 785 | 28.54 | **32.35** |
| Estrés | 1 | 470 | – | – |
| Fatiga | 3 | 141 | **62.41** | 60.99 |
| Fuente | 8 | 416 | 39.18 | **42.78** |
| Grado | 12 | 720 | 46.11 | **65** |
| Imagen | 8 | 231 | **83.12** | 83.11 |
| Material | 2 | 185 | **98.92** | 98.91 |
| Naturaleza | 2 | 691 | **99.28** | 99.27 |
| Pala | 2 | 50 | **88** | **88** |
| Papel | 2 | 648 | **99.54** | 99.53 |
| Partido | 8 | 476 | 44.75 | **56.72** |
| Plan | 2 | 21 | 52.38 | **85.71** |
| Post | 3 | 68 | **66.18** | 66.17 |
| Rendimiento | 5 | 103 | **64.08** | 63.1 |
| Restricción | 2 | 10 | 70 | **90** |
| Sentido | 3 | 84 | 58.33 | **67.85** |
| Silla | 5 | 243 | 52.26 | **57.2** |
| Sort | 2 | 43 | **95.35** | 95.34 |
| AVERAGE | 4.2 | 275 | 72.74 | **77.14** |

## 9.6 Related Work

In word sense disambiguation, the line of work most closely related to ours consists of methods trying to address the sense-tagged data bottleneck problem.

A first set of methods consists of algorithms that generate sense annotated data using words semantically related to a given ambiguous word [2, 24, 34]. Related non-ambiguous words, such as monosemous words or phrases from dictionary definitions, are used to automatically collect examples from the Web. These

**Table 9.6** Word sense disambiguation results on Italian, using Wikipedia 2012 data. In addition to a baseline (MFS = most frequent sense) and the word sense disambiguation system results, the number of senses (#s) and number of examples (#ex) are also indicated

| | | | Baseline | Word sense |
|---|---|---|---|---|
| Word | #s | #ex | MFS(%) | Disambig.(%) |
| Argomento | 2 | 17 | 52.94 | **70.58** |
| Atmosfera | 5 | 922 | **90.56** | 90.45 |
| Banca | 2 | 468 | **99.57** | **99.57** |
| Bar | 7 | 610 | 33.11 | **55.4** |
| Bocca | 2 | 359 | 51.81 | **52.08** |
| Braccio | 6 | 149 | **59.06** | **59.06** |
| Canale | 7 | 331 | 31.72 | **35.34** |
| Carta | 5 | 645 | 89.15 | **89.76** |
| Circuito | 5 | 85 | 48.24 | **58.82** |
| Contenimento | 1 | 5 | – | – |
| Differenza | 2 | 37 | 89.19 | **91.89** |
| Diga | 2 | 618 | **99.35** | **99.35** |
| Disco | 13 | 458 | **44.32** | **44.32** |
| Fatica | 1 | 41 | – | – |
| Fonte | 6 | 68 | 38.24 | **44.11** |
| Grado | 9 | 547 | 48.45 | **58.86** |
| Immagine | 2 | 241 | 58.51 | **70.53** |
| Materiale | 2 | 152 | **96.71** | **96.71** |
| Natura | 2 | 816 | **99.26** | 99.14 |
| Ordinamento | 3 | 72 | **80.56** | 80.55 |
| Partito | 3 | 87 | 59.77 | **71.26** |
| Performance | 3 | 48 | **87.5** | **87.5** |
| Piano | 8 | 746 | 59.38 | **68.36** |
| Posta | 2 | 140 | 86.43 | **87.85** |
| Presa | 2 | 46 | **91.3** | **91.3** |
| Rifugio | 3 | 278 | 89.93 | **91** |
| Sedia | 2 | 35 | **97.14** | **97.14** |
| Senso | 5 | 70 | **67.14** | **67.14** |
| Spada | 6 | 696 | 87.36 | **89.22** |
| Stress | 7 | 337 | **61.42** | **61.42** |
| AVERAGE | 4.39 | 324 | 71.36 | **75.32** |

examples are then turned into sense-tagged data by replacing the non-ambiguous words with their ambiguous equivalents.

Another approach proposed in the past is based on the idea that an ambiguous word tends to have different translations in a second language [49]. Starting with a collection of parallel texts, sense annotations were generated either for one word at a time [11, 42], or for all words in unrestricted text [12], and in both cases the systems trained on these data were found to be competitive with other word sense disambiguation systems.

**Table 9.7** Word sense disambiguation results on German, using Wikipedia 2012 data. In addition to a baseline (MFS = most frequent sense) and the word sense disambiguation system results, the number of senses (#s) and number of examples (#ex) are also indicated

| Word | #s | #ex | Baseline MFS(%) | Word sense Disambig.(%) |
|---|---|---|---|---|
| Argument | 2 | 143 | **96.5** | **96.5** |
| Arm | 3 | 109 | 83.49 | **88.99** |
| Art | 4 | 3,796 | 98.89 | **99.13** |
| Atmosphäre | 5 | 570 | **54.91** | 54.73 |
| Bank | 5 | 563 | **68.03** | 68.02 |
| Bar | 6 | 662 | 53.02 | **67.82** |
| Bild | 7 | 451 | 74.06 | **79.37** |
| Deich | 2 | 520 | **99.81** | 99.8 |
| Differenz | 4 | 96 | **65.63** | 65.62 |
| Ermüdung | 3 | 29 | **58.62** | 51.72 |
| Grad | 5 | 192 | 72.4 | **83.85** |
| Griff | 2 | 26 | **61.54** | 50 |
| Kanal | 8 | 406 | 80.3 | **82.01** |
| Material | 2 | 223 | **97.76** | 97.75 |
| Mund | 3 | 170 | **90** | **90** |
| Natur | 9 | 854 | **71.9** | 71.89 |
| Papier | 3 | 988 | **97.87** | **97.87** |
| Partei | 4 | 539 | **89.05** | **89.05** |
| Performance | 5 | 601 | **81.53** | **81.53** |
| Plan | 6 | 67 | 35.82 | **41.79** |
| Post | 10 | 449 | 84.19 | **85.07** |
| Quelle | 5 | 966 | 80.33 | **83.02** |
| Schaltung | 4 | 48 | **41.67** | 41.66 |
| Scheibe | 2 | 40 | **82.5** | **82.5** |
| Shelter | 2 | 22 | 68.18 | **90.9** |
| Sinn | 5 | 243 | 27.16 | **37.86** |
| Spaten | 2 | 64 | **95.31** | **95.31** |
| Stress | 2 | 495 | 91.52 | **93.93** |
| Stuhl | 3 | 89 | 48.31 | **60.67** |
| Zwang | 3 | 27 | 74.06 | **74.07** |
| AVERAGE | 4.2 | 448 | 74.14 | **76.75** |

The lack of sense-tagged corpora can also be circumvented using bootstrapping algorithms, which start with a few annotated seeds and iteratively generate a large set of disambiguation patterns. This method, initially proposed by Yarowsky [54], was successfully evaluated in the context of the SENSEVAL framework [31].

A series of studies have explored an alternative line of research in which large scale sense annotated corpora are extracted automatically from collaboratively constructed language resource. In an effort related to the Wikipedia collection process, Chklovski and Mihalcea [8] have implemented the Open Mind Word Expert system for collecting sense annotations from volunteer contributors over the Web.

The data generated using this method was then used by the systems participating in several of the SENSEVAL-3 tasks. More recently, Henrich et al. [21] used the web-based dictionary Wiktionary[6] to create a sense annotated corpus for German based on a mapping [19] between GermaNet [23] and Wiktionary. A similar effort [20] has led to a sense tagged corpus for English, by using the mapping between WordNet and Wiktionary created by Meyer and Gurevych [30]. Recent methods for mapping Wikipedia to the WordNet sense repository [43,45] could be used to replace the role of Wiktionary with Wikipedia in the same corpora acquisition approach.

Notably, the method we propose has several advantages over these previous methods. First, our method relies exclusively on monolingual data, thus avoiding the possible constraints imposed by methods that require parallel texts, which may be difficult to find. Second, the Wikipedia-based annotations follow a natural Zipfian sense distribution, unlike the equal distributions typically obtained with the methods that rely on the use of monosemous relatives or bootstrapping methods. Finally, the growth pace of Wikipedia is much faster than other more task-focused and possibly less-engaging activities such as Open Mind Word Expert [8], and therefore has the potential to lead to significantly higher coverage.

Whereas the focus of our work is on disambiguating common nouns, a number of studies have looked at the utility of Wikipedia for proper name disambiguation. The name entity disambiguation algorithm proposed by Bunescu and Pasca [7] maps ambiguous names to their correct interpretation by integrating the information provided by the Wikipedia articles and their categories in an SVM disambiguation kernel. A different approach to proper name disambiguation [10] builds a context out of Wikipedia for each named entity, which is then used to train an automatic disambiguation system. An extensive comparative evaluation of Wikipedia-based named entity disambiguation systems is conducted in [17].

Finally, while Wikipedia has proven its utility as a repository of entities and word senses, a number of approaches have gone beyond this view and created rich semantic networks anchored in Wikipedia concepts and entities. Early approaches such as YAGO [51] and DBPedia [5] distill knowledge bases from the semi-structured information available in Wikipedia and other readily available knowledge repositories. More recently, the scale of the taxonomies and semantic networks automatically extracted from Wikipedia has been greatly expanded by exploiting its extensive multilingual structure [29, 38, 39].

## 9.7 Conclusions

In this paper, we described an approach for using Wikipedia as a source of sense annotations for word sense disambiguation. Starting with the hyperlinks available in Wikipedia, we showed how we can generate a sense annotated corpus that can be

[6]http://www.wiktionary.org

used to train accurate sense classifiers. Through experiments performed on a subset of the SENSEVAL words, we showed that the Wikipedia sense annotations can be used to build a word sense disambiguation system leading to a relative error rate reduction of up to 44 % as compared to simpler baselines. We used the same method to generate sense tagged datasets and train WSD systems for an additional set of three languages (Spanish, Italian, and German), and observed a similar behavior in terms of error reduction with respect to the baselines.

Despite some limitations inherent to this approach – definitions and annotations in Wikipedia are available almost exclusively for nouns, word and sense distributions are sometime skewed, the annotation labels are occasionally inconsistent –, these limitations are overcome by the clear advantage that comes with the use of Wikipedia: large sense tagged data for a large number of words at virtually no cost.

We believe that this approach is particularly promising for two main reasons. First, the size of Wikipedia is growing at a steady pace, which consequently means that the size of the sense tagged corpora that can be generated based on this resource is also continuously growing. While techniques for supervised word sense disambiguation have been repeatedly criticized in the past for their limited coverage, mainly due to the associated sense-tagged data bottleneck, Wikipedia seems a promising resource that could provide the much needed solution for this problem. Second, Wikipedia editions are available for many languages (currently more than 280), which means that this method can be used to generate sense tagged corpora and build accurate word sense classifiers for a large number of languages.

**Acknowledgements** This material is based in part upon work supported by the National Science Foundation IIS awards #1018613 and #1018590 and CAREER award #0747340. Any opinions, findings, and conclusions or recommendations expressed in this material are those of the authors and do not necessarily reflect the views of the National Science Foundation.

## References

1. Agirre E, de Lacalle OL (2009) Supervised domain adaption for WSD. In: Proceedings of the 12th conference of the European chapter of the association for computational linguistics, association for computational linguistics, EACL ’09, Stroudsburg, PA, USA, pp 42–50
2. Agirre E, Martinez D (2004) Unsupervised word sense disambiguation based on automatically retrieved examples: the importance of bias. In: Proceedings of EMNLP 2004, Barcelona, Spain
3. Agirre E, De Lacalle OL, Soroa A (2009) Knowledge-based WSD on specific domains: performing better than generic supervised WSD. In: Proceedings of the 21st international joint conference on artifical intelligence, IJCAI’09. Morgan Kaufmann, San Francisco, pp 1501–1506
4. Ahn D, Jijkoun V, Mishne G, Muller K, de Rijke M, Schlobach S (2004) Using Wikipedia at the TREC QA track. In: Proceedings of the 13th text retrieval conference (TREC 2004), Gaithersburg, MD
5. Bizer C, Lehmann J, Kobilarov G, Auer S, Becker C, Cyganiak R, Hellmann S (2009) DBpedia – a crystallization point for the Web of data. Web Semant 7:154–165
6. Bryl V, Giuliano C, Serafini L, Tymoshenko K (2010) Using background knowledge to support coreference resolution. In: Proceedings of the 2010 conference on ECAI 2010: 19th European conference on artificial intelligence, Amsterdam, The Netherlands, pp 759–764

7. Bunescu R, Pasca M (2006) Using encyclopedic knowledge for named entity disambiguation. In: Proceedings of the European conference of the association for computational linguistics, Trento, Italy
8. Chklovski T, Mihalcea R (2002) Building a sense tagged corpus with open mind word expert. In: Proceedings of the ACL 2002 workshop on word sense disambiguation: recent successes and future directions, Philadelphia
9. Cimiano P, Schultz A, Sizov S, Sorg P, Staab S (2009) Explicit versus latent concept models for cross-language information retrieval. In: International joint conference on artificial intelligence, IJCAI-09, Pasadena, CA, pp 1513–1518
10. Cucerzan S (2007) Large-scale named entity disambiguation based on Wikipedia data. In: Proceedings of the conference on empirical methods in natural language processing, Prague, Czech Republic, pp 708–716
11. Diab M (2004) Relieving the data acquisition bottleneck in word sense disambiguation. In: Proceedings of the 42nd meeting of the association for computational linguistics (ACL 2004), Barcelona, Spain
12. Diab M, Resnik P (2002) An unsupervised method for word sense tagging using parallel corpora. In: Proceedings of the 40st annual meeting of the association for computational linguistics (ACL 2002), Philadelphia, PA
13. Ferrucci DA, Brown EW, Chu-Carroll J, Fan J, Gondek D, Kalyanpur A, Lally A, Murdock JW, Nyberg E, Prager JM, Schlaefer N, Welty CA (2010) Building Watson: an overview of the DeepQA project. AI Mag 31(3):59–79
14. Gabrilovich E, Markovitch S (2006) Overcoming the brittleness bottleneck using Wikipedia: enhancing text categorization with encyclopedic knowledge. In: Proceedings of the national conference on artificial intelligence (AAAI), Boston
15. Gabrilovich E, Markovitch S (2007) Computing semantic relatedness using Wikipedia-based explicit semantic analysis. In: Proceedings of the international joint conference on artificial intelligence, Hyderabad, pp 1606–1611
16. Galley M, McKeown K (2003) Improving word sense disambiguation in lexical chaining. In: Proceedings of the 18th international joint conference on artificial intelligence (IJCAI 2003), Acapulco, Mexico
17. Hachey B, Radford W, Nothman J, Honnibal M, Curran JR (2013) Evaluating entity linking with Wikipedia. Artif Intell 194:130–150
18. Haghighi A, Klein D (2009) Simple coreference resolution with rich syntactic and semantic features. In: Proceedings of the 2009 conference on empirical methods in natural language processing, Singapore, pp 1152–1161
19. Henrich V, Hinrichs EW, Vodolazova T (2011) Semi-automatic extension of GermaNet with sense definitions from Wiktionary. In: Proceedings of the 5th language and technology conference: human language technologies as a challenge for computer science and linguistics, Poznań, Poland pp 126–130
20. Henrich V, Hinrichs EW, Vodolazova T (2012) An automatic method for creating a sense-annotated corpus harvested from the Web. In: 13th international conference on intelligent text processing and computational linguistics, CICLing-2012, New Delhi, India
21. Henrich V, Hinrichs EW, Vodolazova T (2012) Webcage – a Web-harvested corpus annotated with GermaNet senses. In: 13th conference of the European chapter of the association for computational linguistics, EACL '12, Avignon, France, pp 387–396
22. Kaisser M (2008) The QuALiM question answering demo: supplementing answers with paragraphs drawn from Wikipedia. In: Proceedings of the ACL-08 human language technology demo session, Columbus, Ohio, pp 32–35
23. Kunze C, Lemnitzer L (2002) GermaNet – Representation, visualization, application. In: 3rd international conference on language resources and evaluation, LREC'02, Las Palmas, Spain, pp 1485–1491
24. Leacock C, Chodorow M, Miller G (1998) Using corpus statistics and WordNet relations for sense identification. Comput Linguist 24(1):147–165

25. Lee Y, Ng H (2002) An empirical evaluation of knowledge sources and learning algorithms for word sense disambiguation. In: Proceedings of the 2002 conference on empirical methods in natural language processing (EMNLP 2002), Philadelphia
26. Lesk M (1986) Automatic sense disambiguation using machine readable dictionaries: how to tell a pine cone from an ice cream cone. In: Proceedings of the SIGDOC conference 1986, Toronto
27. Li Y, Luk R, Ho E, Chung K (2007) Improving weak ad-hoc queries using Wikipedia as external corpus. In: proceedings of the 30th annual international ACM SIGIR conference on research and development in information retrieval, Amsterdam, Netherlands, pp 797–798
28. Medelyan O, Milne D, Legg C, Witten IH (2009) Mining meaning from Wikipedia. Inter J Human Comput Stud 67(9):716–754
29. de Melo G, Weikum G (2010) Menta: inducing multilingual taxonomies from Wikipedia. In: Proceedings of the 19th ACM international conference on information and knowledge management, CIKM '10. ACM, New York, pp 1099–1108
30. Meyer CM, Gurevych I (2011) What psycholinguists know about chemistry: aligning Wiktionary and WordNet for increased domain coverage. In: Proceedings of the 5th international joint conference on natural language processing (IJCNLP), pp 883–892
31. Mihalcea R (2002) Bootstrapping large sense tagged corpora. In: Proceedings of the third international conference on language resources and evaluation LREC 2002, Canary Islands, Spain, pp 1407–1411
32. Mihalcea R (2007) Using Wikipedia for automatic word sense disambiguation. In: Human language technologies 2007: the conference of the North American chapter of the association for computational linguistics, Rochester, New York
33. Mihalcea R, Csomai A (2007) Wikify!: linking documents to encyclopedic knowledge. In: Proceedings of the sixteenth ACM conference on information and knowledge management, Lisbon, Portugal
34. Mihalcea R, Moldovan D (1999) An automatic method for generating sense tagged corpora. In: Proceedings of AAAI-99, Orlando, FL, pp 461–466
35. Miller G (1995) Wordnet: A lexical database for English. Commun ACM 38(11):39–41
36. Milne D (2007) Computing semantic relatedness using Wikipedia link structure. In: Proceedings of the New Zealand computer science research student conference, Hamilton, New Zealand
37. Milne D, Witten I (2008) Learning to link with Wikipedia. In: Proceedings of the seventeenth ACM conference on information and knowledge management, Napa Valley, CA
38. Nastase V, Strube M, Boerschinger B, Zirn C, Elghafari A (2010) WikiNet: a very large scale multi-lingual concept network. In: 7th international conference on language resources and evaluation, LREC'10, Valletta
39. Navigli R, Ponzetto S (2010) BabelNet: Building a very large multilingual semantic network. In: Proceedings of the 48th annual meeting of the association for computational linguistics, Uppsala, Sweden
40. Navigli R, Velardi P (2005) Structural semantic interconnections: a knowledge-based approach to word sense disambiguation. IEEE Trans Pattern Anal Mach Intell (PAMI) 27:1075–1086
41. Ng H, Lee H (1996) Integrating multiple knowledge sources to disambiguate word sense: an examplar-based approach. In: Proceedings of the 34th annual meeting of the association for computational linguistics (ACL 1996), Santa Cruz
42. Ng H, Wang B, Chan Y (2003) Exploiting parallel texts for word sense disambiguation: an empirical study. In: Proceedings of the 41st annual meeting of the association for computational linguistics (ACL 2003), Sapporo, Japan
43. Niemann E, Gurevych I (2011) The people's Web meets linguistic knowledge: automatic sense alignment of Wikipedia and Wordnet. In: Proceedings of the ninth international conference on computational semantics, association for computational linguistics, IWCS '11, Stroudsburg, PA, USA, pp 205–214
44. Pedersen T (2001) A decision tree of bigrams is an accurate predictor of word sense. In: Proceedings of the North American chapter of the association for computational linguistics (NAACL 2001), Pittsburgh, pp 79–86

45. Ponzetto SP, Navigli R (2009) Large-scale taxonomy mapping for restructuring and integrating Wikipedia. In: Proceedings of the 21th international joint conference on artificial intelligence, Pasadena, CA
46. Ponzetto SP, Navigli R (2010) Knowledge-rich word sense disambiguation rivaling supervised systems. In: Proceedings of the 48th annual meeting of the association for computational linguistics, association for computational linguistics, Stroudsburg, PA, USA, pp 1522–1531
47. Potthast M, Stein B, Anderka MA (2008) Wikipedia-based multilingual retrieval model. In: Proceedings of the 30th European conference on IR research, Glasgow, United Kingdom
48. Rahman A, Ng V (2011) Coreference resolution with world knowledge. In: Proceedings of the 49th annual meeting of the association for computational linguistics: human language technologies – volume 1, association for computational linguistics, Stroudsburg, PA, USA, pp 814–824
49. Resnik P, Yarowsky D (1999) Distinguishing systems and distinguishing senses: new evaluation methods for word sense disambiguation. Nat Lang Eng 5(2):113–134
50. Strube M, Ponzetto SP (2006) Wikirelate! computing semantic relatedeness using Wikipedia. In: Proceedings of the American association for artificial intelligence, Boston, MA
51. Suchanek FM, Kasneci G, Weikum G (2007) Yago: A core of semantic knowledge. In: Proceedings of the 16th World Wide Web conference, Banff, Alberta, Canada
52. Wu F, Weld D (2007) Autonomously semantifying Wikipedia. In: Proceedings of the 16th ACM conference on information and knowledge management, Lisbon, Portugal
53. Wu F, Weld D (2008) Automatically refining the Wikipedia Infobox ontology. In: Proceedings of the 17th international World Wide Web conference, Beijing, China
54. Yarowsky D (1995) Unsupervised word sense disambiguation rivaling supervised methods. In: Proceedings of the 33rd annual meeting of the association for computational linguistics (ACL 1995), Cambridge, MA

# Part III
# Interconnecting and Managing Collaboratively Constructed Language Resources

# Chapter 10
# An Open Linguistic Infrastructure for Annotated Corpora

**Nancy Ide**

**Abstract** One means to offset the high cost of corpus creation is to distribute effort among members of the research community, and thereby distribute the cost as well. To this end, the American National Corpus (ANC) project undertook to provide data and linguistic annotations to serve as the base for a collaborative, community-wide resource development effort (the ANC Open Linguistic Infrastructure, ANC-OLI). The fundamental premises of the effort are, first, that all data and annotations must be freely available to all members of the community, without restriction on use or redistribution, and second, that once a base of data and annotation was established, the resources would grow as community members contributed their enhancements and derived data. To ensure maximum flexibility and usability, the project has also developed an infrastructure for representing linguistically annotated resources intended to solve some of the usability problems for annotations produced at different sites by harmonizing their representation formats. We describe here the resources and infrastructure developed to support this collaborative community development and the efforts to ensure full community engagement.

## 10.1 Introduction

Annotated corpora are a fundamental resource for research and development in the field of natural language processing (NLP). Although unannotated corpora (for example, Gigaword, Wikipedia, etc.) are often used to build language models, annotations for linguistic phenomena provide a richer set of features and hence, potentially better models in the long run. It is widely accepted that a first step in the pursuit of NLP applications for any language is to develop a high quality

N. Ide (✉)
Vassar College, Poughkeepsie, NY, USA
e-mail: ide@cs.vassar.edu

I. Gurevych and J. Kim (eds.), *The People's Web Meets NLP*, Theory and Applications of Natural Language Processing, DOI 10.1007/978-3-642-35085-6_10,

annotated corpus with at least a basic set of annotations for phenomena such as part of speech and shallow syntax, while corpora for languages such as English, for which substantial annotated resources already exist, are increasingly being enhanced to include additional annotations for semantic and discourse phenomena (e.g., semantic roles, sense annotations, coreference, named entities, discourse structure). This is occurring for at least two reasons: first, more and deeper linguistic information, together with study of intra-level interactions, may lead to insights that can improve NLP applications; and second, in order to handle more subtle and difficult aspects of language understanding, there is a trend away from purely statistical approaches and (back) toward symbolic or rule-based approaches. Richly annotated corpora provide the raw materials for this kind of development. As a result, there is an increased demand for high quality linguistic annotations of corpora representing a wide range of phenomena, especially at the semantic level, to support machine learning and computational linguistics research in general. At the same time, there is a demand for annotated corpora representing a broad range of genres, due to the impact of domain on both syntactic and semantic characteristics. Finally, there is a keen awareness of the need for annotated corpora that are both easily accessible and available for use by anyone.

Despite the need, there are very few richly annotated corpora, even for major languages such as English. This lack is most directly attributable to the high cost of producing such corpora. First, appropriate and, above all, available language data must be identified and acquired, often after lengthy copyright negotiations or painstaking web search for data unfettered by licensing limitations. Preparation of the data for annotation is notoriously difficult, especially when data come in a variety of formats, each of which must be cleaned to remove formatting information or, in the case of web data, extensive amounts of interspersed HTML (even more difficult if the format needs to be preserved); differences in character sets also have to be resolved in this step. Once prepared, annotation software may be applied to provide a base for manual validation, or annotations may be performed manually from the start; in either case, some environment for accomplishing the manual work must be provided. To be maximally useful, manual validation or annotation must be performed by multiple annotators and under controlled circumstances. For annotations at the semantic or discourse level, such as sense tagging or coreference, considerable effort to ensure the quality of the manual work must be expended, for example, by computing inter-annotator agreement metrics. Thus, corpus development can require several man-years of labor-intensive effort and, correspondingly, substantial funding. But while there has been some support for corpus creation and development over the past two decades, especially in Europe, in general the substantial funding required to produce high quality, richly annotated corpora, can be relatively difficult to acquire. Furthermore, the production and annotation of corpora, even when they involve significant scientific research, often do not, *per se*, lead to publishable research results. It is therefore understandable that many researchers are unwilling to get involved in such a massive undertaking for relatively little reward.

One means to offset the high cost of corpus creation is to distribute effort among members of the research community, and thereby distribute the cost as well. To this end, the American National Corpus (ANC) project[1] undertook to provide data and linguistic annotations to serve as the base for a collaborative, community-wide resource development effort (the ANC Open Linguistic Infrastructure, ANC-OLI) [12]. The fundamental premises of the effort are, first, that all data and annotations must be freely available to all members of the community, without restriction on use or redistribution, and second, that once a base of data and annotation was established, the resources would grow as community members contributed their enhancements and derived data. To ensure maximum flexibility and usability, the project has also developed an infrastructure for representing linguistically annotated resources intended to solve some of the usability problems for annotations produced at different sites by harmonizing their representation formats. We describe here the resources and infrastructure developed to support this collaborative community development and the efforts to ensure full community engagement.

## 10.2 Requirements for a Collaborative Annotation Effort

To be successful, an effort to involve the language processing community in collaborative resource development must meet several requirements so that the resources meet community needs and contribution of data and annotations as well as use of the available resources is easy for community members. Building on discussions held at a U.S. National Science Foundation (NSF)-sponsored workshop held in Fall, 2006,[2] we identified the following general criteria for a collaborative community annotation effort for the field.

### *10.2.1 Open Data*

In order to ensure that the entire community, including large teams as well as individual researchers, has access and means to use the resources in their work as well as the ability to redistribute the data with their enhancements, all data and annotations included in the ANC-OLI should be either in the public domain or under a license that does not restrict redistribution of the data or its use for any purpose, including commercial use. (e.g., the Creative Commons Attribution

[1] www.anc.org

[2] The NSF workshop, held October 29–30, 2006, included the following participants: Collin Baker, Hans Boas, Branimir Bogureav, Nicoletta Calzolari, Christopher Cieri, Christiane Fellbaum, Charles Fillmore, Sanda Harabagiu, Rebecca Hwa, Nancy Ide, Judith Klavans, Adam Meyers, Martha Palmer, Rebecca Passonneau, James Pustejovsky, Janyce Wiebe, and funding organization representatives Tatiana Korelsky (NSF) and Joseph Olive (DARPA). A report summarizing the consensus of the workshop participants is available at http://anc.org/nsf-workshop-2006.

(CC-BY) license.[3] Data under licenses such as GNU General Public License[4] or Creative Commons Attribution-ShareAlike[5] should be avoided because of the potential obstacle to commercial use imposed by the requirement to redistribute under the same terms.

### *10.2.2 Data Diversity*

The lack of diverse data to support NLP research and development is well-known within the community. Even today, the corpora most frequently used by the community are the Penn Treebank corpus, the Chinese Treebank, EuroParl, and Wikipedia,[6] all of which are either very skewed for genre and/or unannotated. This is a result, of course, of the labor required to obtain large amounts of broad genre open data that can be annotated *and* redistributed with its annotations. The ANC-OLI should therefore include data from a range of different written and spoken genres, including but not limited to the genres in "representative" corpora such as the Brown Corpus and the British National Corpus. It should also include topic-specific data and newer genres unrepresented in older language data collections, such as tweets, blogs, wikis, email, etc. Although modalities other than text should be the focus at the start, in principle the ANC-OLI should include and support audio, image, and video as well.

### *10.2.3 Annotation Types*

The ANC-OLI should include automatically-produced annotations, especially annotations of the same phenomenon, which are valuable for comparison and development of heuristics that can improve the performance of automatic annotation software. In addition, there is a critical need for data that is manually annotated for a broad range of linguistic phenomena, in order to provide much needed training data to improve automatic annotation software and machine learning. The ANC-OLI should seek support for manual validation of a (possibly small) sub-component of its holdings, but we expect to rely heavily on community contributions to provide high quality, manual annotations. In general, the production of annotations should be application-driven (e.g., discourse level annotations useful to Question Answering).

---

[3]creativecommons.org/licenses/by/3.0/

[4]www.gnu.org/licenses/gpl.html

[5]creativecommons.org/licenses/by-sa/2.5/

[6]Based on entries in the LRE Map, http://www.resourcebook.eu/LreMap/faces/views/resourceMap.xhtml

Whether automatically or manually produced, annotations should represent different (possibly competing) theoretical approaches, for example, syntactic annotation using phrase structure and dependency syntax, in order to support research that compares the various approaches to show both how they relate and which are more appropriate for a down-stream use. Manual annotations over the same data utilizing widely used lexical and semantic resources such as WordNet senses and FrameNet frames are valuable as a step toward harmonizing such resources, which is a critical need for the field. Use of WordNet and FrameNet has the further advantage that it provides links to wordnets and framenets in other languages; for example, a WordNet sense-tagged lexical unit is automatically associated with its translations in the over fifty existing wordnets in other languages.

Ensuring the compatibility of annotation semantics–i.e., the linguistic categories used to describe the data–is still an area for research, and no attempt should be made by the ANC-OLI to resolve it. Rather, the ANC-OLI should encourage and contribute to efforts to devise means to harmonize linguistic annotation categories such as the ISO TC37/SC4 Data Category Registry (ISOCat [17]), General Ontology of Linguistic Description (GOLD, [7] and Ontologies of Linguistic Annotation (OLiA [3]), and in general foster the movement toward "semantics by reference" wherein the definition of a linguistic category used in an annotation is provided by referencing the URI of the category in question.

### *10.2.4 Format*

To be successful, an effort to involve the community in a collaborative resource development effort must ensure that it is easy for community members to contribute and that the resulting resources are easy for community members to use, both individually and together. In the past, widely used corpora have been enhanced by community members, and in some cases the added resources have been made publicly available, but the lack of consistency among formats has prevented combined use of the existing and added annotations. The most obvious case in point is the one million word *Wall Street Journal* corpus known as the Penn Treebank [19], which over the years has been fully or partially annotated for several phenomena over and above the original part-of-speech tagging and phrase structure annotation. The usability of these annotations is limited, however, by the fact that most of them were produced by independent projects using their own tools and formats, making it difficult to combine them in order to study their inter-relations.

The obvious obstacle to the combined use of annotations produced at different sites is the lack of standards for representing linguistically annotated language data, including not only annotated corpora but also lexicons, treebanks, propbanks, etc. The ANC-OLI should address this obstacle as broadly as possible, by seeking a solution that would cover the greatest number of situations in the short term, and at the same time serve over the long term as a viable approach to the multiple formats problem.

Transducing among different formats, especially complex formats such as the Penn Treebank syntax and PropBank semantic role annotations that depend on it, is often non-trivial. Therefore, ANC-OLI contributors cannot be expected to expend resources to provide their annotations in any format other than the one their in-house tools produce, and users cannot be expected to adapt ANC-OLI annotations for use with either in-house or off-the-shelf tools. However, to be usable together, both internally-produced and contributed annotations must be represented in a single, usable format. This format must be both powerful and generic enough to allow annotations in any representation (e.g. LISP structures, XML) and with any internal structure (e.g., tree, graph) to be readily *mapped* to it without information loss, and flexible and standardized enough to enable linking to resource efforts in other areas of the world. For ease of use, ANC-OLI data and annotations should also be made available not only in a common format, but also in formats compatible with widely used tools such as the Natural Language Tool Kit (NLTK), GATE, and UIMA, as well as other commonly used formats such as the Resource Description Format (RDF) and the IOB format used in CoNLL shared tasks.

### 10.2.5 Access

Access should be easy and open via the web. Selective access should also be provided, so that users can choose to download only the annotations and data of interest to them, in a format that is convenient for their purposes. In addition, there should be tool support for the data and annotations in the common format.

### 10.2.6 Maintenance

There must be provision for maintenance and sustainability. There is a history in both the US and Europe of resource development that is not followed up with funding to maintain and, where necessary, update the resource. This has led to a situation where resources have become obsolete, or, more often, become unavailable because developers have no support for distribution. Therefore, to ensure sustainability of the resource, the resources should be made available through a major data center such as the LDC, which can guarantee long term availability of the resource. In the short term, availability through a major data center will increase the visibility and accessibility of the resources.

### 10.2.7 Coverage

The ANC-OLI is based on the American National Corpus, which by definition contains only American English data. The ANC-OLI should be expanded to include other languages and media such as audio, video, image, etc. at the earliest possible time.

### *10.2.8 Fostering Community Involvement*

The idea of an annotated resource deliberately intended for collaborative development is a relatively new one in the field. Until recently, the addition of annotations to common data by different individuals or groups was done in an *ad hoc*, uncoordinated way, and there was never a clear intention to use the annotations together or even share them with the rest of the community. The growing promotion of sharable, "open" resources over the past few years (largely engendered by the open software movement) has created a major shift in community perspective concerning the need to accommodate more universal usability of resources and tools, but in general, the *de facto* scenario in people's minds does not include giving resources to another individual or group for their use. Therefore, there is, as yet, no collective mentality fostering collaborative resource development, although this is clearly on the horizon. In the meantime, to engage the community and perhaps move them more quickly toward adoption of the collaborative model, it is necessary to familiarize researchers and developers with the premises behind collaborative development and promote its adoption.

## 10.3 ANC-OLI

### *10.3.1 History*

The American National Corpus project was launched in 1998 [9], motivated by developers of major linguistic resources such as FrameNet[7] and Nomlex,[8] who found that usage examples extracted from the 100 million word British National Corpus (BNC), the largest corpus of English across several genres available at the time, were often unusable or misrepresentative for developing templates for the description of semantic arguments and the like, due to significant syntactic differences between British and American English. The ANC project was originally conceived as a near-identical twin to its British cousin: the ANC would include the same amount of data (100 million words), balanced over the same range of genres and including 10 % spoken transcripts just like the BNC.

[7] www.icsi.berkeley.edu/~framenet

[8] nlp.cs.nyu.edu/nomlex/index.html

The BNC was substantially funded by the British government, together with a group of publishers who provided both financial support and contributed a majority of the data that would appear in the corpus. Based on this model, the ANC looked to similar sources, but gained the support of only a very few U.S. publishers and a handful of major software developers, who provided about $400,000 to support the first 4 years of ANC development, an order of magnitude less funding than that which supported development of the BNC.

British publishers provided the bulk of the data in the 100 million word BNC. The plan for the ANC was that the sponsoring publishers and software vendors would do the same for the ANC. However, only a very few of the ANC supporters eventually contributed data to the corpus.[9] As a result, it was necessary to attempt to find data from other sources, including existing corpora such as the Indiana Center for Intercultural Communication (ICIC) Corpus of Philanthropic Fundraising Discourse, and the Charlotte Narrative and Conversation Collection (CNCC), together with government documents, biomedical articles, and other public domain material on the web.

In 2003, the ANC produced its first release of 11 million words of data, which included a wide range of genres of both spoken and written data. Annotations included word and sentence boundaries and part-of-speech annotation produced by two different taggers in standoff form, that is, provided as separate files with links into the data.[10] To our knowledge, the ANC First Release was the first large, publicly available corpus to be published with standoff annotations. In 2005, the ANC released an additional 11 million words, bringing the size of the corpus to 22 million words. The Second Release includes data from additional genres, most notably a sizable sub-corpus of blog data, biomedical and technical reports, and the *9/11 Report* prepared by the U.S. Government. The Second Release was issued with standoff annotations for the same phenomena as in the First Release, as well as annotations for shallow parse (noun chunks and verb chunks). Notably, the ANC Second release also included the first community contributed annotations of the corpus: manually produced coreference annotation of about 100,000 words of Slate magazine articles contributed by University of Alberta, and two additional part of speech annotations using the CLAWS 5 and 7 tags used in the BNC contributed by University of Lancaster.

In 2006, the project made 15 million of the ANC's 22 million words that were not restricted for any use available for download as the "Open ANC" (OANC) from the ANC website.[11] The fully open distribution model pioneered by the OANC has now been adopted for all future releases of data and annotations.[12] It was at this point that the ANC-OLI was conceived [12], thus creating the first collaborative, community-wide resource development effort in the field. Since then, three syntactic parses of

[9]The consortium members who contributed texts to the ANC are Oxford University Press, Cambridge University Press, Langenscheidt Publishers, and the Microsoft Corporation.

[10]The contents of the ANC First Release are described at http://www.anc.org/FirstRelease/

[11]www.anc.org/OANC/index.html

[12]However, since 2005 the ANC project had no funding for production of additional data.

11 million words of the OANC (using the Charniak and Johnson parser, MaltParser, and LHT dependency converter, respectively) and named entity annotations of the entire OANC produced by the BBN tagger [20], have been contributed.

The next year, the ANC project received a substantial grant from the U.S. National Science Foundation[13] to produce a half-million word Manually Annotated Sub-Corpus (MASC) of the OANC that would include automatically-produced annotations for logical structure (paragraph, section, headings, etc.), word and sentence boundaries, part of speech and lemma, shallow parse, and named entities, and to manually add annotations for WordNet senses and FrameNet frames to portions of the corpus. From the outset, the project was designed to serve as a centerpiece for the ANC-OLI, and so to facilitate initial community contribution, materials for the MASC were drawn from sources that have already been heavily annotated by others (where licensing permitted). MASC currently includes a 50 K subset consisting of OANC data that has been previously annotated for Penn Treebank syntax, PropBank predicate argument structures, Pittsburgh Opinion annotation (opinions, evaluations, sentiments, etc.), TimeML time and events and several other linguistic phenomena. It also includes a handful of small texts from the so-called Language Understanding (LU) Corpus[14] that was annotated by multiple groups for a wide variety of phenomena, including events and committed belief; and 5.5 K words of Wall Street Journal texts that have been annotated by several projects, including Penn Treebank, PropBank, Penn Discourse Treebank, TimeML, and the Pittsburgh Opinion project. All of these annotations, apart from 420 K of annotations for Penn Treebank syntax,[15] were contributed to the project.

The first full version of the corpus was released in 2012, including a separate sentence corpus [23] that provides sense-tags for approximately 1,000 occurrences of each of 114 words chosen by the WordNet and FrameNet teams (ca. 114,000 annotated occurrences).

## *10.3.2 Meeting the Requirements for Community Collaboration*

### 10.3.2.1 Open Data and Data Diversity

The requirement for *open data* imposes severe limits on what can be included in the corpora distributed by the ANC-OLI, making data acquisition the major issue for ANC-OLI development. Over the past 5 years we have gathered approximately

[13] NSF CRI 0708952

[14] MASC contains about 4 K words of the 10 K LU corpus, eliminating non-English and translated LU texts as well as texts that are not free of usage and redistribution restrictions.

[15] The MASC project commissioned the remainder of the annotation from the Penn Treebank project.

**Table 10.1** Genre distribution in MASC

| Genre | No. files | No. words | Pct corpus |
|---|---|---|---|
| Court transcript | 2 | 30,052 | 6 % |
| Debate transcript | 2 | 32,325 | 6 % |
| Email | 78 | 27,642 | 6 % |
| Essay | 7 | 25,590 | 5 % |
| Fiction | 5 | 31,518 | 6 % |
| Gov't documents | 5 | 24,578 | 5 % |
| Journal | 10 | 25,635 | 5 % |
| Letters | 40 | 23,325 | 5 % |
| Newspaper | 41 | 23,545 | 5 % |
| Non-fiction | 4 | 25,182 | 5 % |
| Spoken | 11 | 25,783 | 5 % |
| Technical | 8 | 27,895 | 6 % |
| Travel guides | 7 | 26,708 | 5 % |
| Twitter | 2 | 24,180 | 5 % |
| Blog | 21 | 28,199 | 6 % |
| Ficlets | 5 | 26,299 | 5 % |
| Movie script | 2 | 28,240 | 6 % |
| Spam | 110 | 23,490 | 5 % |
| Jokes | 16 | 26,582 | 5 % |
| TOTAL | 376 | 506,768 | |

50 million words of open data,[16] not including public domain data that can be acquired from government sites and web archives of technical documents. While these latter sources can provide virtually limitless amounts of data, the requirement for *data diversity* means that acquisition efforts must focus on other data types, especially those that are rarely published as open data such as fiction, tweets, etc. The OANC contains about three million words of spoken data spoken data (face to face, telephone conversations, academic discourse), and over 11 million words of written texts (government documents, technical articles, travel guides, fiction, letters, non-fiction). The contents of the MASC corpus are given in Table 10.1.

To date, the ANC-OLI has gathered open data from the following sources:

(a) Contributions from publishers who are willing to provide data under a non-restrictive license, including non-fiction materials donated to the ANC by Oxford University Press and Cambridge University Press, travel guides from Langenscheidt, and SLATE magazine articles from Microsoft. To protect their interests, publishers sometimes provide only a subset of a complete book or collection.
(b) Web materials in the public domain or licensed under non-viral licenses such as CC-BY. Government documents and debate and court transcripts, as well as technical articles in collections such as Biomed Central[17] and the Public Library

[16]Lack of funding for processing the data currently prevents its publication.

[17]www.biomedcentral.com

of Science,[18] are typically in the public domain. Although more difficult to track down, blogs, fiction, and other writing such as essays are very often distributed over the web under licenses such as CC-BY.

(c) Contributions from college students of class essays and other writing. College students produce considerable volumes of prose during their academic careers, and very often this data is discarded or forgotten once handed in to satisfy an assignment. The ANC-OLI provides a web interface for contributions of this kind that includes a grant of permission to use the contributed materials.[19]

(d) Direct solicitation for use of web materials. We have on occasion identified a web site containing interesting or substantial materials and contacted the relevant parties directly to explain our use of the data and ask for permission to use it. We have also contacted providers whose data are freely available for access to the materials in a form more manageable for processing purposes. So far, none of our requests has been turned down.

(e) Contributions from colleagues in the field and data centers such as the Linguistic Data Consortium (LDC).[20] We have received data contributions, including significant amounts of spoken data, from several NLP and linguistics projects, including the Indiana Center for Intercultural Communication (ICIC) Corpus of Philanthropic Fundraising Discourse,[21] Project MUSE's Charlotte Narrative and Conversation Collection (CNCC),[22] the Michigan Corpus of Academic Spoken English (MICASE),[23] and the International Computer Science Institute (ICSI) Meeting Corpus [16]. We have also received contributed annotations from the Penn Treebank project, the PropBank project, the Pittsburg Opinion annotation project, TimeBank, and several others.

Acquisition of almost all of these data was non-trivial, requiring substantial time and effort to solicit contributions from publishers, projects, and even college students, and to identify suitably open materials on the web. Contributions from the research community at large have also so far been relatively meagre, typically due to licensing constraints. As awareness of the nature of and need for open data increases, these contributions are more and more readily forthcoming.

#### 10.3.2.2 Annotations

The 15 million word OANC includes automatically-produced annotations for logical structure, sentence and token boundaries, part of speech and lemma (four

[18] www.plos.org

[19] www.anc.org/contribute.html

[20] www.ldc.upenn.edu

[21] liberalarts.iupui.edu/icic/research/corpus_of_philanthropic_fundraising_discourse

[22] newsouthvoices.uncc.edu/

[23] http://quod.lib.umich.edu/m/micase/

**Table 10.2** Summary of MASC annotations

| Annotation type | No. words |
|---|---|
| Logical | 506,659 |
| Token | 506,659 |
| Sentence | 506,659 |
| POS/lemma (GATE) | 506,659 |
| POS (Penn) | 506,659 |
| Noun chunks | 506,659 |
| Verb chunks | 506,659 |
| Named entities | 506,659 |
| FrameNet | 39,160 |
| Penn Treebank syntax | 506,659 |
| PropBank | 55,599 |
| Opinion | 51,243 |
| TimeBank | 55,599 |
| Committed belief | 4,614 |
| Event | 4,614 |
| Dependency treebank | 5,434 |
| Coreference | 506,659 |
| Discourse segments | 506,659 |

different taggers and tag sets), noun chunks, verb chunks, and named entities. Eleven million words are automatically annotated for two dependency parses and one phrase structure parse. MASC contains a richer set of annotations, all manually produced or hand validated, over all or parts of the corpus as shown in Table 10.2. The MASC Sentence Corpus consists of approximately 110,000 sentences with WordNet sense annotations for 114 words. The sentences include every occurrence of each of the 114 words in MASC together with occurrences drawn from the OANC to fill out the balance of 1,000 sentences per word.

While every effort has been made to include as diverse a set of annotations, including multiple annotations of the same type representing different theoretical approaches, the ANC-OLI does not have the resources to produce the full range of possible types, especially for the MASC data which requires manual validation. One particular lack is a dependency parse of MASC, which would provide a complement to the Penn Treebank phrase structure analysis. Discourse-level annotation of a variety of types would also be desirable for MASC. We will rely on community collaboration and contribution to fill these gaps.

Automatic annotation of OANC data is easier to produce but still requires programming effort to render into GrAF. Also, the accuracy of automatically produced annotations over OANC data tends to degrade severely, since most annotation software is trained on a single or relatively constrained set of genres, whereas the OANC data is far more varied. Hopefully, the availability of diverse data will spark experimentation with the impact of domain and genre on the performance of automatic annotation software.

#### 10.3.2.3 Format

The representation format of the ANC-OLI annotations must serve two purposes: it must be possible to transduce from formats of contributed annotations to the ANC-OLI format without loss of information, and the format must be interoperable with diverse tools and frameworks for searching, processing, and enhancing the corpus. For this reason, the representation of all ANC-OLI annotations follows the specifications of the International Standards Organization (ISO) Linguistic Annotation Framework (LAF) [15], which provides a framework for representing annotations based on an abstract model consisting of a graph of features structures, two very powerful and general data structures that have been widely used, either directly or as an underlying model, to represent linguistic information [11]. A fundamental tenet of the LAF model is that all annotations are in stand-off format, with references to primary data or other annotations.[24] The Graph Annotation Format (GrAF) [13, 14], the XML serialization of the model, is intended to function in much the same way as an interlingua in machine translation, that is, as a "pivot" representation into and out of which user- and tool-specific formats are transduced, so that a transduction of any specific format into and out of GrAF accomplishes the transduction between it and any number of other GrAF-conformant formats. The rendering of ANC-OLI data and annotations in GrAF thus satisfies the criteria outlined above: it is powerful enough to represent annotations contributed in any format, easy to transduce ANC-OLI annotations to other formats, and it conforms to a widely adopted international standard. The graph-based format also enables trivial merging of annotations rendered in GrAF. Furthermore, the generality of the abstract model makes mapping to formats such as the Resource Description Format (RDF) [18], which is the format used in the Semantic Web, and the UIMA Common Analysis System (CAS) [8].

The generic graph model underlying GrAF is isomorphic to that of emerging Semantic Web standards, notably RDF/OWL, thus making conversion between GrAF and RDF/OWL representations trivial. The GrAF representation of MASC has recently been rendered into POWLA [4], an RDF/OWL linearization of PAULA, a generic data model for the representation of annotated corpora [5, 6]. This representation includes linkage of its WordNet and FrameNet annotations to RDF instantiations of these resources, as well as linkage of linguistic categories used in MASC's other annotation layers to types in the POWLA OWL/DL ontology.[25] The RDF/OWL instantiation opens up the potential to formulate queries that combine information from the linked versions of WordNet, FrameNet, and MASC using an

[24]Allowing annotations to reference other annotations differentiates GrAF from other representation formats, such as Annotation Graphs [2]

[25]For more details, see Chiarcos, et al., in this volume.

RDF query language such as SPARQL [25]. The RDF/OWL version of MASC is publicly available as a part of the Linguistic Linked Open Data cloud.[26]

The ANC-OLI project is committed to rendering contributed annotations into GrAF. To date, all of the annotation types in Table 10.2, which came to us in a variety of both stand-off and embedded (in-line) formats, have been rendered into GrAF without information loss. The transduction process is not always trivial; for example, to be transduced to GrAF standoff form, in-line annotations must first be extracted from the text and then realigned to refer to the primary data document. Another problem results from variations in tokenization among the different annotations; this is solved by GrAF's provision for a segmentation document that defines minimally granular regions over a primary resource, which may then be combined (if necessary) and referenced by different tokenizations. The difficulties encountered in the transduction process typically arise from inconsistencies or omissions in the original format, which must be rectified in the GrAF representation. The problems are at any rate informative for the development of best practice annotation guidelines.

#### 10.3.2.4 Access and Maintenance

All ANC-OLI data are freely downloadable from the web, without the need to sign a license or provide any information. In addition, to ensure sustainability of the resources, all data and annotations are held and distributed by the Linguistic Data Consortium for no cost.

For use of the available resources, the ANC-OLI has developed an open source GrAF API[27] for reading and writing GrAF files and also provides a web application, called ANC2Go, that enables a user to choose any portion or all of MASC and the OANC together with any of their annotations to create a "customized corpus". The customized corpus can be delivered in any of several formats, including inline XML (input to any XML-aware program, including BNC's XIARA which then allows for comparative studies), token/pos (input to commonly used concordancing software), CONLL IOB format, tagged input for the Natural Language Toolkit (NLTK), and RDF. The project also provides modules to import/export from the widely used annotation and analysis frameworks GATE[28] and UIMA, so that ANC-OLI annotations are directly usable in these systems. However, in addition to being readily transduced to other formats, the GrAF format is useful in itself: one of the most salient features of the graph representation for linguistic annotations is the ability to exploit the wealth of graph-analytic algorithms for information extraction and analysis. For example, it is trivial to merge independently-produced annotations

[26]linguistics.okfn.org/llod

[27]http://sourceforge.net/projects/iso-graf/

[28]General Architecture for Text Engineering; http://gate.ac.uk

of the same data in GrAF form, as well as to apply algorithms to find common sub-graphs that reflect relations among different annotations.

#### 10.3.2.5 Coverage

Because the ANC-OLI grew out of the American National Corpus project, the included corpus resources currently include only American English spoken transcripts and written texts. Ideally, the project should expand to cover other modalities, including speech (audio), video, and image, as well as other languages. To address the lack of multilingual data in the ANC-OLI, we have recently launched MultiMASC [10], which builds upon MASC by extending it to include comparable corpora in other languages. Here, "comparable" means not only representing the same genres and styles, but also including similar types and number of annotations represented in a common format. Like MASC, MultiMASC will contain only completely open data and expand the collaboration effort upon which it depends. The eventual result is envisaged to be a massive, multi-lingual, multi-genre corpus with comparable multi-layered annotations that are inter-linked via reference to the original MASC or, perhaps more interestingly, to the RDF/OWL instantiation of MASC and associated resources described in Chiarcos et al. (this volume).

#### 10.3.2.6 Fostering Community Involvement

Following the familiar quote "build it and they will come",[29] by virtue of their existence and availability, community use of the OANC and MASC has been immediate and substantial. Contribution of annotations, on the other hand, has been slower to develop but is now beginning to gain momentum. In the first years of OANC availability (2005 onward), only a handful of annotations were contributed, including the output of three different parsers[30] and named entity annotation produced by the BBN Tagger [21]. MASC has enjoyed better success, in large part because it is both a newer resource and one that has been more widely publicized within the community via conference papers and workshops. The first release of 82 K includes a 50 K subcomponent for which several annotation layers were contributed, including Penn Treebank syntax, PropBank semantic roles, TimeML time and event annotation, and Pittsburgh opinion annotations. Additional annotations of MASC data for spatial information, PropBank semantic roles, discourse (Penn Discourse Treebank), and "deep semantics" (Groningen Meaning Bank), among others, are underway. We also expect that MASC–either the corpus

[29]Taken from *Field of Dreams*; see http://en.wikipedia.org/wiki/Field_of_Dreams

[30]The Charniak and Johnson (2005) parser, MaltParser, and LHT dependency converter.

and some or all annotations, or the sense-tagged sentence corpus–will be used in upcoming SemEval exercises.[31]

Collaborative community development goes beyond the contribution of annotations. Such development crucially relies on the community to identify errors in order to continually improve the resources, together with contribution of derived data such as frequency lists, ngrams, statistics reflecting the distribution of various phenomena, etc. Another important development activity involves the incorporation of ANC-OLI data and annotations into platforms and frameworks that enable others to work with them, beyond those already provided by the ANC project itself. Currently, community members have spontaneously taken up incorporation of MASC into the OpenNLP machine learning toolkit[32] and development of a corpus reader for ANC-OLI data and annotations for the Natural Language Toolkit (NLTK), two important frameworks for NLP research and education.

As noted earlier, collaborative development is not yet in the mainstream of activity within the language processing community, and so it is still necessary to promote community involvement through publicity at conferences and workshops, together with the use of OANC and, in particular, MASC, in shared tasks such as CONLL, SemEval, and *SEM. It will require a significant shift in the community mindset before its members reflexively contribute their annotations of ANC-OLI data and derived information, given the established practice in the field of *consuming* resources with no expectation of return, a practice most evident in the procedures of resource repositories such as LDC and ELRA.[33] Widespread acceptance of the collaborative resource development model is exacerbated by the fact that preparing annotations and derived data for use by others can require additional and sometimes considerable effort. Nonetheless, recognition that the need for richly annotated and inter-linked language resources can be most efficiently met through a collaborative community development effort is increasingly widespread and motivates numerous national and international funding programs aimed at infrastructure development for NLP research.

## 10.4 ANC-OLI in Context

The ANC-OLI corpora provide a unique resource in terms of both their content and configuration, as well as the collaborative aspect of their development. For example, the two standard broad genre corpora for English, the Brown Corpus and the British National Corpus (BNC), provide only part of speech annotations, in contrast to the richer set of annotations in the OANC and particularly in MASC. In addition,

---

[31] http://aclweb.org/aclwiki/index.php?title=SemEval_Portal

[32] http://opennlp.apache.org

[33] Such repositories were set up to answer the call for resource reusability which, no doubt in large part because information added to these resources was until recently unlikely to be usable by others, always referred to the consumer-only model.

Brown and BNC include only data produced prior to widespread use of the web, which has radically affected lexical and syntactic usage and fostered the emergence of new genres. The one million word *Wall Street Journal* corpus known as the Penn Treebank [19] has been fully or partially annotated for several phenomena beyond the original part-of-speech tagging and phrase structure annotation over the years, but most were produced by independent projects using their own tools and formats, making it difficult to use these annotations together. Of course, the lack of genre diversity of this corpus, which contains texts from a single domain that have been edited to conform to a consistent "Wall Street Journal style", is well known as a major drawback for its use in training language models for broad-range syntactic and semantic phenomena.

The corpus closest to ANC-OLI in terms of richness of annotation and currency of language is the one million word English OntoNotes corpus [24], which includes annotations for Penn Treebank syntax, sense annotations using an in-house sense inventory, PropBank predicate argument structures, coreference, and named entities represented in a "normal form". As in MASC, all annotations have been hand-validated. However, the OntoNotes corpus represents a limited set of genres (newswire, broadcast news, and broadcast conversation), and, because of the need to compile annotations into the internal OntoNotes database, annotations produced by others cannot be added to the corpus. Also, unlike ANC-OLI data, OntoNotes is restricted for research use only and requires licensing through the LDC.

Very recently, two collaborative annotation efforts have been initiated that share some aspects of ANC-OLI development. One, the *Language Library*,[34] asks community members to apply their software to provided data and contribute the results. The Language Library data are for the most part freely available, although inclusion of large amounts of multi-lingual Wikipedia data imposes the "share-alike" restriction that typically prevents its use for commercial purposes. The inspiration for the Language Library came directly from the ANC-OLI collaborative model, with the intent to expand the coverage to multiple languages. The new MultiMASC effort will extend the ANC-OLI to other languages, but will differ from the Language Library because the data will represent a broad range of genres, include only manually produced or validated annotations, ensure all annotations are represented in a harmonized format, and, by virtue of the common format, enable inter-linkage of linguistic phenomena at all levels across languages. Thus MultiMASC is far more ambitious and, correspondingly, more labor-intensive collaborative project than the Language Library, but promises to deliver resources that can be used to train learning algorithms and provide new insights about relationships across linguistic levels and languages that the Language Library cannot provide.

The second new collaborative project, the Groningen Meaning Bank (GMB) [1], has established an effort to provide manual validations of automatically-produced annotations for several linguistic layers from part of speech through discourse structure. Validation is done by volunteer linguists. The data are chosen to be in

[34] http://www.languagelibrary.eu

the public domain; interestingly, the project has chosen the MASC data as a part of its corpus. However, the MASC annotations for phenomena included in the GMB are not used but rather re-generated and hand validated, thus effectively duplicating the work done for MASC. Given that one goal of collaborative annotation is to avoid duplication of effort, it is somewhat tautological for a collaborative project to (in part) discard and re-do the same work as another collaborative effort. The GMB does not use the MASC annotations because of differences in tokenization and (some) annotation categories that are incompatible with their annotation tools.

The GMB has recently established a "game with a purpose" called *Wordrobe* that enables collecting validations from non-experts, which, if enough redundant validations are collected, can provide reliable results by majority vote (see, for example, [22]). The success of *Wordrobe* and *PhraseDetectives*[35] for co-reference annotation (which is also annotating MASC data), together with crowdsourcing in general, suggest the possibility to exploit these strategies for development of ANC-OLI annotations. However, although crowdsourcing dramatically reduces the overhead for gathering validated annotations on a relatively large scale, it is not without some cost for setting up, collecting, evaluating, and preparing the results. The purely collaborative development model of the ANC-OLI requires considerably less investment, since the only requirement is conversion of annotations in different formats to GrAF for compatibility. As time goes on, fewer and fewer annotations are contributed in a format for which a converter has not already developed, if they are not contributed in GrAF itself, thus further reducing the overhead. As a result, of the foreseeable future the ANC-OLI will likely not pursue this development option.

## 10.5 Looking Forward

The eventual vision for the ANC-OLI is to expand to include additional resources – not only corpora but also lexicons, lists, etc. – not only in English but also in multiple languages. As mentioned earlier in Sect. 10.3.2.5, the multi-lingual effort starts with MultiMASC, which will immediately expand MASC and the collaboration effort upon which it depends by exploiting the infrastructure and expertise established in the ANC-OLI to support development in other languages. Although this effort has only been recently launched, it has already drawn substantial interest within the community. Development of MultiMASC will expand the collaborative activity of the ANC-OLI to include the creation of comparable corpora, for which we have published a first set of guidelines, together with an incremental process for developing a fully inter-linked multi-lingual network of linguistic annotations [10].

We envision linkage across hundreds of languages among linguistic phenomena at many levels, e.g., part-of-speech categories, syntactic structures, paraphrases, semantic roles, named entities, events, etc. For example, Fig. 10.1 depicts linkage

[35]http://anawiki.essex.ac.uk/phrasedetectives/

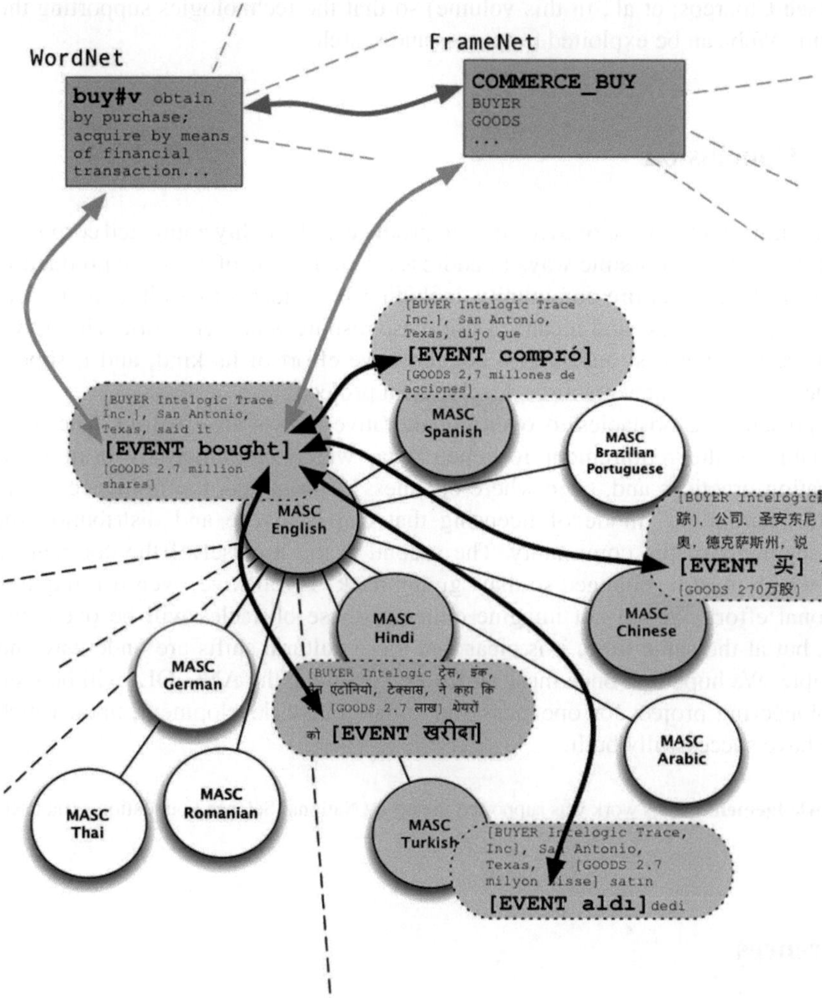


**Fig. 10.1** Overview of MultiMASC

among several languages for lexical units representing a common semantic role, in this case the EVENT of "buying". Such inter-linkage would utilize a reference set of categories residing in a data category registry such as ISOCat or OLiA that provides information about the annotation content and, more importantly, cross-references linguistic annotations using the same *conceptual* categories, regardless of physical label, within all of the inter-linked resources. Additional linkage to resources such as WordNet and FrameNet, which are themselves linked to wordnets and framenets in other languages, would add another dimension to this resource network, which would in turn enable cross-linguistic and inter-layer studies on a scale that is currently impossible. Ideally, this network would ultimately be available as Linked

Data (see Chiarcos, et al., in this volume) so that the technologies supporting the Semantic Web can be exploited for access and search.

## 10.6 Conclusion

A community-wide, collaborative effort to produce high quality annotated corpora is one of the very few possible ways to address the high costs of resource production and ensure that the entire community, including large teams as well as individual researchers, has access and means to use these resources in their work. The ANC-OLI represents the first and largest collaborative effort of its kind, and it should provide a model for new resource development projects.

At present, the obstacles to open collaborative efforts are twofold. The most formidable is the requirement for open data, which is limited by established publication practices and, even where openness is promoted, the influence of the default "share-alike" mode of licensing that can limit use and distribution for some segments of the community. The second is the mindset of the community itself, which must be changed so that "giving back" is reflexive, even if it requires additional effort. We do not imagine either of these obstacles will be overcome easily, but at the same time, it is clear that these cultural shifts are underway and inevitable. We hope that once these shifts are complete, the ANC-OLI will be seen as a pioneering project for openness and collaborative development, upon which others have successfully built.

**Acknowledgements** This work was supported in part by National Science Foundation grant CRI-0708952.

## References

1. Basile V, Bos J, Evang K, Venhuizen N (2012) Developing a large semantically annotated corpus. In: Proceedings of the eighth international conference on language resources and evaluation (LREC 2012), Istanbul, Turkey, pp 3196–3200
2. Bird S, Liberman M (2001) A formal framework for linguistic annotation. Speech Commu 33(1–2):23–60
3. Chiarcos C (2008) An ontology of linguistic annotations. LDV Forum 23(1):1–16
4. Chiarcos C (2012) Ontologies of linguistic annotation: survey and perspectives. In: Proceedings of the eighth international conference on language resources and evaluation (LREC), Istanbul, Turkey
5. Chiarcos C, Ritz J, Stede M (2012) By all these lovely tokens... merging conflicting tokenizations. Lang Resour Eval 46(1):53–74
6. Dipper S (2005) XML-based stand-off representation and exploitation of multi-level linguistic annotation. In: Eckstein R, Tolksdorf R (eds) Berliner XML Tage, pp 39–50
7. Farrar S, Langendoen DT (2010) An OWL-DL implementation of GOLD: an ontology for the semantic web. In: Witt A, Metzing D (eds) Linguistic modeling of information and markup languages. Springer, Dordrecht

8. Ferrucci D, Lally A (2004) UIMA: an architectural approach to unstructured information processing in the corporate research environment. J Nat Lang Eng 10(3–4):327–348
9. Fillmore CJ, Jurafsky D, Ide N, Macleod C (1998) An American national corpus: a proposal. In: Proceedings of the first annual conference on language resources and evaluation. European Language Resources Association, Paris, pp 965–969
10. Ide N (2012) MultiMASC: An open linguistic infrastructure for language research. In: Proceedings of the fifth workshop on building and using comparable corpora, Istanbul, Turkey
11. Ide N, Romary L (2004) International standard for a linguistic annotation framework. J Nat Lang Eng 10(3–4):211–225
12. Ide N, Suderman K (2006) An open linguistic infrastructure for American English. In: Proceedings of the fifth language resources and evaluation conference (LREC). European Language Resources Association, Paris, Genoa, Italy
13. Ide N, Suderman K (2007) GrAF: a graph-based format for linguistic annotations. In: Proceedings of the first linguistic annotation workshop, Prague, Czech Republic, pp 1–8
14. Ide N, Suderman K (Submitted) The linguistic annotation framework: a standard for annotation interchange and merging. Lang Resour Eval, in press
15. ISO 24612 (2012) Language resource management – linguistic annotation framework. International Standard ISO 24612
16. Janin A, Baron D, Edwards J, Ellis D, Gelbart D, Morgan N, Peskin B, Pfau T, Shriberg E, Stolcke A, Wooters C (2003) The ICSI meeting corpus. In: Proceedings of ICASSP-03, Hong Kong, pp 364–367
17. Kemps-Snijders M, Windhouwer M, Wittenburg P, Wright SE (2008) ISOcat: corralling data categories in the wild. In: Proceedings of the sixth international conference on language resources and evaluation (LREC'08), Marrakech, Morocco. European Language Resources Association (ELRA)
18. Klyne G, Carroll JJ (2004) Resource description framework (RDF): concepts and abstract syntax. World Wide Web Consortium, Recommendation REC-RDF-Concepts-20040210
19. Marcus MP, Marcinkiewicz MA, Santorini B (1993) Building a large annotated corpus of English: the Penn Treebank. Comput Linguist 19(2):313–330
20. Miller S, Guinness J, Zamanian A (2004) Name tagging with word clusters and discriminative training. In: Susan Dumais DM, Roukos S (eds) HLT-NAACL 2004: main proceedings, association for computational linguistics, Boston, MA, USA, pp 337–342
21. Miller S, Guinness J, Zamanian A (2004) Name tagging with word clusters and discriminative training. In: Proceedings of human language technologies, Boston, MA, USA, pp 337–342
22. Nowak S, Rüger S (2010) How reliable are annotations via crowdsourcing: a study about inter-annotator agreement for multi-label image annotation. In: Proceedings of the international conference on multimedia information retrieval. ACM, New York. MIR '10, pp 557–566. doi:10.1145/1743384.1743478, http://doi.acm.org/10.1145/1743384.1743478
23. Passonneau RJ, Baker CF, Fellbaum C, Ide N (2012) The MASC word sense corpus. In: Proceedings of the eighth international conference on language resources and evaluation (LREC'12), Istanbul, Turkey. European Language Resources Association (ELRA)
24. Pradhan SS, Hovy E, Marcus M, Palmer M, Ramshaw L, Weischedel R (2007) OntoNotes: a unified relational semantic representation. In: ICSC '07: Proceedings of the international conference on semantic computing. IEEE Computer Society, Washington, DC, pp 517–526
25. Prud'hommeaux E, Seaborne A (2007) SPARQL query language for rdf (working draft). Technical report, W3C. http://www.w3.org/TR/2007/WD-rdf-sparql-query-20070326/

# Chapter 11
# Towards Web-Scale Collaborative Knowledge Extraction

**Sebastian Hellmann and Sören Auer**

**Abstract** While the Web of Data, the Web of Documents and Natural Language Processing are well researched individual fields, approaches to combine all three are fragmented and not yet well aligned. This chapter analyzes current efforts in collaborative knowledge extraction to uncover connection points between the three fields. The special focus is on three prominent RDF data sets (DBpedia, LinkedGeoData and Wiktionary2RDF), which allow users to influence the knowledge extraction process by adding another crowd-sourced layer on top. The recently published NLP Interchange Format (NIF) provides a way to annotate textual resources on the Web through the assignment of URIs with fragment identifiers. We will show how this formalism can easily be extended to encompass new annotation layers and vocabularies.

## 11.1 Introduction

The vision of the Giant Global Graph[1] was conceived by Tim Berners-Lee aiming at connecting all data on the Web and allowing to discover new relations between the data. This vision has been pursued by the Linked Open Data (LOD) community, where the cloud of published datasets now comprises 295 data repositories and more than 30 billion RDF triples.[2] Although it is difficult to precisely identify the reasons for the success of the LOD effort, advocates generally argue that open licenses as well as open access are key enablers for the growth of such a network as they provide

[1]http://dig.csail.mit.edu/breadcrumbs/node/215

[2]http://www4.wiwiss.fu-berlin.de/lodcloud/state/

S. Hellmann (✉) · S. Auer (✉)
AKSW, Universität Leipzig, Leipzig, Germany
e-mail: hellmann@informatik.uni-leipzig.de; auer@informatik.uni-leipzig.de

I. Gurevych and J. Kim (eds.), *The People's Web Meets NLP*, Theory and Applications of Natural Language Processing, DOI 10.1007/978-3-642-35085-6_11,

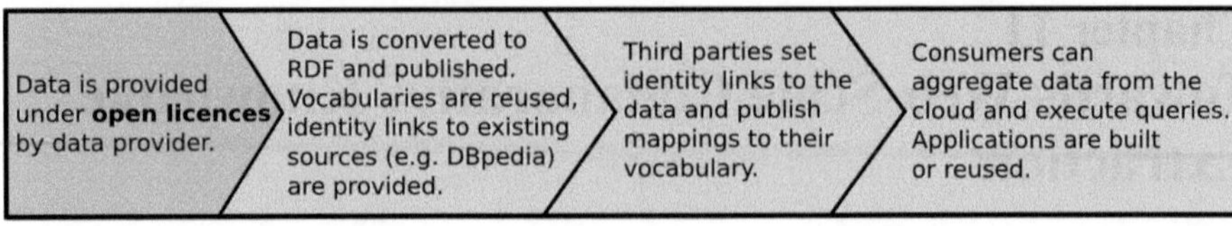


**Fig. 11.1** Summary of the above-mentioned methodologies for publishing and exploiting Linked Data [10]. The data provider is only required to make data available under an open license (*left-most step*). The remaining, data integration steps can be contributed by third parties and data consumers

a strong incentive for collaboration and contribution by third parties. Bizer [5] argues that with RDF the overall data integration effort can be "split between data publishers, third parties, and the data consumer", a claim that can be substantiated by looking at the evolution of many large data sets constituting the LOD cloud. We outline some stages of the linked data publication and refinement (cf. [1, 4, 5]) in Fig. 11.1 and discuss these in more detail throughout this article.

### Natural Language Processing

In addition to the increasing availability of open, structured and interlinked data, we are currently observing a plethora of *Natural Language Processing* (NLP) tools and services being made available and new ones appearing almost on a weekly basis. Some examples of web services providing just *Named Entity Recognition* (NER) services are *Zemanta*,[3] *OpenCalais*,[4] *Ontos*,[5] *Enrycher*,[6] *Extractiv*,[7] *Alchemy API*.[8] Similarly, there are tools and services for language detection, part-of-speech (POS) tagging, text classification, morphological analysis, relationship extraction, sentiment analysis and many other NLP tasks. Each of the tools and services has its particular strengths and weaknesses, but exploiting the strengths and synergistically combining different tools is currently an extremely cumbersome and time consuming task. The programming interfaces and result formats of the tools have to be analyzed and differ often to a great extend. Also, once a particular set of tools is integrated this integration is *not reusable* by others.

We argue that simplifying the interoperability of different NLP tools performing similar but also complementary tasks will facilitate the comparability of results, the building of sophisticated NLP applications as well as the synergistic combination

[3]http://www.zemanta.com/
[4]http://www.opencalais.com/
[5]http://www.ontos.com/
[6]http://enrycher.ijs.si/
[7]http://extractiv.com/
[8]http://www.alchemyapi.com/

of tools. Ultimately, this might yield a boost in precision and recall for common NLP tasks. Some first evidence in that direction is provided by tools such as *RDFaCE* [20], *Spotlight* and *Fox*,[9] which already combine the output from several backend services and achieve superior results.

Another important factor for improving the quality of NLP tools is the availability of large quantities of qualitative background knowledge on the currently emerging Web of Linked Data [1]. Many NLP tasks can greatly benefit from making use of this wealth of knowledge being available on the Web in structured form as *Linked Open Data* (LOD). The precision and recall of Named Entity Recognition, for example, can be boosted when using background knowledge from *DBpedia*, *Geonames* or other LOD sources as crowdsourced and community-reviewed and timely-updated gazetteers. Of course the use of gazetteers is a common practice in NLP. However, before the arrival of large amounts of Linked Open Data their creation, curation and maintenance in particular for multi-domain NLP applications was often impractical.

The use of LOD background knowledge in NLP applications poses some particular challenges. These include: *identification* – uniquely identifying and reusing identifiers for (parts of) text, entities, relationships, NLP concepts and annotations etc.; *provenance* – tracking the lineage of text and annotations across tools, domains and applications; *semantic alignment* – tackle the semantic heterogeneity of background knowledge as well as concepts used by different NLP tools and tasks.

### NLP Interchange Format

In order to simplify the combination of tools, improve their interoperability and facilitating the use of Linked Data, we developed the NLP Interchange Format (NIF). NIF addresses the interoperability problem on three layers: the *structural*, *conceptual* and *access* layer. NIF is based on a Linked Data enabled URI scheme for identifying elements in (hyper-)texts (structural layer) and a comprehensive ontology for describing common NLP terms and concepts (conceptual layer). NIF-aware applications will produce output (and possibly also consume input) adhering to the NIF ontology as REST services (access layer). Other than more centralized solutions such as *UIMA* and *GATE*, NIF enables the creation of heterogeneous, distributed and loosely coupled NLP applications, which use the Web as an integration platform. Another benefit is, that a NIF wrapper has to be only created once for a particular tool, but enables the tool to interoperate with a potentially large number of other tools without additional adaptations. Ultimately, we envision an ecosystem of NLP tools and services to emerge using NIF for exchanging and integrating rich annotations.

The remainder of this article is structured as follows: In the next section, we will take up the cudgels on behalf of open licenses and RDF and give relevant

[9]http://aksw.org/Projects/FOX

background information and facts about the used technologies and the current state of the Web of Data. We will especially elaborate on the following aspects: The importance of open licenses and open access as an enabler for collaboration; the ability to interlink data on the Web as a key feature of RDF; a discussion about scalability and decentralization; as well as an introduction on how conceptual interoperability can be achieved by (1) re-using vocabularies and (2) agile ontology development (3) meetings to refine and adapt ontologies (4) tool support to enrich ontologies and match schemata. In Sect. 11.3, we will describe three data sets that were created by a knowledge extraction process and maintained collaboratively by a community of stakeholders. Especially, we will focus on DBpedia's[10] *Mappings Wiki*[11] (which governs the extraction from Wikipedia), the mapping approach of *LinkedGeoData*[12] (extracted from OpenStreetMaps) and the configurable extraction of RDF from Wiktionary. While Sect. 11.4 introduces key concepts of the NLP Interchange Format (NIF), Sect. 11.5 shows how to achieve interoperability between NIF and existing annotation ontologies which are modelling different layers of NLP annotations. Section 11.5, also shows how extensions of NIF have the potential to connect the Giant Global Graph (especially the resources introduced in Sect. 11.3), the Web of Documents and NLP tool output. The article concludes with a short discussion and an outlook on future work in Sect. 11.6.

## 11.2 Background

### *11.2.1 Open Licenses, Open Access and Collaboration*

DBpedia, FlickrWrappr, 2000 U.S. Census, LinkedGeoData, LinkedMDB are some prominent examples of LOD data sets, where the conversion, interlinking, as well as the hosting of the links and the converted RDF data has been completely provided by third parties with no effort and cost for the original data providers.[13] DBpedia [23], for example, was initially converted to RDF solely from the openly licensed database dumps provided by Wikipedia. With Openlink Software a company supported the project by providing hosting infrastructure and a community evolved, which created links and applications. Although it is difficult to determine whether open licenses are a necessary or sufficient condition for the collaborative evolution of a data set, the opposite is quite obvious: *Closed* licenses or *unclearly licensed* data are an impediment to an architecture which is focused on (re-)publishing and linking of data. Several data sets, which were converted to RDF could not be re-published due to licensing issues. Especially, these include the Leipzig Corpora Collection (LCC) [28] and the RDF data used in the TIGER Corpus

[10] http://dbpedia.org

[11] http://mappings.dbpedia.org/

[12] http://linkedgeodata.org

[13] More data sets can be explored here: http://thedatahub.org/tag/published-by-third-party

Navigator [13]. Very often (as it is the case for the previous two examples), the reason for closed licenses is the strict copyright of the primary data (such as newspaper texts) and researchers are unable to publish their annotations and resulting data. The open part of the American National Corpus (OANC[14]) on the other hand has been converted to RDF and was re-published successfully using the POWLA ontology [9]. Thus, the work contributed to OANC was directly reusable by other scientists and likewise the same accounts for the RDF conversion.

Note that the *Open* in Linked Open Data refers mainly to *open access*, i.e. retrievable using the HTTP protocol.[15] Only around 18 % of the data sets of the LOD cloud provide clear licensing information at all.[16] Of these 18 % an even smaller amount is considered *open* in the sense of the open definition[17] coined by the Open Knowledge Foundation. One further important criteria for the success of a collaboration chain is whether the data set explicitly allows to redistribute data. Very often self-made licenses allow scientific and non-commercial use, but do not specify how redistribution is handled.

### *11.2.2 RDF as a Data Model*

RDF as a data model has distinctive features, when compared to its alternatives. Conceptually, RDF is close to the widely used Entity-Relationship Diagrams (ERD) or the Unified Modeling Language (UML) and allows to model entities and their relationships. XML is a serialization format, that is useful to (de-)serialize data models such as RDF. Major drawbacks of XML and relational databases are the lack of (1) global identifiers such as URIs, (2) standardized formalisms to explicitly express links and mappings between these entities and (3) mechanisms to publicly access, query and aggregate data. Note that (2) can not be supplemented by transformations such as XSLT, because the linking and mappings are implicit. All three aspects are important to enable ad-hoc collaboration. The resulting technology mix provided by RDF allows any collaborator to join her data into the decentralized data network employing the HTTP protocol which immediate benefits herself and others. In addition, features of OWL can be used for inferencing and consistency checking. OWL – as a modelling language – allows, for example, to model transitive properties, which can be queried on demand, without expanding the size of the data via backward-chaining reasoning. While XML can only check for validity, i.e. the occurrence and order of data items (elements and attributes), consistency checking allows to verify, whether a data set adheres to the semantics imposed by the formal definitions of the used ontologies.

---

[14]http://www.anc.org/OANC/

[15]http://richard.cyganiak.de/2007/10/lod/#open

[16]http://www4.wiwiss.fu-berlin.de/lodcloud/state/#license

[17]http://opendefinition.org/

### *11.2.3 Performance and Scalability*

RDF, its query language SPARQL and its logical extension OWL provide features and expressivity that go beyond relational databases and simple graph-based representation strategies. This expressivity poses a performance challenge to query answering by RDF triples stores, inferencing by OWL reasoners and of course the combination thereof. Although the scalability is a constant focus of RDF data management research,[18] the primary strength of RDF is its flexibility and suitability for data integration and not superior performance for specific use cases. Many RDF-based systems are designed to be deployed in parallel to existing high-performance systems and not as a replacement. An overview over approaches that provide Linked Data and SPARQL on top of relational database systems, for example, can be found in [2]. The NLP Interchange Format (cf. Sect. 11.4) allows to express the output of highly optimized NLP systems (e.g. UIMA) as RDF/OWL. The architecture of the Data Web, however, is able to scale in the same manner as the traditional WWW as the nodes are kept in a de-centralized way and new nodes can join the network any time and establish links to existing data. Data Web search engines such as *Swoogle*[19] or *Sindice*[20] index the available structured data in a similar way as Google does with the text documents on the Web and provide keyword-based query interfaces.

### *11.2.4 Conceptual Interoperability*

While RDF and OWL as a standard for a common data format provide structural (or syntactical) interoperability, conceptual interoperability is achieved by globally unique identifiers for entities, properties and classes, that have a fixed meaning. These unique identifiers can be interlinked via `owl:sameAs` on the entity-level, re-used as properties on the vocabulary level and extended or set equivalent via `rdfs:subClassOf` or `owl:equivalentClass` on the schema-level. Following the ontology definition of [12], the aspect that ontologies are a "shared conceptualization" stresses the need to collaborate to achieve agreement. On the class and property level RDF and OWL give users the freedom to reuse, extend and relate to other work in their own conceptualization. Very often, however, it is the case that groups of stakeholders actively discuss and collaborate in order to form some kind of agreement on the meaning of identifiers as has been described in [16]. In the following, we will give four examples to elaborate how conceptual interoperability is achieved:

---

[18]http://factforge.net or http://lod.openlinksw.com provide SPARQL interfaces to query billions of aggregated facts.

[19]http://swoogle.umbc.edu

[20]http://sindice.com

- In a knowledge extraction process (e.g. when converting relational databases to RDF) vocabulary identifiers can be reused during the extraction process. Especially community-accepted vocabularies such as FOAF, SIOC, Dublin Core and the DBpedia Ontology are suitable candidates for reuse as this leads to conceptual interoperability with all applications and databases that also use the same vocabularies. This aspect was the rationale for designing Triplify [2], where the SQL syntax was extended to map query results to existing RDF vocabularies.
- During the creation process of ontologies, direct collaboration can be facilitated with tools that allow agile ontology development such as *OntoWiki*, *Semantic Mediawiki* or the *DBpedia Mappings Wiki*.[21] This way, conceptual interoperability is achieved by a distributed group of stakeholders, who work together over the Internet. The created ontology can be published and new collaborators can register and get involved to further improve the ontology and tailor it to their needs.
- In some cases, real life meetings are established, e.g. in the form of Vo(cabulary) Camps, where interested people meet to discuss and refine vocabularies. VoCamps can be found and registered on http://vocamp.org.
- A variety of RDF tools exists, which aid users in creating links between individual data records as well as in mapping ontologies.
- Semi-automatic enrichment tools such as ORE [7] allow to extend ontologies based on the entity-level data.

## 11.3 Collaborative Knowledge Extraction

Knowledge Extraction is the creation of knowledge from structured (relational databases, XML) and unstructured (text, documents, images) sources. The resulting knowledge needs to be in a machine-readable and machine-interpretable format and must represent knowledge in a manner that unambiguously defines its meaning and facilitates inferencing [31]. By this definition, almost all RDF/OWL knowledge bases that where created from "legacy" sources can be considered as being created by a knowledge extraction process. In this section, we will focus on three prominent knowledge bases that fall in this category: *DBpedia*, *LinkedGeoData* and *Wiktionary2RDF*. The crowd-sourcing process that yielded these knowledge bases stretched over different stages of their development process:

- All three knowledge bases originate from crowd-sourced wiki approaches, i.e. *Wikipedia*, *OpenStreetMaps* and *Wiktionary*.
- The knowledge extraction process itself is crowd-sourced: (1) DBpedia provides a mappings wiki, which allows to define extraction rules on Wikipedia's infoboxes; (2) LinkedGeoData provides a mapping XML file from terms

[21] http://mappings.dbpedia.org

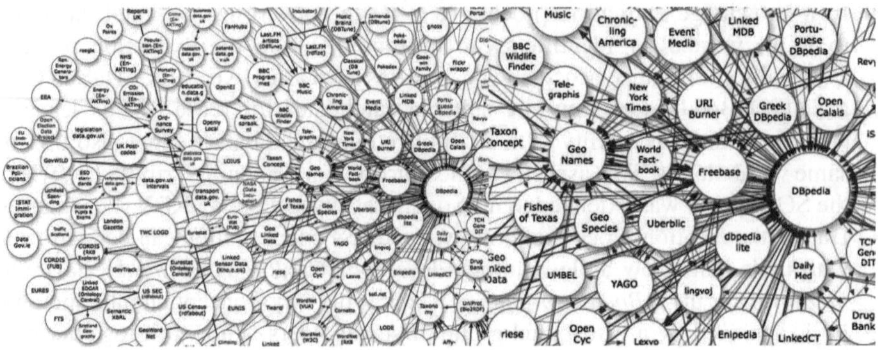


**Fig. 11.2** Excerpt of the data sets interlinked with DBpedia (Source: http://lod-cloud.net with kind permission of Anja Jentzsch and Richard Cyganiak)

occurring in OpenStreetMaps to RDF properties; (3) Wiktionary2RDF allows domain experts to create and maintain wrappers for language-specific Wiktionary editions

- Each project has a mailing list and a bug tracker, where data consumers can report bugs and discuss modelling issues. Occasionally, patches are directly provided by the community.
- Third parties have provided link sets for inclusion into the data set itself (e.g. DBpedia contains links to Yago, WordNet, Umbel).
- Third parties publish links into one of the projects alongside their own data sets, as can be seen on the LOD cloud image.

Due to continuous reviewing by a large community of stakeholders, DBpedia has evolved into a paragon of best practices for linked data. The same accounts to a lesser extent for LinkedGeoData and Wiktionary2RDF as both projects are much younger.

## *11.3.1 DBpedia*

DBpedia [23] is a community effort to extract structured information from Wikipedia and to make this information available on the Web. The main output of the DBpedia project is a data pool that (1) is widely used in academics as well as industrial environments, that (2) is curated by the community of Wikipedia and DBpedia editors, and that (3) has become a major crystallization point and a vital infrastructure for the Web of Data. DBpedia is one of the most prominent Linked Data examples and presently the largest hub in the Web of Linked Data (Fig. 11.2). The extracted RDF knowledge from the English Wikipedia is published and interlinked according to the Linked Data principles and made available under the same license as Wikipedia (cc-by-sa).

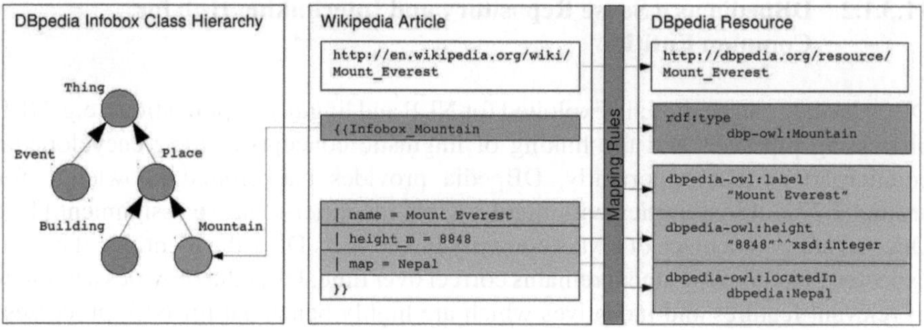


**Fig. 11.3** Rule-based manipulation of extracted data in DBpedia Mappings Wiki [15]

In its current version 3.8 DBpedia contains more than 3.77 million things, of which 2.35 million are classified in a consistent ontology, including 764,000 persons, 573,000 places, 112,000 music albums, 72,000 films, 18,000 video games, 192,000 organizations, 202,000 species and 5,500 diseases. The DBpedia data set features labels and abstracts in up to 111 different languages; 8.0 million links to images and 24.4 million links to external Web pages; 27.2 million data links into other RDF datasets, and 55.8 million Wikipedia categories. The dataset consists of 1.89 billion RDF triples out of which 400 million were extracted from the English edition of Wikipedia and 1.46 billion were extracted from other Wikipedia language editions and around 27 million links to external datasets [6].

Currently, the DBpedia Ontology is maintained in a crowd-sourcing approach and thus freely editable on a *Mappings Wiki*[22]: each OWL class can be modeled on a Wiki page and the `subClassOf` axioms (shown on the left side of Fig. 11.3) are created manually. The classification of articles according to the ontology classes is based on rules. In Fig. 11.3, the article is classified as `dbp-owl:Mountain`, because it contains the Infobox "Infobox_Mountain" in its source.

#### 11.3.1.1 Internationalization of DBpedia

While early versions of the DBpedia Information Extraction Framework (DIEF) used only the English Wikipedia as their sole source, its focus later shifted integrate information from many different Wikipedia editions. During the fusion process, however, language-specific information was lost or ignored. The aim of the current research in internationalization [21,22] is to establish best practices (complemented by software) that allow the DBpedia community to easily generate, maintain and properly interlink language-specific DBpedia editions. In a first step, we realized a language-specific DBpedia version using the Greek Wikipedia [21]. Soon, the approach was generalized and applied to 15 other Wikipedia language editions [6]

[22]http://mappings.dbpedia.org

#### 11.3.1.2 DBpedia as a Sense Repository and Interlinking Hub for Common Entities

DBpedia data can be directly exploited for NLP and linguistic applications, e.g. NLP processing pipelines and the linking of linguistic concepts to their encyclopedic counterparts. Most importantly, DBpedia provides background knowledge for around 3.77 million entities with highly stable identifier-to-sense assignment [17]: Once an entity or a piece of text is correctly linked to its DBpedia identifier, it can be expected that this assignment remains correct over time. DBpedia provides a number of relevant features and incentives which are highly beneficial for NLP processes: (1) the senses are curated in a crowd-sourced community process and remain stable; (2) Wikipedia is available in multiple languages; (3) data in Wikipedia and DBpedia[23] remains up-to-date and users can influence the knowledge extraction process in the Mappings Wiki; (4) the open licensing model allows all contributors to freely exploit their work.

Note that most of the above-mentioned properties are inherited from Wikipedia. The additional benefit added by DBpedia is the standardization and re-usability of the data for NLP developers. Especially, the community around DBpedia Spotlight has specialized in providing datasets refined from DBpedia that are directly tailored towards NLP processes [26].

#### 11.3.1.3 DBpedia Spotlight

The band-width of applications of DBpedia data in NLP research is immense, but here, we focus on a single example application, DBpedia Spotlight by Mendes et al. [25], a tool for annotating mentions of DBpedia resources in text, providing a solution for linking unstructured information sources to the Linked Open Data cloud through DBpedia. DBpedia Spotlight performs named-entity extraction, including entity detection and Name Resolution. Several strategies are used to generate candidate sets and automatically select a resource based on the context of the input text.

The most basic candidate generation strategy in DBpedia Spotlight is based on a dictionary of known DBpedia resource names extracted from page titles, redirects and disambiguation pages. These names are shared in the DBpedia Lexicalization dataset.[24] The graph of labels, redirects and disambiguations in DBpedia is used to extract a lexicon that associates multiple surface forms to a resource and interconnects multiple resources to an ambiguous name. One recent development is the internationalization of DBpedia Spotlight, and the development of entity disambiguation services for German and Korean has begun. Other languages will follow soon including the evaluation of the performance of the algorithms in other languages.

---

[23]For DBpedia Live see http://live.dbpedia.org/

[24]http://wiki.dbpedia.org/Lexicalizations

### *11.3.2 LinkedGeoData*

With the *OpenStreetMap* (OSM)[25] project, a rich source of spatial data is freely available. It is currently used primarily for rendering various map visualizations, but has the potential to evolve into a crystallization point for spatial Web data integration (e.g. as gazetteer for NLP applications focusing on recognition of spatial entities). The goal of the *LinkedGeoData* (LGD) [30] project is to lift OSM's data into the Semantic Web infrastructure. This simplifies real-life information integration and aggregation tasks that require comprehensive background knowledge related to spatial features. Such tasks might include, for example, to locally depict the offerings of the bakery shop next door, to map distributed branches of a company, or to integrate information about historical sights along a bicycle track.

The majority of LGD data, which comprises 15 billion spatial facts, is obtained by converting data from the popular OpenStreetMap community project to RDF and deriving a lightweight ontology from it. Furthermore, interlinking is performed with *DBpedia*, *GeoNames* and other datasets as well as the integration of icons and multilingual class labels from various sources. As a side effect, LGD is striving for the establishment of an OWL vocabulary with the purpose of simplifying exchange and reuse of geographic data. Besides coarse-grained spatial entities such as countries, cities and roads LGD also contains millions of buildings, parking lots, hamlets, restaurants, schools, fountains or recycling trash bins. Since the initial LGD release in [3], a substantial effort was invested in maintaining and improving LinkedGeoData, which includes improvements of the project infrastructure, the generated ontology, and data quality in general. To date, the LinkedGeoData project comprises in particular:

- A flexible system for mapping OpenStreetMap data to RDF including support for nice URIs (camel case), typed literals, language tags, and a mapping of the OSM data to classes and properties.
- Support for ways: Ways are OpenStreetMap entities used for modelling things such as streets but also areas. The geometry of a way (a line or a polygon) is stored in a literal of the corresponding RDF resource, which makes it easy to e.g. display such a resource on a map. Furthermore, all nodes referenced by a way are available both via the Linked Data interface and the SPARQL endpoints.
- A *REST interface* with integrated search functions as well as a publicly accessible *live SPARQL endpoint* that is being interactively updated with the minutely changesets that OpenStreetMap publishes.
- A simple *republication method* of the corresponding RDF changesets so that LinkedGeoData data consumers can replicate the LinkedGeoData store.
- Direct *interlinking* with *DBpedia*, *GeoNames* and the *UN FAO* data. Integration of appropriate *icons and multilingual labels* for LinkedGeoData ontology elements from external sources.

[25] http://openstreetmap.org

- The spatial-semantic user interface *LinkedGeoData browser* as well as the *Vicibit* application to facilitate the integration of LGD facet views in external web pages.

In essence, the transformation and publication of the OpenStreetMap data according to the Linked Data principles in LinkedGeoData adds a new dimension to the Data Web: spatial data can be retrieved and interlinked on an unprecedented level of granularity. For NLP applications, the LinkedGeoData resource opens possibilities previously hardly thinkable. For example, entity references in text such as 'the bakery on Broad Street' can possibly be resolved by using the vast knowledge comprised in LGD's 15 billion spatial facts.

### *11.3.3 Wiktionary2RDF*

*Wiktionary* is one of the biggest collaboratively created lexical-semantic and linguistic resources available, written in 171 languages (of which approximately 147 can be considered active,[26]) containing information about hundreds of spoken and even ancient languages. For example, the English *Wiktionary* contains nearly three million words.[27] A *Wiktionary* page provides for a lexical word a hierarchical disambiguation to its language, part of speech, sometimes etymologies and most prominently senses. Within this tree numerous kinds of linguistic properties are given, including synonyms, hyponyms, hyperonyms, example sentences, links to Wikipedia and many more. Meyer and Gurevych [27] gave a comprehensive overview on why this dataset is so promising and how the extracted data can be automatically enriched and consolidated. Aside from building an upper-level ontology, one can use the data to improve NLP solutions, using it as comprehensive background knowledge. The noise should be lower when compared to other automatic generated text corpora (e.g. by web crawling) as all information in *Wiktionary* is entered and curated by humans. Opposed to expert-built resources, the openness attracts a huge number of editors and thus enables a faster adaption to changes within the language.

The fast changing nature together with the fragmentation of the project into *Wiktionary* language editions (*WLE*) with independent layout rules (ELE) poses the biggest problem to the automated transformation into a structured knowledge base. We identified this as a serious problem: Although the value of *Wiktionary* is known and usage scenarios are obvious, only some rudimentary tools exist to extract data from it. Either they focus on a specific subset of the data or they only cover one or two WLE. The development of a flexible and powerful tool is challenging to be accommodated in a mature software architecture and has been neglected in the past. Existing tools can be seen as adapters to single WLE – they are hard to maintain and there are too many languages, that constantly change. Each change in the *Wiktionary* layout requires a programmer to refactor complex code. The last years showed, that

[26] http://s23.org/wikistats/wiktionaries_html.php

[27] See http://en.wiktionary.org/wiki/semantic for a simple example page

only a fraction of the available data is extracted and there is no comprehensive RDF dataset available yet. The key question is: Can the lessons learned by the successful DBpedia project be applied to *Wiktionary*, although it is fundamentally different from Wikipedia? The critical difference is that only word forms are formatted in infobox-like structures (e.g. tables). Most information is formatted covering the complete page with custom headings and often lists. Even the infoboxes itself are not easily extractable by default DBpedia mechanisms, because in contrast to DBpedias *one entity per page* paradigm, *Wiktionary* pages contain information about *several* entities forming a complex graph, i.e. the pages describe the lexical word, which occurs in several languages with different senses per part of speech and most properties are defined *in context* of such child entities. Opposed to the currently employed classic and straight-forward approach (implementing software adapters for scraping), Wiktionary2RDF employs a declarative mediator/wrapper pattern. The aim is to enable non-programmers (the community of adopters and domain experts) to tailor and maintain the WLE wrappers themselves. We created a simple XML dialect to encode the "entry layout explained" (ELE) guidelines and declare triple patterns, that define how the resulting RDF should be built. This configuration is interpreted and run against *Wiktionary* dumps. The resulting dataset is open in every aspect and hosted as linked data.[28] Furthermore the presented approach can be extended easily to interpret (or *triplify*) other MediaWiki installations or even general document collections, if they follow a global layout.

In order to conceive a flexible, effective and efficient solution, we survey in this section the challenges associated with Wiki syntax, *Wiktionary* and large-scale extraction.

#### 11.3.3.1 Processing Wiki Syntax

Pages in *Wiktionary* are formatted using the *wikitext* markup language.[29] Operating on the parsed HTML pages, rendered by the *MediaWiki engine*, does not provide any significant benefit, because the rendered HTML does not add any valuable information for extraction. Processing the database backup XML dumps[30] instead, is convenient as we could reuse the DBpedia extraction framework[31] in our implementation. The framework mainly provides input and output handling and also has built-in multi-threading by design. Actual features of the wikitext syntax are not notably relevant for the extraction approach, but we will give a brief introduction to the reader, to get familiar with the topic. A wiki page is formatted using the lightweight (easy to learn, quick to write) markup language *wikitext*. Upon request of a page, the MediaWiki engine renders this to an HTML page and sends it to the

[28] http://wiktionary.dbpedia.org/

[29] http://www.mediawiki.org/wiki/Markup_spec

[30] http://dumps.wikimedia.org/backup-index.html

[31] http://wiki.dbpedia.org/Documentation

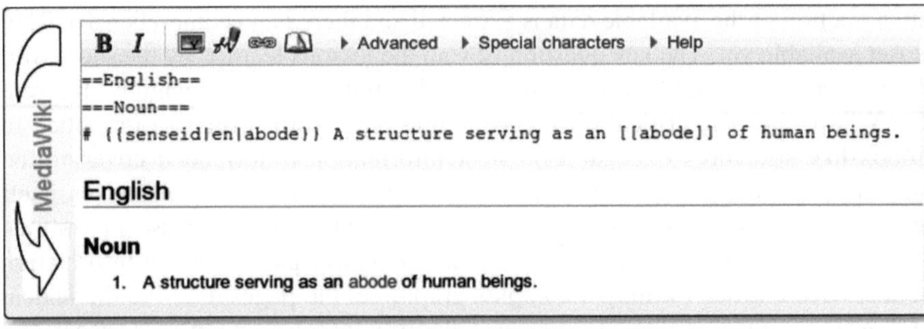


**Fig. 11.4** An excerpt of the *Wiktionary* page *house* with the rendered HTML

user's browser. An excerpt of the *Wiktionary* page *house* and the resulting rendered page are shown in Fig. 11.4.

The markup == is used to denote headings, # denotes a numbered list (∗ for bullets), `[[link label]]` denotes links and {{}} calls a template. Templates are user-defined rendering functions that provide shortcuts aiming to simplify manual editing and ensuring consistency among similarly structured content elements. In MediaWiki, they are defined on special pages in the `Template:` namespace. Templates can contain any wikitext expansion, HTML rendering instructions and placeholders for arguments. In the example page in Fig. 11.4, the `senseid` template[32] is used, which does nothing being visible on the rendered page, but adds an id attribute to the HTML `li`-tag (which is created by using #). If the English *Wiktionary* community decides to change the layout of senseid definitions at some point in the future, only a single change to the template definition is required. Templates are used heavily throughout *Wiktionary*, because they substantially increase maintainability and consistency. But they also pose a problem to extraction: on the unparsed page only the template name and its arguments are available. Mostly this is sufficient, but if the template adds static information or conducts complex operations on the arguments (which is fortunately rare), the template result can only be obtained by a running MediaWiki installation hosting the pages. The resolution of template calls at extraction time slows the process down notably and adds additional uncertainty.

#### 11.3.3.2 Wiktionary

*Wiktionary* has some unique and valuable properties:

- **Crowd-sourced**. *Wiktionary* is community edited, instead of expert-built or automatically generated from text corpora. Depending on the activeness of its

[32] http://en.wiktionary.org/wiki/Template:senseid

community, it is up-to-date to recent changes in the language, changing perspectives or new research. The editors are mostly semi-professionals (or guided by one) and enforce a strict editing policy. Vandalism is reverted quickly and bots support editors by fixing simple mistakes and adding automatically generated content. The community is smaller than Wikipedia's but still quite vital (between 50 and 80 very active editors with more than 100 edits per month for the English *Wiktionary* in 2012.[33])

- **Multilingual**. The data is split into different Wiktionary Language Editions (WLE, one for each language). This enables the independent administration by communities and leaves the possibility to have different perspectives, focus and localization. Simultaneously one WLE describes multiple languages; only the representation language is restricted. For example, the German *Wiktionary* contains German description of German words as well as German descriptions for English, Spanish or Chinese words. Particularly the linking across languages shapes the unique value of *Wiktionary* as a rich multi-lingual linguistic resource. Especially the WLE for not widely spread languages are valuable, as corpora might be rare and experts are hard to find.
- **Feature rich**. As stated before, *Wiktionary* contains for each lexical word (A lexical word is just a string of characters and has no disambiguated meaning yet) a disambiguation regarding language, part of speech, etymology and senses. Numerous additional linguistic properties exist normally for each part of speech. Such properties include word forms, taxonomies (hyponyms, hyperonyms, synonyms, antonyms) and translations. Well maintained pages (e.g. frequent words) often have more sophisticated properties such as derived terms, related terms and anagrams.
- **Open license**. All the content is dual-licensed under both the *Creative Commons CC-BY-SA 3.0 Unported License*[34] as well as the *GNU Free Documentation License (GFDL)*.[35] All the data extracted by our approach falls under the same licenses.
- **Big and growing**. English contains 2.9 M pages, French 2.1 M, Chinese 1.2 M, German 0.2 M. The overall size (12 M pages) of *Wiktionary* is in the same order of magnitude as Wikipedia's size (20 M pages).[36] The number of edits per month in the English *Wiktionary* varies between 100 k and 1M – with an average of 200 k for 2012 so far. The number of pages grows – in the English *Wiktionary* with approx. 1 k per day in 2012.[37]

---

[33]http://stats.wikimedia.org/wiktionary/EN/TablesWikipediaEN.htm

[34]http://en.wiktionary.org/wiki/Wiktionary:Text_of_Creative_Commons_Attribution-ShareAlike_3.0_Unported_License

[35]http://en.wiktionary.org/wiki/Wiktionary:GNU_Free_Documentation_License

[36]http://meta.wikimedia.org/wiki/Template:Wikimedia_Growth

[37]http://stats.wikimedia.org/wiktionary/EN/TablesWikipediaEN.htm

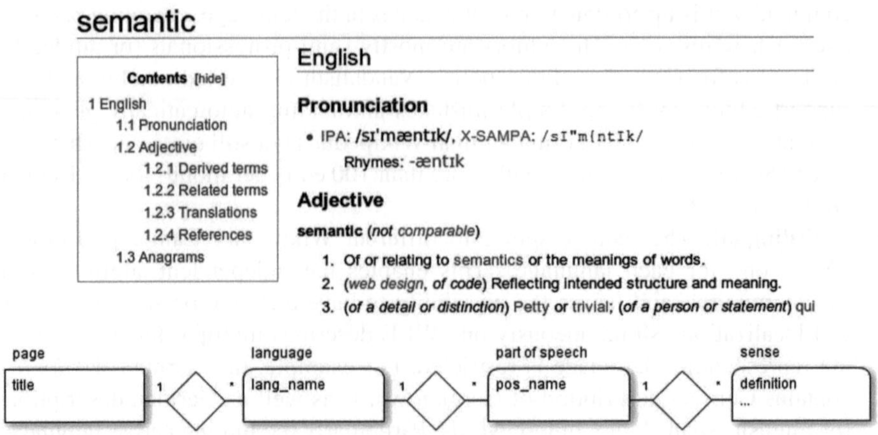


**Fig. 11.5** Example page http://en.wiktionary.org/wiki/semantic and underlying schema (only valid for the English *Wiktionary*, other WLE might look very different)

The most important resource to understand how *Wiktionary* is organized are the *Entry Layout Explained* (ELE) help pages. As described above, a page is divided into sections that separate languages, part of speech etc. The table of content on the top of each page also gives an overview of the hierarchical structure. This hierarchy is already very valuable as it can be used to disambiguate a lexical word. The schema for this tree is restricted by the ELE guidelines.[38] The entities illustrated in Fig. 11.5 of the ER diagram will be called *block* from now on. The schema can differ between WLEs and normally evolves over time.

#### 11.3.3.3 Wiki-Scale Data Extraction

The above listed properties that make *Wiktionary* so valuable, unfortunately pose a serious challenge to extraction and data integration efforts. Conducting an extraction for specific languages at a fixed point in time is indeed easy, but it eliminates some of the main features of the source. To fully synchronize a knowledge base with a community-driven source, one needs to make distinct design choices to fully capture all desired benefits. MediaWiki was designed to appeal to non-technical editors and abstains from intensive error checking as well as formally following a grammar – the community gives itself just layout guidelines. One will encounter fuzzy modelling and unexpected information. Editors often see no problem with such "noise" as long as the page's visual rendering is acceptable. Overall, the main challenges can be summed up as (1) the constant and frequent changes to data *and schema*, (2) the heterogeneity in WLE schemas and (3) the human-centric nature of a wiki.

[38]For English see http://en.wiktionary.org/wiki/Wiktionary:ELE

**Table 11.1** Statistical comparison of extractions for different languages. XML lines measures the number of lines of the XML configuration files

| Language | *#words* | *#triples* | *#resources* | *#predicates* | *#senses* | *XML lines* |
|---|---|---|---|---|---|---|
| *en* | 2,142,237 | 28,593,364 | 11,804,039 | 28 | 424,386 | 930 |
| *fr* | 4,657,817 | 35,032,121 | 20,462,349 | 22 | 592,351 | 490 |
| *ru* | 1,080,156 | 12,813,437 | 5,994,560 | 17 | 149,859 | 1,449 |
| *de* | 701,739 | 5,618,508 | 2,966,867 | 16 | 122,362 | 671 |

#### 11.3.3.4 Resulting Data

The extraction has been conducted as a proof-of-concept on four major WLE: The English, French, German and Russian *Wiktionary*. The datasets combined contain more than 80 million facts. The data is available as N-Triples dumps,[39] Linked Data,[40] via the *Virtuoso Faceted Browser*[41] or a SPARQL endpoint.[42] Table 11.1 compares the size of the datasets from a quantitative perspective.

The statistics show, that the extraction produces a vast amount of data with broad coverage, thus resulting in one of the largest lexical linked data resource. There might be partially data quality issues with regard to missing information (for example the number of *words with senses* seems to be relatively low intuitively), but detailed quality analysis has yet to be done.

**Community Process.** For each of the languages, a configuration XML file was created, which describes how the Wiktionary2RDF framework should transform the Wiki syntax into triples. Existing configuration files are public and can be altered by everybody without touching the source code of the project and patches can be submitted back into the project. Additionally, they serve as templates to aid creation of config files for more languages by a community. We can identify three sources for low data quality during the extraction process: (1) An error or missing feature in the extraction algorithm of the software framework (2) An erroneous or incomplete configuration file (3) a *Wiktionary* page that does not adhere to the ELE guidelines. While the Wiktionary2RDF project requires a developer for the first point, two and three can be fixed by domain experts and *Wiktionary* users. Providing a live extraction, similar to DBpedia also has the potential to become a great supportive resource to help editors of *Wiktionary* in spotting inconsistencies.

[39]http://downloads.dbpedia.org/wiktionary

[40]for example http://wiktionary.dbpedia.org/resource/dog

[41]http://wiktionary.dbpedia.org/fct

[42]http://wiktionary.dbpedia.org/sparql

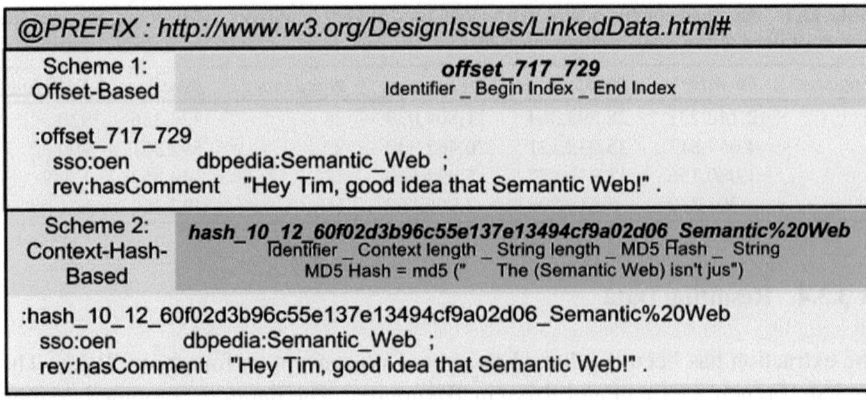


**Fig. 11.6** NIF URI schemes: Offset (*top*) and context-hashes (*bottom*) are used to create identifiers for strings [14]

## 11.4 The NLP Interchange Format

The motivation behind NIF is to allow NLP tools to exchange annotations about documents in RDF. Hence, the main prerequisite is that parts of the documents (i.e. strings) are referenceable by URIs, so that they can be used as subjects in RDF statements. We call an algorithm to create such identifiers *URI Scheme*: For a given text $t$ (a sequence of characters) of length $|t|$ (number of characters), we are looking for a *URI Scheme* to create a URI, that can serve as a *unique* identifier for a substring $s$ of $t$ (i.e. $|s| \leq |t|$). Such a substring can (1) consist of adjacent characters only and it is therefore a unique character sequence within the text, if we account for parameters such as context and position or (2) derived by a function which points to several substrings as defined in (1).

NIF provides two URI schemes, which can be used to represent strings as RDF resources. In this section, we focus on the first scheme using offsets. In the top part of Fig. 11.6, two triples are given that use the following URI as subject:
http://www.w3.org/DesignIssues/LinkedData.html#offset_717_729

According to the above definition, the URI points to a substring of a given text $t$, which starts at index 717 until the index 729.

For the URI creation scheme, there are three basic requirements – *uniqueness*, *ease of implementation* and *URI stability* during document changes. Since these three conflicting requirements can not be easily addressed by a single URI creation scheme, NIF defines two URI schemes, which can be chosen depending on which requirement is more important in a certain usage scenario. Naturally further schemes for more specific use cases can be developed easily. After discussing some guidelines on the selection of URI namespaces, we explain in this section how stable URIs can be minted for parts of documents by using *offset-based* and *context-hash* based schemes (see Fig. 11.6 for examples).

### 11.4.1 Namespace Prefixes

A NIF URI is constructed from a namespace prefix and the actual *identifier* (e.g. "offset_717_729"). Depending on the selected context, different prefixes can be chosen. For practical reasons, it is recommended that the following guidelines should be met for NIF URIs: If we want to annotate a (web) resources, the whole content of the document is considered as `str:Context`, as explained in the next section, and it is straightforward to use the existing document URL as the basis for the prefix. The prefix should then either end with slash ('/') or hash ('#').[43]

Recommended prefixes for http://www.w3.org/DesignIssues/LinkedData.html are:

- http://www.w3.org/DesignIssues/LinkedData.html/
- http://www.w3.org/DesignIssues/LinkedData.html#

### 11.4.2 Offset-Based URIs

The offset-based URI scheme focuses on ease of implementation and is compatible with the position and range definition of RFC 5147 by Wilde and Duerst [32] (esp. Sect. 2.1.1) and builds upon it in terms of encoding and counting character positions (See [14] for a discussion). Offset-based URIs are constructed of three parts separated by an underscore '_': (1) a *scheme identifier*, in this case the string 'offset', (2) *start index*, (3) the *end index*. The indexes are counting the gaps between the characters starting from 0 as specified in RFC 5147 with the exception that the encoding is defined to be Unicode Normal Form C (NFC)[44] and counting is fixed on Unicode Code Units.[45] This scheme is easy and efficient to implement and the addressed string can be referenced unambiguously. Due to its dependency on start and end indexes, however, a substantial disadvantage of offset-based URIs is the *instability* with regard to changes in the document. In case of a document change (i.e. insertion or deletion of characters), all offset-based URIs after the position the change occurred become invalid. The context-hash-based scheme is explained in more detail by Hellmann et al. [14].

### 11.4.3 Usage of Identifiers in the String Ontology

We are able to fix the referent of NIF URIs in the following manner: To avoid ambiguity, NIF requires that the whole string of the document has to be included in the

[43]Note that with '/' the identifier is sent to the server during a request (e.g. Linked Data), while everything after '#' can only be processed by the client.

[44]http://www.unicode.org/reports/tr15/#Norm_Forms

[45]http://unicode.org/faq/char_combmark.html#7

RDF output as an `rdf:Literal` to serve as the reference point, which we will call *inside context* formalized using an OWL class called `str:Context`.[46] By typing NIF URIs as `str:Context` we are referring to the content only, i.e. an arbitrary grouping of characters forming a unit. The term *document* would be inappropriate to capture the real intention of this concept as `str:Context` could also be applied to a *paragraph* or a *sentence* and is **absolutely independent** upon the *wider context* in which the string is actually used such as a Web document reachable via HTTP.

We will distinguish between the notion of outside and inside context of a piece of text. The *inside context* is easy to explain and formalize, as it is the text itself and therefore it provides a *reference context* for each substring contained in the text (i.e. the characters before or after the substring). The *outside context* is more vague and is given by an outside observer, who might arbitrarily interpret the text as a "book chapter" or a "book section".

The class `str:Context` now provides a clear reference point for all other relative URIs used in this context and blocks the addition of information from a larger (outside) context. `str:Context` is therefore disjoint with `foaf:Document`, because labeling a context resource as a document is an information, which is not contained within the context (i.e. the text) itself. It is legal, however, to say that the string of the context occurs in (`str:occursIn`) a `foaf:Document`. Additionally, `str:Context` is a subclass of `str:String` and therefore its instances denote textual strings as well.

```
@prefix : <http://www.w3.org/DesignIssues/LinkedData.html#> .
@prefix str: <http://nlp2rdf.lod2.eu/schema/string/> .
:offset_0_26546
    rdf:type str:Context ;
# the exact retrieval method is left underspecified
    str:occursIn <http://www.w3.org/DesignIssues/LinkedData.html> ;
# [...] are all 26547 characters as rdf:Literal
    str:isString "[...]" .
:offset_717_729
    rdf:type str:String ;
    str:referenceContext :offset_0_26546 .
```

As mentioned in Sect. 11.4, NIF URIs are grounded on Unicode Characters using Unicode Normalization Form C counted in Code Units. For all resources of type `str:String`, the universe of discourse will then be the **words over the alphabet of Unicode characters** (sometimes called $\Sigma^*$). According to the "RDF Semantics W3C Recommendation", such an interpretation is considered a "semantic extension"[47] of RDF, because "extra semantic conditions" are "imposed on the meanings of terms".[48] This "semantic extension" allows – per definitionem – for an unambiguous interpretation of NIF by machines. In particular, the `str:isString` term points to the string that fixes the referent of the context. The meaning of

---

[46]for the resolution of prefixes, we refer the reader to http://prefix.cc

[47]http://www.w3.org/TR/2004/REC-rdf-mt-20040210/#urisandlit

[48]http://www.w3.org/TR/2004/REC-rdf-mt-20040210/#intro

a `str:Context` NIF URI is then exactly the string contained in the object of `str:isString`. Note that Notation 3 even permits literals as subjects of statements, a feature, which might even be adopted to RDF.[49]

## 11.5 Interoperability Between Different Layers of Annotations

In this section, we describe the extension mechanisms used to achieve interoperability between different annotation layers using RDF and the NIF URI schemes. Several vocabularies (or ontologies) were developed and published by the Semantic Web community, where each one describes one or more layers of annotations. The current best practice to achieve interoperability on the Semantic Web is to re-use the provided identifiers. Therefore, it is straightforward to generate one or more RDF properties for each vocabulary and thus connect the identifiers to NIF. We call such an extension a *Vocabulary Module*.

We introduce three generic properties called `annotation` (for URIs as object), `literalAnnotation` (for literals as object) and `classAnnotation` (for OWL classes as object), which are made available in the NIF namespace. The third one is typed as OWL annotation property in order to stay within the OWL DL language profile. All further properties used for annotation should be either modelled as a subproperty (via `rdfs:subPropertyOf`) of `annotation`, `literalAnnotation` or `classAnnotation` or left underspecified by using the `annotation`, `literalAnnotation` or `classAnnotation` property directly. This guarantees that on the one hand conventions are followed for uniform processing, while on the other hand developers can still use their own annotations using the extension mechanism. The distinction between `annotation`, `literalAnnotation` and `classAnnotation` guarantees that each vocabulary module will still be valid OWL/DL, which is essential for standard OWL reasoners.

When modeling an extension of NIF via a vocabulary module, vocabulary providers can use the full expressiveness of OWL. In the following, we will present several vocabulary modules, including design choices, so they can serve as templates for adaption and further extensions.

### *11.5.1 OLiA*

The *Ontologies of Linguistic Annotation* (OLiA) [8][50] provide stable identifiers for morpho-syntactical annotation tag sets, so that NLP applications can use these

[49]http://lists.w3.org/Archives/Public/www-rdf-comments/2002JanMar/0127.html

[50]http://purl.org/olia

identifiers as an interface for interoperability. OLiA provides *Annotation Models* for the most frequently used tag sets, such as Penn.[51] These annotation models are then linked to a *Reference Model*, which provides the interface for applications. Consequently, queries such as 'Return all Strings that are annotated (i.e. typed) as `olia:PersonalPronoun` are possible, regardless of the underlying tag set. In the following example, we show how *Penn Tag Set*[52] identifiers are combined with NIF:

```
@prefix sso: <http://nlp2rdf.lod2.eu/schema/sso/> .
# POS tags produced by Stanford Parser online demo
# http://nlp.stanford.edu:8080/parser/index.jsp
:offset_713_716
     str:anchorOf "The" ;
     str:referenceContext :offset_0_26546 ;
     sso:oliaIndividual <http://purl.org/olia/penn.owl#DT> ;
     sso:oliaCategory <http://purl.org/olia/olia.owl#Determiner>  .
:offset_717_725
     str:anchorOf "Semantic" ;
     str:referenceContext :offset_0_26546 ;
     sso:oliaIndividual <http://purl.org/olia/penn.owl#NNP> ;
     sso:oliaCategory <http://purl.org/olia/olia.owl#ProperNoun>  .
```

`oliaIndividual` and `oliaCategory` are subproperties of `annotation` and `classAnnotation` respectively and link to the tag set specific annotation model of OLiA as well as to the tag set independent reference ontology. The main purpose of OLiA is not the modelling of linguistic features, but to provide a mapping for data integration. Thus OLiA can be extended by third-parties easily to accommodate more tag sets currently not included. Furthermore, all the ontologies are available under an open license.[53]

### 11.5.2 ITS 2.0 and NERD

At the time of writing the *MultilingualWeb-LT Working Group*[54] is working on a new specification for the *Internationalization Tag Set* (ITS) Version 2.0,[55] which will allow to include coarse-grained NLP annotation into XML and HTML via custom attributes. Because attributes can only occur once per element, a corresponding NIF vocabulary module would require to reflect that in its design. Complementary to the ITS standardization effort, the *Named Entity Recognition and Disambiguation* (NERD) project [29] has created mappings between different existing entity type hierarchies to normalize named entity recognition tags. In this case, a vocabulary module can be composed of (1) DBpedia identifiers, (2) the functional OWL

[51] http://purl.org/olia/penn.owl

[52] http://www.comp.leeds.ac.uk/amalgam/tagsets/upenn.html

[53] http://sourceforge.net/projects/olia/

[54] http://www.w3.org/International/multilingualweb/lt/

[55] http://www.w3.org/TR/2012/WD-its20-20120829/

property `disambigIdentRef` to connect NIF with DBpedia (3) and additional type attachment to the included DBpedia identifier (`nerd:Organisation` in this case):

```
@prefix itsrdf: <http://www.w3.org/2005/11/its/rdf#> .
:offset_23107_23110
    str:anchorOf "W3C" ;
    itsrdf:disambigIdentRef <http://dbpedia.org/resource/World_Wide_Web_Consortium> ;
    str:referenceContext :offset_0_26546 .
<http://dbpedia.org/resource/World_Wide_Web_Consortium>
    rdf:type  <http:/nerd.eurecom.fr/ontology#Organisation> .
```

Note that the functionality of OWL properties allows to infer that, if the same subject has two different objects, then these are the same:

```
:offset_23107_23110
    itsrdf:disambigIdentRef <http://dbpedia.org/resource/World_Wide_Web_Consortium> ;
    itsrdf:disambigIdentRef <http://rdf.freebase.com/ns/m.082bb> .
# entails that
<http://dbpedia.org/resource/dbpedia:World_Wide_Web_Consortium>
    owl:sameAs <http://rdf.freebase.com/ns/m.082bb> .
```

### 11.5.3 Lemon and Wiktionary2RDF

URIs of RDF datasets using lemon [24] can be attached to NIF URIs employing two properties, which link to lexical entries and senses contained in a lemon lexicon.

```
@prefix wiktionary: <http://wiktionary.dbpedia.org/resource/> .
:offset_717_725
    str:anchorOf "Semantic" ;
    str:referenceContext :offset_0_26546 ;
    sso:lexicalEntry wiktionary:semantic ;
    sso:lexicalSense wiktionary:semantic-English-Adjective-1en .
```

### 11.5.4 Apache Stanbol

*Apache Stanbol*[56] is a Java framework, that provides a set of reusable components for semantic content management. One component is the content enhancer that serves as an abstraction for entity linking engines. For Stanbol's use case, it is necessary to keep provenance, confidence of annotations as well as full information about alternative annotations (often ranked by confidence) and not only the best estimate. In this case the vocabulary module uses an extra RDF node with a uniform resource name (urn).[57]

[56] http://stanbol.apache.org

[57] http://tools.ietf.org/html/rfc1737

```
@prefix fise: <http://fise.iks-project.eu/ontology/> .
@prefix dcterms: <http://purl.org/dc/terms/> .
@prefix dbo: <http://dbpedia.org/ontology/> .
:offset_23107_23110
    str:anchorOf "W3C" ;
    str:referenceContext :offset_0_26546 ;
    sso:annotation <urn:enhancement-3f794cd6-11d1-3cae-f514-154d4e6a3b59>
.
<urn:enhancement-3f794cd6-11d1-3cae-f514-154d4e6a3b59>
    fise:confidence 0.9464704504529554 ;
    fise:entity-label "W3C"@en ;
    fise:entity-reference <http://dbpedia.org/resource/World_Wide_Web_Consortium> ;
    fise:entity-type <http:/nerd.eurecom.fr/ontology#Organisation> ;
    fise:entity-type dbo:Organisation, owl:Thing,
                     <http://schema.org/Organization> ;
    dcterms:created "2012-07-25T09:02:38.703Z"^^xsd:dateTime ;
    dcterms:creator "stanbol.enhancer.NamedEntityTaggingEngine"^^xsd:string ;
    dcterms:relation <urn:enhancement-c5377650-41af-7ea2-8ac8-44356007821a> ;
    rdf:type fise:Enhancement ;
    rdf:type fise:EntityAnnotation .
```

## 11.6 Discussion and Outlook

In recent years, the interoperability of linguistic resources and NLP tools has become a major topic in the fields of computational linguistics and Natural Language Processing [18]. The technologies developed in the Semantic Web during the last decade have produced formalisms and methods that push the envelop further in terms of expressivity and features, while still trying to have implementations that scale on large data. Some of the major current projects in the NLP area seem to follow the same approach such as the graph-based formalism GrAF developed in the ISO TC37/SC4 group [19] and the ISOcat data registry [33], which can benefit directly by the widely available tool support, once converted to RDF. Note that it is the declared goal of GrAF to be a pivot format for supporting conversion between other formats and not designed to be used directly and the ISOcat project already provides a Linked Data interface. In addition, other data sets have already converted to RDF such as the typological data in Glottolog/Langdoc [10]. An overview can be found in [11].

One important factor for improving the quality of NLP tools is the availability of large quantities of qualitative background knowledge on the currently emerging Web of Linked Data [1]. Many NLP tasks can greatly benefit from making use of this wealth of knowledge being available on the Web in structured form as *Linked Open Data* (LOD). The precision and recall of Named Entity Recognition, for example, can potentially be boosted when using background knowledge from LinkedGeoData, Wiktionary2RDF, DBpedia, Geonames or other LOD sources as crowd-sourced and community-reviewed and timely-updated gazetteers. Of course the use of gazetteers is a common practice in NLP. However, before the arrival of large amounts of Linked Open Data their creation and maintenance in particular for multi-domain NLP applications was often impractical.

In this article, we have:

- Described challenges and benefits of RDF for NLP.
- Investigated the collaborative nature of three large data sets, which were created by a knowledge extraction process from crowd-sourced community projects.

- Provided the extension mechanism of the NLP Interchange Format as a proof of concept, that NLP tool output can be represented in RDF as well as connected with existing LOD data sets.

**Acknowledgements** We would like to thank our colleagues from AKSW research group and the LOD2 project for their helpful comments during the development of NIF. Especially, we would like to thank Christian Chiarcos for his support while using OLiA and Jonas Brekle for his work on Wiktionary2RDF. This work was partially supported by a grant from the European Union's 7th Framework Programme provided for the project LOD2 (GA no. 257943).

## References

1. Auer S, Lehmann J (2010) Making the web a data washing machine – creating knowledge out of interlinked data. Semant Web J 1:97–104
2. Auer S, Dietzold S, Lehmann J, Hellmann S, Aumueller D (2009) Triplify: light-weight linked data publication from relational databases. In: Proceedings of the 18th international conference on world wide web, WWW 2009, Madrid, Spain, 20–24 April 2009. ACM, pp 621–630
3. Auer S, Lehmann J, Hellmann S (2009) LinkedGeoData – adding a spatial dimension to the web of data. In: Proceedings of 8th international semantic web Conference (ISWC), Chantilly, VA, USA
4. Berners-Lee T (2006) Design issues: linked data. http://www.w3.org/DesignIssues/LinkedData.html
5. Bizer C (2011) Evolving the web into a global data space. http://www.wiwiss.fu-berlin.de/en/institute/pwo/bizer/research/publications/Bizer-GlobalDataSpace-Talk-BNCOD2011.pdf, keynote at 28th British National Conference on Databases (BNCOD2011)
6. Bizer C (2012) Dbpedia 3.8 released, including enlarged ontology and additional localized versions. http://tinyurl.com/dbpedia-3-8
7. Bühmann L, Lehmann J (2012) Universal owl axiom enrichment for large knowledge bases. In: Proceedings of EKAW 2012, Galway, Ireland. http://jens-lehmann.org/files/2012/ekaw_enrichment.pdf
8. Chiarcos C (2012) Ontologies of linguistic annotation: survey and perspectives. In: Proceedings of the eight international conference on language resources and evaluation (LREC'12), Istanbul, Turkey
9. Chiarcos C (2012) Powla: modeling linguistic corpora in owl/dl. In: Proceedings of 9th extended semantic web conference (ESWC2012), Heraklion, Crete, Greece
10. Chiarcos C, Hellmann S, Nordhoff S (2011) Towards a linguistic linked open data cloud: the open linguistics working group. TAL 52(3):245–275. http://www.atala.org/Towards-a-Linguistic-Linked-Open
11. Chiarcos C, Nordhoff S, Hellmann S (eds) (2012) Linked data in linguistics. Representing language data and metadata. Springer, Heidelberg. (ISBN 978-3-642-28248-5). http://www.springer.com/computer/ai/book/978-3-642-28248-5
12. Gruber TR (1993) A translation approach to portable ontology specifications. Knowl Acquis 5(2):199–220
13. Hellmann S, Unbehauen J, Chiarcos C, Ngonga Ngomo AC (2010) The TIGER corpus navigator. In: 9th international workshop on treebanks and linguistic theories (TLT-9), Tartu, Estonia, pp 91–102
14. Hellmann S, Lehmann J, Auer S (2012) Linked-data aware uri schemes for referencing text fragments. In: EKAW 2012, Galway, Ireland. Lecture notes in artificial intelligence (LNAI). Springer,

15. Hellmann S, Stadler C, Lehmann J (2012) The German DBpedia: a sense repository for linking entities. In: Chiarcos C, Nordhoff S, Hellmann S (eds) (2012) Linked data in linguistics. Representing language data and metadata. Springer, Berlin/New York, pp 181–190
16. Hepp M, Bachlechner D, Siorpaes K (2006) Harvesting wiki consensus – using wikipedia entries as ontology elements. In: Völkel M, Schaffert S (eds) Proceedings of the first workshop on semantic wikis – from wiki to semantics, co-located with the 3rd annual european semantic web conference (ESWC 2006), Budva, Montenegro. http://www.eswc2006.org/
17. Hepp M, Siorpaes K, Bachlechner D (2007) Harvesting wiki consensus: using wikipedia entries as vocabulary for knowledge management. IEEE Internet Comput 11(5):54–65
18. Ide N, Pustejovsky J (2010) What does interoperability mean, anyway? Toward an operational definition of interoperability. In: Proceedings of the second international conference on global interoperability for language resources (ICGL 2010), Hong Kong, China
19. Ide N, Suderman K (2007) GrAF: a graph-based format for linguistic annotations. In: Proceedings of the linguistic annotation workshop (LAW 2007), Prague, Czech Republic, pp 1–8
20. Khalili A, Auer S, Hladky D (2012) The rdfa content editor – from wysiwyg to wysiwym. In: Proceedings of COMPSAC 2012 – trustworthy software systems for the digital society, 16–20 July 2012, Izmir, Turkey. Best paper award
21. Kontokostas D, Bratsas C, Auer S, Hellmann S, Antoniou I, Metakides G (2011) Towards linked data internationalization – realizing the greek dbpedia. In: Proceedings of the ACM WebSci'11, Koblenz, Germany
22. Kontokostas D, Bratsas C, Auer S, Hellmann S, Antoniou I, Metakides G (2012) Internationalization of linked data: the case of the Greek DBpedia edition. J Web Semant 15:51–61
23. Lehmann J, Bizer C, Kobilarov G, Auer S, Becker C, Cyganiak R, Hellmann S (2009) DBpedia – a crystallization point for the web of data. J Web Semant 7(3):154–165
24. McCrae J, Cimiano P, Montiel-Ponsoda E (2012) Integrating WordNet and Wiktionary with lemon. In: Chiarcos C, Nordhoff S, Hellmann S (eds) Linked data in linguistics, Springer, Heidelberg. (ISBN 978-3-642-28248-5). http://www.springer.com/computer/ai/book/978-3-642-28248-5
25. Mendes PN, Jakob M, García-Silva A, Bizer C (2011) Dbpedia spotlight: shedding light on the web of documents. In: Proceedings of the 7th international conference on semantic systems (I-Semantics), Graz, Austria
26. Mendes PN, Jakob M, Bizer C (2012) Dbpedia for nlp: a multilingual cross-domain knowledge base. In: Proceedings of the eight international conference on language resources and evaluation (LREC'12), Istanbul, Turkey
27. Meyer CM, Gurevych I (2011) OntoWiktionary – constructing an ontology from the collaborative online dictionary wiktionary. In: Pazienza M, Stellato A (eds) Semi-automatic ontology development: processes and resources. IGI Global, Hershey, PA, USA. http://www.ukp.tudarmstadt.de/publications/details/?no_cache=1&tx_bibtex_pi1[pub_id]=TUD-CS-2011-0202&type=99&tx_bibtex_pi1[bibtex]=yes
28. Quasthoff M, Hellmann S, Höffner K (2009) Standardized multilingual language resources for the web of data: http://corpora.uni-leipzig.de/rdf. In: 3rd prize at the LOD triplification challenge, Graz. http://triplify.org/files/challenge_2009/languageresources.pdf
29. Rizzo G, Troncy R, Hellmann S, Brümmer M (2012) NERD meets NIF: lifting NLP extraction results to the LinkedData cloud. In: Proceedings of linked data on the web workshop (WWW), Lyon, France
30. Stadler C, Lehmann J, Höffner K, Auer S (2011) Linkedgeodata: a core for a web of spatial open data. Semant Web J 3(4):333–354. http://iospress.metapress.com/content/141w054666871326
31. Unbehauen J, Hellmann S, Auer S, Stadler C (2012) Knowledge extraction from structured sources. In: Search computing – broadening web search. Lecture Notes in Computer Science, vol 7538. Springer, Berlin/Heidelberg. http://link.springer.com/chapter/10.1007/978-3-642-34213-4_3

32. Wilde E, Duerst M (2008) URI fragment identifiers for the text/plain media type. http://tools.ietf.org/html/rfc5147, [Online; Accessed 13-April-2011]
33. Windhouwer M, Wright SE (2012) Linking to linguistic data categories in isocat. In: Chiarcos C, Nordhoff S, Hellmann S (eds) (2012) Linked data in linguistics. Representing language data and metadata. Springer, Berlin/New York

# Chapter 12
# Building a Linked Open Data Cloud of Linguistic Resources: Motivations and Developments

**Christian Chiarcos, Steven Moran, Pablo N. Mendes, Sebastian Nordhoff, and Richard Littauer**

**Abstract** We describe on going community-efforts to create a Linked Open Data (sub-)cloud of linguistic resources, with an emphasis on resources that are specific to linguistic research, namely annotated corpora and linguistic databases. We argue that for both types of resources, the application of the Linked Open Data paradigm and the representation in RDF represents a promising approach to address interoperability problems, and to integrate information from different repositories. This is illustrated with example studies for different kinds of linguistic resources.

The efforts described in this chapter are conducted in the context of the Open Linguistics Working Group (OWLG) of the Open Knowledge Foundation. The OWLG is a network of researchers interested in linguistic resources and/or their publication under open licenses, and a number of its members are engaged in the application of the Linked Open Data paradigm to their resources. Under the umbrella of the OWLG, these efforts will eventually emerge in the creation of a Linguistic Linked Open Data cloud (LLOD).

---

C. Chiarcos (✉)
Goethe University Frankfurt am Main, Germany
e-mail: chiarcos@em.uni-frankfurt.de

S. Moran
Ludwig-Maximilians-Universität, München, Germany
e-mail: steve.moran@lmu.de

P.N. Mendes
Freie Universität Berlin, Berlin, Germany
e-mail: pablo.mendes@fu-berlin.de

S. Nordhoff
Max Planck Institute for Evolutionary Anthropology, Leipzig, Germany
e-mail: sebastian_nordhoff@eva.mpg.de

R. Littauer
Universität des Saarlandes, Saarbrücken, Germany
e-mail: littauer@coli.uni-saarland.de

I. Gurevych and J. Kim (eds.), *The People's Web Meets NLP*, Theory and Applications of Natural Language Processing, DOI 10.1007/978-3-642-35085-6_12, 

## 12.1 Background and Motivation

In recent years, the limited interoperability between linguistic resources has been recognized as a major obstacle for data use and re-use within and across discipline boundaries. After half a century of computational linguistics [24], quantitative typology [33], empirical, corpus-based study of language [29], and computational lexicography [63], researchers in Computational Linguistics, Natural Language Processing (NLP) or Information Technology, as well as in Digital Humanities, are confronted with an immense wealth of linguistic resources, that are not only growing in number, but also in their heterogeneity.

Interoperability involves two aspects [40]:

**Structural** ('syntactic') interoperability: Resources use comparable formalisms to represent and to access data (formats, protocols, query languages, etc.), so that they can be accessed in a uniform way and that their information can be integrated with each other.

**Conceptual** ('semantic') interoperability: Resources share a common vocabulary, so that linguistic information from one resource can be resolved against information from another resource, e.g., grammatical descriptions can be linked to a terminology repository.

With the rise of the Semantic Web, new representation formalisms and novel technologies have become available, and, independently from each other, researchers in different communities have recognized the potential of these developments with respect to the challenges posited by the heterogeneity and multitude of linguistic resources available today. Many of these approaches follow the **Linked (Open) Data paradigm** [6] that postulates four rules for the publication and representation of Web resources: (1) Referred entities should be designated by using URIs, (2) these URIs should be resolvable over HTTP, (3) data should be represented by means of specific W3C standards (such as RDF), (4) and a resource should include links to other resources. These rules facilitate information integration, and thus, interoperability, in that they require that entities can be addressed in a globally unambiguous way (1), that they can be accessed (2) and interpreted (3), and that entities that are associated on a conceptual level are also physically associated with each other (4).

In the definition of Linked Data, the **Resource Description Framework (RDF)** receives special attention. RDF was designed as a language to provide metadata about resources that are available both offline (e.g., books in a library) and online (e.g., eBooks in a store). RDF provides a data model that is based on labeled directed (multi-)graphs, which can be serialized in different formats. In RDF, information is expressed in terms of *triples* – consisting of a *property* (relation, in graph-theoretical terms a labeled edge) that connects a *subject* (a resource, in graph-theoretical terms a labeled node) with its *object* (another resource, or a literal, e.g., a string). RDF resources (nodes)[1] are represented by *Uniform Resource Identifiers (URIs)*. They are thus globally unambiguous in the web of data. This allows resources hosted at different locations to refer to each other, and thereby to create a network of data collections whose elements are densely interwoven.

Several data base implementations for RDF data are available, and these can be accessed using **SPARQL** [67], a standardized query language for RDF data. At its very core, SPARQL uses a triple notation similar to RDF, only that properties and RDF resources can be replaced by variables. SPARQL is inspired by SQL, variables can be introduced in a separate `SELECT` block, and constraints on these variables are expressed in a `WHERE` block in a triple notation (see Sect. 2.4 for an example). SPARQL does not only support running queries against individual RDF data bases that are accessible over HTTP (so-called 'SPARQL end points'), but also, it allows us to combine information from multiple repositories (federation, see Sect. 6.1). RDF can thus not only be used to *establish* a network, or cloud, of data collections, but also, to *query* this network directly.[2]

RDF has been applied for various purposes beyond its original field of application. In particular, it evolved into a generic format for knowledge representation. It was readily adapted by disciplines as different as biomedicine and bibliography, and eventually it became one of the building stones of the **Semantic Web**. Due to its application across discipline boundaries, RDF is maintained by a large and active community of users and developers, and it comes with a rich infrastructure of APIs, tools, databases, query languages, and multiple sub-languages that have been developed to define data structures that are more specialized than the graphs represented by RDF. These sub-languages can be used to define *reserved vocabularies* and *structural constraints* for RDF data. For example, the Web Ontology Language (OWL) introduces datatypes necessary for the representation of ontologies as an extension of RDF, i.e., *classes* (concepts), *instances* (individuals) and *properties* (relations). OWL/DL is an OWL dialect that is restricted such that the language corresponds to a description logic, i.e., a decidable fragment of first-order predicate logic. Exploiting this restriction, a number of reasoners have been developed that allow the verification of consistency constraints (*axioms*) as well as methods to draw inferences from logical relations

[1] The term 'resource' is ambiguous here. As understood in this chapter, resources are structured collections of data which can be represented, for example, using RDF. In RDF, however, 'resource' is the conventional name of a node in the graph, because, historically, these nodes were meant to represent objects that are described by metadata. Hence, we use the terms 'node' or 'concept' whenever *RDF resources* are meant.

[2] Federation is possible with SPARQL, although not necessarily very performant with state-of-the-art implementations. A more efficient way than federation is thus to retrieve the content necessary for a particular application from another end point and to query it locally. SPARQL end points provide this functionality, and publishing data under open licenses (see below) warantees that the necessary legal preconditions for this practice are met.

in the ontology. If modeled as ontologies, the semantic consistency of linguistic resources can be validated and implicit information can be inferred.

The concept of Linked Data is closely coupled with the idea of **openness** (otherwise, the linking is only reproducible under certain conditions, see Sect. 12.6.2 for the definition of openness applied here and its ramifications), and in 2010, the original definition of Linked Open Data has been extended with a 5 star rating system for data on the Web.[3] The first star is achieved by publishing data on the Web (in any format) under an open license, the second, third and fourth star require machine-readable data, a non-proprietary format, and using standards like RDF, respectively. The fifth star is achieved by linking the data to other people's data to provide context. If (linguistic) resources are published in accordance with these rules, it is possible to follow links between existing resources to find other, related data and exploit network effects.

In this chapter, we formulate and substantiate the claim that publishing linguistic resources as Linked Data helps to overcome both critical challenges identified above, i.e., the interoperability of language resources and the integration of information from different sources. The application of Linked Data principles is an established technique for lexical-semantic resources [31] and terminology repositories for linguistic concepts [27]. For other types of linguistic resources, however, its potential has not been recognized to the full extent possible so far.[4] This chapter focuses on these types of resources, and specifically deals with linguistic corpora and linguistic databases as those compiled in typology and language documentation. The linking of lexical-semantic resources with other linguistic resources has been described elsewhere [20].

For **linguistic corpora**, the potential of the Linked Data paradigm for modeling, processing and querying of corpora is immense. RDF provides a graph-based data model as required for the interoperable representation of arbitrary kinds of annotation [8, 44], and this flexibility makes it a promising candidate for a general means of representation for corpora with complex and heterogeneous annotations. Section 12.2 describes the application of the Linked Data paradigm to the MASC corpus, an open, multi-layer corpus of American English. RDF does not only establish interoperability between annotations within a corpus, but also between corpora and other linguistic resources, as illustrated in Sect. 12.3.

RDF also provides suitable means to represent **linguistic databases**. Linguistic databases are a particularly heterogeneous group of linguistic resources, they contain complex and heterogeneous types of information, e.g., feature structures that represent typologically relevant phenomena, along with examples for their

---

[3] http://www.w3.org/DesignIssues/LinkedData.html, paragraph 'Is your Linked Open Data 5 Star?'

[4] Although the application of RDF to linguistic resources as described here has been occasionally suggested, see [11, 13] for linguistic corpora, but these approaches focused on the RDF representation of individual resources rather than linking them with other types of linguistic resources. As opposed to this, the focus of this chapter is not on modeling linguistic resources, but rather, on the potential to linking these with each other.

illustration and annotations (glosses) and translations applied to these examples (structurally comparable to corpus data), or word lists (structurally comparable to lexical-semantic resources). Section 12.4 describes several such databases that evolved in the context of typological research, as well as currently on going efforts to harmonize them on the basis of RDF and Linked Data.

Modeled as Linked Data, both corpora and typological data collections can be fully integrated into a Linked Open Data (sub-)cloud of linguistic resources, along with lexical-semantic resources and knowledge bases of information about languages and linguistic terminology. These examples show how the Linked Data paradigm provides a holistic approach to the problem of **structural interoperability**, i.e., that the *same* formalism can be applied to achieve interoperable representations of these types of linguistic resources along with more established resource types like lexical-semantic resources, metadata repositories and bibliographies. Accordingly, integration from different resources is substantially enhanced, and by using resolvable URIs and openly available resources, stable links between these resources can be established. These links can then be employed to formulate queries over multiple, distributed resources, to enrich and to verify the information from one resource with information from another.

Beyond this, Linked Data paradigm facilitates the establishment of **conceptual interoperability**, i.e., that resource-specific annotations or abbreviations are expanded into references to repositories of linguistic terminology and/or metadata categories. Section 12.5 illustrates the establishment of conceptual interoperability between linguistic databases and annotated corpora in this way, and suggests that the use of shared terminology repositories may help researchers from one particular community, say, NLP, to access and to re-use resources built up in the context of another linguistic sub-discipline, say, typology.

Finally, Sect. 12.6 summarizes these and other benefits of the application of the Linked Data paradigm to linguistic resources, and describes the Open Linguistics Working Group (OWLG), an interdisciplinary network of individual researchers interested in open linguistic resources that provides the broader context for the aforementioned efforts to apply the Linked Data paradigm to linguistic resources. In particular, it introduces the idea of a Linguistic Linked Open Data (LLOD) cloud that covers the linguistic resources described in this chapter as well as various lexical-semantic resources, together with their respective linking.

The goal of the paper is to provide an overview over recent activities with respect to the application of the Linked Open Data paradigm to linguistic resources, focusing on integration of these efforts. As compared to other chapters in this book, the focus of this chapter is thus not so much to describe how people collaborate in the *creation* of resources, but rather, how independently created resources can be *integrated* with each other, to provide examples for these integration efforts, and to describe collaboration efforts of the participating communities. This novel form of collaboration is heavily based on using Semantic Web formalisms that support the integration of linguistic resources physically distributed in different HTTP-accessible RDF data bases ('SPARQL end points'), thereby forming a cloud, or network of resources. This chapter focuses on exemplary resources, it shows the applicability of the Linked Data paradigm to different types of resources and the community efforts to integrate them. One goal of this chapter is to familiarize researchers coming from linguistics or NLP with these technologies, with their potential and recent developments in the communities, while a more detailed description of technical aspects and evaluation results for specific resources and their respective linking can be found in the literature referred to in the project descriptions.

## 12.2 Structural Interoperability for Annotated Corpora

Generally speaking, a **linguistic corpus** can be defined as "a collection of texts when considered as an object of language or literary study" [47, p. 334]. More specifically, we are interested in *annotated* corpora, where one or multiple layers of transcription, transliteration, translation, or linguistic analysis are attached to the primary data, which may be textual or multi-modal content. The multitude of possible annotations raises the issue of structural interoperability, i.e., how and whether these different types of information can be integrated with each other such that they can be queried, and evaluated.

The application of the Linked Data paradigm to an annotated corpus is illustrated here for the Manually Annotated (Sub-)Corpus of American English (MASC), a corpus of 500 K tokens of contemporary American English text drawn from the Open American National Corpus [45, also see Chap. 10, this volume].[5] Annotations were gathered by crowdsourcing volunteers, aggregated from other, pre-annotated corpora, or created in research projects that employed MASC data. The MASC project is committed to a fully open model of distribution, without restriction, for all data and annotations produced or contributed. The corpus has become increasingly popular in different projects because of its openness. It comprises various layers of annotations, including parts-of-speech, nominal and verbal chunks, constituent syntax, annotations of WordNet senses, frame-semantic annotations, document structure, illocutionary structure, as well as other layers; it is thus a representative example for a **multi-layer corpus**, where different types of annotations are applied to the same stretch of data, and thus where interoperability issues are particularly likely to arise.

### *12.2.1 Interoperability Challenges*

In accordance with the definition given in Sect. 12.1, structural interoperability of annotated corpora or different annotation layers in a corpus can be said to be successfully achieved if different corpora and annotations are represented within the same data format. Then they can be stored within the same database, they can be accessed with a single and uniform query language, and they can actually be

[5] http://www.anc.org/MASC

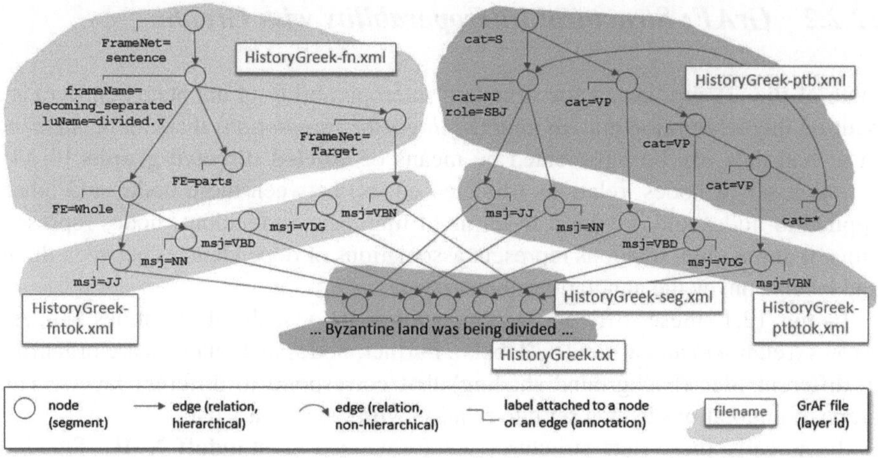


**Fig. 12.1** Representing and integrating annotations for syntax and frame-semantics in GrAF [14]

physically merged. As querying different corpora for the same type of annotations is trivial, structural interoperability can thus be shown by formulating queries across different layers of annotation (Sect. 12.2.4).

A minimal requirement for structural interoperability is that different linguistic resources can be integrated without complicated conversion routines. For this purpose, the NLP and linguistics communities have developed a number of representation formalisms that address this problem either in a fully generic approach [1, 8,12,17,44], or for the full band-width of annotations for one particular phenomenon (e.g., syntax annotation, [22, 69]). Under the umbrella of the Linguistic Annotation Framework developed by ISO TC37/SC4, these approaches gradually converge towards the establishment of standard data models and formalisms [41,43], including, for example, the recently published Graph Annotation Format (GrAF, ISO 24612, see below), for whose development the MASC corpus served as a test-bed.

Figure 12.1 shows an example clause drawn from the sentence, *"While Byzantine land was being divided, there was no one in control of the seas, so pirates raided towns on many of the islands."* (taken from the file `HistoryGreek`, written section of MASC v.1.0.3, http://www.anc.org/MASC), with annotations for syntax [7, right] and frame semantics [4, left].

Originally, both 'branches' were stored in different files with different formats, but the figure shows what is meant by structural interoperability: both annotations are represented using the same elementary representation formalism (nodes, edges and labels). Furthermore, a relationship between different annotations has been established: both syntactic and semantic annotations refer to the same stretch of primary data. Therefore they make use of a common segmentation of the underlying text. Thus it is possible to formulate queries across both annotation layers. The formalism applied to the MASC is the GrAF data model, described in the following section.

### 12.2.2 GrAF: Structural Interoperability with Graphs

State-of-the-art approaches for structural interoperability of annotated corpora are built on the assumption that *all kinds of linguistic annotations* that can be attached to textual data can be represented by means of **labeled directed graphs** [8, 44], i.e., as a set of nodes, relations (directed edges) between these nodes and labels applied to nodes and/or edges. In terms of linguistic annotations, nodes represent units of annotation, relations represent associations or dependencies between them, and labels convey the annotations themselves.

In Fig. 12.1, these different data structures are visualized as circles (nodes), arrows (relations) and attributes (labels). Further, nodes and relations are organized in different files (background shading) that correspond to different layers. This graph of annotations is then set into relation with the primary data.

Technically these data structures are formalized as **standoff XML**. Standoff formats are based on the physical separation between primary data and different annotation layers, usually in independent files, which are interconnected with XLink/XPointer references. Figure 12.1 illustrates annotations for syntax [7] and frame semantics [4], as represented in GrAF, the Graph Annotation Format developed by the ISO TC37/SC4 [44, ISO 24612]. Because of the heavy use of XLink/XPointer, the efforts to parse, validate and process standoff annotations are relatively high. As XLink/XPointer references are untyped, no off-the-shelf validation mechanisms are available. Even worse, there are no efficient means for storing and querying general standoff XML data [26]. Therefore, it is necessary to convert standoff XML to other representations (e.g., tables for a relational DB system) in order to process it efficiently.

But standoff XML is not the only option to encode graph-based data structures. For example, communities working with graph-based data structures have developed their own tools (e.g., databases) based on their own representational standards (e.g., GraphML [9], whose application to linguistic data collections has also been suggested [51]). It should be noted, however, that graphs do not provide a sufficiently restrictive data model to represent linguistic annotations, but that additional constraints apply that need to be captured in GraphML in the form of *naming conventions* for specific labels. Thus, off-the-shelf tools cannot be used to verify the consistency of annotations represented as directed graphs in this formalism, and additional means of validation are required.

### 12.2.3 Towards Formally Defined Data Structures with OWL/DL

Although representing corpora in GraphML already establishes interoperability in the sense that the same formalism is used, such general formalisms are underspecified with respect to specific constraints on corpus data. RDF is another graph-based formalism, but it may be more suitable to the modeling of structured resources, because it comes with a number of sub-languages that can be used to define and to validate formal axioms, e.g., OWL/DL, the description logic dialect of the Web Ontology Language (OWL) that makes it possible to formulate axioms that capture consistency conditions.

Chiarcos [14, 16] described how a generic, graph-based data model for annotated corpora (comparable to GrAF) can be reconstructed as an OWL/DL ontology, and how this data model, **POWLA** (see below), defines data types for corpus data – that can then be represented in RDF. As compared to traditional approaches, using RDF as a representation formalism relieves us from developing a special-purpose XML standoff format that requires the development of its own infrastructure to parse, manipulate, store and query linguistic resources. Instead, for the processing of RDF, a rich ecosystem of APIs, databases and query languages is already available. Similar approaches to interoperability have been proposed before, e.g., [13], whose representation of GrAF in RDF, however, did not provide formal (OWL/DL) definitions of data types, and it did not cover information about the primary data, but only preserved the original GrAF pointers to the annotated text. It was thus not possible to use a standard RDF query language to gain information about both annotations and the original strings that these annotations were applied to.

Other approaches [11, 36] describe proof-of-principle implementations that illustrate the applicability of RDF and/or OWL to modeling of language resources, but they do not provide a generic model for linguistic annotations in general. Neither do recent approaches to develop RDF-based NLP pipelines (e.g., the NLP Interchange Format NIF [37], cf. Chap. 11, this volume) focus on genericity, but rather, economic considerations play a role here, resulting in more compact representations that are often specific to a tool at hand or the tasks that the NLP architecture is intended for. NIF, for example, does not support the concept of annotation layers which is essential for annotated corpora. However, efforts to establish ties and mappers between NLP-oriented and corpus-oriented RDF formats are underway. In fact, this is an example for closer collaboration between different OWLG members (Sect. 12.6.2) – one of the ultimate goals of the working group.

### *12.2.4 POWLA: Establishing Structural Interoperability of Multi-layer Corpora Using RDF, OWL and SPARQL*

The idea underlying POWLA is to represent linguistic annotations by means of RDF, to employ OWL/DL to define data types and consistency constraints for these RDF data, and to adopt these data types and constraints from an existing XML standoff formalism capable to represent arbitrary kinds of text-oriented linguistic annotation in a generic and lossless fashion. Consequently, all annotations currently representable by the underlying data model (i.e., any text-oriented linguistic annotation) can be represented as Linked Data. A converter from GrAF to POWLA, applied to data from the MASC, can be found under http://purl.org/powla. The converter provides an isomorphic mapping between GrAF data structures and

POWLA concepts. An important difference as compared to related research is that POWLA is not developed with one specific corpus or application in mind, but that it originates from an *established* generic data model, PAULA [17], and thus preserves the genericity of this model. To our best knowledge, POWLA is the first approach that provides an exhaustive formalization of GrAF-style corpus data in RDF/OWL.

The primary data structures of POWLA are the concepts `Node` and `Relation` that correspond to nodes and edges in labeled directed acyclic graphs. Every `Relation` has a `hasSource` and a `hasTarget` property that link it to one `Node`, respectively. For reasons of space, we restrict ourselves here to an informal introduction by stating that GrAF nodes and edges as depicted in Fig. 12.1 can be isomorphically expressed in POWLA. Annotations are represented by the property `hasAnnotation` that assigns a `Node` or a `Relation` a string, such that, for a given attribute-value pair (label) in the original annotation, say `cat="NP"`, a subproperty of `hasAnnotation` is created that corresponds to the attribute name (e.g., `has_cat`) and that assigns the annotated structures the attribute value as a string value. Further concepts include `Layer`, `Document` and `Corpus`. Concepts and properties are arranged in hierarchies, and complemented with OWL/DL axioms that express, for example, cardinality or type restrictions, e.g., that a `Relation` needs to have exactly one source and one target, or that source and target of a `Relation` can only be `Node`s.

With POWLA specifications for both syntactic and frame annotations, we can now query both layers of annotation simultaneously, and combine their information. Using the original GrAF data from Fig. 12.1, it would not be possible to realize this with out-of-the-box tools, say XQuery in an XML data base. For RDF, a standardized query language and numerous data base implementations are available. As an example, one may be interested to find out which grammatical role the `Whole` argument of the frame `Becoming_separated` is assigned in a corpus. Using POWLA, this can be expressed in SPARQL as follows:

```
SELECT ?gr
WHERE {
  ?frame a powla:Node.
  ?frame has_frameName "Becoming_separated".
  ?frame powla:hasChild ?wholeArg.
  ?wholeArg has_FE "Whole".
  ?wholeArg powla:firstTerminal ?startTerm.
  ?phrase powla:firstTerminal ?startTerm.
  ?wholeArg powla:lastTerminal ?endTerm.
  ?phrase powla:lastTerminal ?endTerm.
  ?phrase has_cat "NP".
  ?phrase has_role ?gr.
}
```

In this query, the `powla` namespace indicates concepts and properties defined in the POWLA ontology. The variable `?frame` is instantiated as a `Node` that is assigned the original annotation `frameName="Becoming_separated"` (attribute names and string values are preserved in the transformation from GrAF to POWLA whenever a new sub-property of `hasAnnotation` is instantiated, e.g., `has_frameName`). The frame argument we are interested in depends on `?frame`, and it is annotated as `FE="Whole"`. Aside from properties and concepts that are motivated from the underlying data model (`Node`, `hasChild`), POWLA provides a set of properties that serve to simplify querying and to facilitate access to information. For example, every Node can be assigned a `firstTerminal` and `lastTerminal` that represent terminal nodes (nodes without dependent nodes) that are (by definition) anchored to the primary data. Both properties are inferrable from positional information and `hasChild` properties. These properties can then be used to find a `?phrase Node` in the syntax annotation that covers the same stretch of primary data (i.e., the same `firstTerminal` and `lastTerminal`) as `?wholeArg`. If `?phrase` is a noun phrase (`cat="NP"`), the query returns the argument of the grammatical role (`role`) annotation as a result.

This example query shows how RDF not only preserves the structural interoperability between different annotation layers that was established by GrAF, but that it renders this data such that it can be seamlessly queried with standard query languages (a functionality that GrAF currently does not provide). The verbosity of such queries can be reduced by introducing macros that expand into more complex expressions, e.g., `?wholeArg _=_ ?phrase` as a shorthand to check the identity of first and last terminal [14]. One important difference as compared to the original GrAF is that data structure can be queried and stored with off-the-shelf tools. While data base implementations for multi-layer corpora have been developed [17], these are specific to the linguistic domain, and thus used and maintained by a relatively small community. RDF and SPARQL, however, come with larger communities and a well-developed technological infrastructure which includes APIs, querying engines, and databases, that can be used to process, to store and to merge data that is represented in standoff XML for other applications. With POWLA as an implementation of a data model that can also be linearized as standoff XML, lossless conversion between standoff XML and RDF can be assured, and technologies developed for either representation can be combined using intermediate conversion modules.

Aside from facilitating the interoperability between different annotation layers, an additional benefit of an RDF representation that corpora can be easily linked with other linguistic resources, and for the specific case of lexical-semantic resources, this is shown in the following section.

## 12.3 Structural Interoperability Between Corpora and Lexical-Semantic Resources

Existing representation standards for corpora and lexical-semantic resources are capable to establish interoperability *among* resources of the *same* type, e.g., GrAF for annotated corpora, and the Lexical Markup Framework [30, LMF] for electronic dictionaries, but it is not clear how these formalisms can be applied to establish interoperability *between* linguistic resources of *different* types.

RDF and the application of the Linked Data paradigm represent one possible approach to address this problem, as GrAF data can be transformed into RDF (see Sect. 12.2) as well as LMF data [54]. This section illustrates the linking of both types of resources for the MASC corpus and the general-purpose semantic knowledge base DBpedia, and further, how this linking allows to use the existing WordNet sense annotations of the MASC corpus for the evaluation and improvement of the DBpedia linking.

### 12.3.1 DBpedia Spotlight: Linking MASC with the DBpedia

DBpedia is a general-purpose knowledge base for the Semantic Web encompassing facts from a wide range of domains of knowledge.[6] It was created through a community effort aiming at extracting information from Wikipedia and making it available as structured data on the Web [49]. Considering the example sentence from Fig. 12.1, DBpedia can provide, for example, a detailed description for the term *Byzantine Empire*[7] as a formal and machine-readable definition of the adjective *Byzantine* in this context.

DBpedia represents a data pool that (1) is widely used in academics as well as industrial environments, that (2) is curated by the community of Wikipedia and DBpedia editors, and that (3) has become a major crystallization point and a vital infrastructure for the Web of Data. The extracted RDF knowledge from the Wikipedia is published and interlinked according to the Linked Data principles and made available under the same license as Wikipedia (CC-BY-SA). In its current version, DBpedia contains labels and abstracts for 3.64 million things in up to 97 different languages, of which 1.83 million are classified in a consistent ontology, including 416,000 persons, 526,000 places, 169,000 organizations, 183,000 species, 106,000 music albums, etc. The data set consists of 1 billion RDF triples out of which 385 million were extracted from the English edition of Wikipedia and roughly 665 million comprise data extracted from other language editions and links to external data sets. DBpedia is a general purpose knowledge base for the Semantic Web, but it can also be fruitfully applied to NLP applications and linguistic research.

DBpedia Spotlight [55] is a tool for annotating mentions of DBpedia concepts (entities and other concepts that are the subject of Wikipedia pages) in natural language. It performs term extraction, maps terms to DBpedia concepts and automatically selects the most likely DBpedia concept based on the context of the input text.

The most basic term extraction strategy in DBpedia Spotlight is based on a dictionary of known DBpedia concept names extracted from Wikipedia page

[6] http://dbpedia.org/

[7] http://live.dbpedia.org/resource/Byzantine_Empire

titles, redirects and disambiguation pages [57]. It operates along the following rationale: (1) Wikipedia page titles can be seen as community-approved names, (2) redirects to URIs indicate synonyms or alternative surface forms, including common misspellings and acronyms, (3) disambiguations provide ambiguous names that are "confusable" with all pages they link to (their labels become names for all target concepts in the disambiguation page). In order to score the association between names and DBpedia concepts, page links in Wikipedia are used. For each page link, one association between a name in the anchor text with the concept represented by the target page is counted. Based on these statistics, a number of scores have been derived. Aside from this vanilla implementation, more advanced term extraction techniques have also been implemented [56].

The disambiguation strategy used in DBpedia Spotlight models each DBpedia concept in a vector space model of words extracted from Wikipedia paragraphs containing page links. A paragraph is added to the context of a DBpedia concept if the corresponding Wikipedia page is the target of a page link in that paragraph. The words in the paragraph are further scored based on the TF*ICF (Term Frequency – Inverse Candidate Frequency) measure [55], a variation of TF*IDF that scores words on their ability to distinguish between known senses of a given term.[8]

Figure 12.2 shows the DBpedia Spotlight Web demo running on a snippet of text from MASC.[9] For each term, DBpedia Spotlight adds a link to the DBpedia concept that has been detected as the most likely concept expressed by that term. If the 'Show n-best candidates' option has been enabled, hovering the mouse pointer over an annotation will reveal a list of candidates and corresponding disambiguation scores – as shown for *Byzantine* in Fig. 12.2.

DBpedia Spotlight is also provided as a Web Service that is able to produce, through content negotiation, other output formats such as XML (Extensible Markup Language), JSON (JavaScript Object Notation) and XHTML+RDFa (Resource Description Framework – in – attributes). Through the usage of a small wrapper around the service, it can also produce output in the NLP Interchange Format (NIF). The system is also provided as open source software (Apache V2 license), allowing its use as a Java/Scala component in third party systems.

Linking MASC to DBpedia provides a first step to evaluate the aforementioned tasks against manual annotations (see below). Moreover, it enables queries that permeate different levels of annotation, in combination with domain knowledge from DBpedia.

[8]Note that [68] have also defined a TF-ICF measure (where "C" stands for "corpus") with the objective to generate vectors of streaming documents in linear time. In DBpedia Spotlight's TF*ICF (where "C" stands for "candidate") the objective is to give more weight to words that are rare among confusable entities.

[9]http://spotlight.dbpedia.org/demo

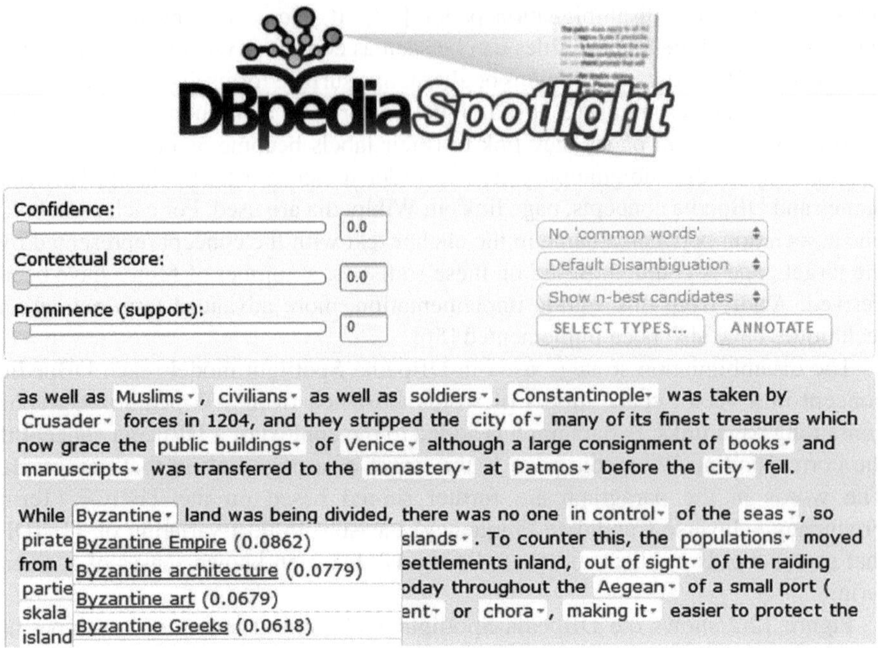


**Fig. 12.2** DBpedia Spotlight web application showing an annotated snippet from MASC

## *12.3.2 MASC+WordNet: Linking Corpora to Lexical-Semantic Resources*

Lexical-semantic resources represent first-class citizens of the Semantic Web world, and aside from the newly created resources described in Sects. 12.2 and 12.4, a number of resources are already available in the Linked Open Data cloud, including several instantiations of WordNet. With corpora being represented in RDF, existing annotations for WordNet senses can thus be transformed into links between these resources.

The MASC corpus includes WordNet sense annotations [2]. Within GrAF, such annotations can only be represented as string values and processed as such. Corpus-oriented representation formalisms would not allow formulating queries that *access* the WordNet specifications for the annotated WordNet senses (or for any other information from outside the corpus). With corpora and WordNets in RDF, this can be easily accomplished, and given a mapping between WordNet sense keys and the URIs of an RDF instantiation of WordNet, the necessary links can be generated automatically.

As one example, the word *Byzantine*, highlighted in Fig. 12.2, is assigned the WordNet sense key `byzantine%3:01:00::` in MASC. Well-documented naming conventions allow us to convert this to an URI in an RDF edition of WordNet.[10] WordNet distinguishes different senses for *Byzantine*, and this particular sense pertains to the Byzantine Empire. DBpedia Spotlight also identifies the Byzantine Empire as the most likely sense of *Byzantine* here, but other DBpedia senses are almost as likely. Given an alignment between WordNet senses and DBpedia concepts (i.e., Wikipedia pages [66]), the manual WordNet annotations of the MASC can be used to develop improved disambiguation routines for DBpedia Spotlight.

This illustrates how the interoperability between corpora and lexical-semantic annotations can be used to improve existing applications, and possibly, to find new applications for existing data sets.

## 12.4 Structural Interoperability of Linguistic Databases

The Linked Data paradigm can not only be applied to establish structural interoperability for annotated corpora and lexical-semantic resources, but also for linguistic databases. In this section we discuss the Linked Data paradigm as applied to selected databases from typology and language documentation, as well as the on going efforts to establish interoperability between them.

### *12.4.1 Glottolog/Langdoc: A Global Database of Language Identifiers and Language Resources*

The Glottolog/Langdoc project[11] maintains an extensive list of hierarchically organized 'languoids' (languages, dialects and families), the genealogical relations between them, and the references treating them [32, 64]. Every languoid and every reference has its own URI. All information is available as XHTML and RDF. The bibliographical data make use of the ontologies of the Dublin Core Metadata Initiative [77] metadata and the Bibliographic Ontology (BIBO).[12] In the domain of languoids and genealogical relations, it was not possible to draw on an existing, rich infrastructure of ontologies, so a special ontology had to be developed. Glottolog/Langdoc links to other projects with a related scope, such as the Open Language Archives Community (OLAC),[13] Ethnologue,[14] Multitree,[15]

[10] http://purl.org/vocabularies/princeton/wn30/synset-Byzantine-adjective-2

[11] http://glottolog.livingsources.org

[12] http://bibliontology.com/

[13] http://language-archives.org

[14] http://www.ethnologue.com

[15] http://multitree.org/

and the World Atlas of Language Structures (WALS).[16] From these, only WALS provides its data as Linked Data in RDF at the moment.

Next to standard bibliographical data such as author and title, references are also tagged for document type (grammar, dictionary, etc.) and languages covered. The combination of this extra information with genealogical data means that very complex queries can be formulated, e.g., "Give me a dictionary of a Semitic language spoken in Eurasia that was published after 1950".

### *12.4.2 PHOIBLE: A Typological Knowledge Base of Segment Inventories*

PHOIBLE (PHOnetics Information Base and LExicon)[17] is a repository of cross-linguistic phonological segment inventory data that encompasses several legacy segment inventory databases and contains additional linguistic (e.g., distinctive features, genealogical information) and non-linguistic information (e.g., population figures, geographic data) about a large number of languages. PHOIBLE is published as Linked Data in RDF, it uses a graph-based model to represent segment inventories and their distinctive features, and it can be used to investigate descriptive universals of phonological inventories [60, 61]. Additional linguistic information about languages (e.g., genealogical classification including language stock and genus) and non-linguistic information (e.g., geographic information, population figures) is linked via other resources by ISO 639-3 language name identifiers, so that data can be queried and extracted for various statistical analyses [62].

The linking of languages and bibliographic references in PHOIBLE to ISO 639-3 URIs provides interoperability of its contents with other metadata resources and typological data sets that also use ISO 639-3 codes, such as Glottolog/Langdoc and the typological data sets mentioned in the next section. When different resources share the same URIs for identifying languages, or different URIs schemes are identified as being equivalent (e.g., through the OWL property `owl:sameAs`), then different data sets in RDF can be merged so that users can query across the various Linked Data sets. For example, with PHOIBLE users can access detailed information about the phonological system of numerous languages. However, by merging its graph with an RDF graph of a lexical resource like the Intercontinental Dictionary Series (IDS),[18] then users can identify both detailed phonological descriptions and lexicons with which to undertake typological analysis. Structural interoperability thus allows queries across linguistic data sets, so that users can locate and combine disparate data.

[16] http://wals.info

[17] http://phoible.org

[18] http://lingweb.eva.mpg.de/ids/

### 12.4.3 Other Typological Data Sets

There are a number of other large typological data sets that are being converted into RDF. Several of these come from the Max Planck Institute of Evolutionary Anthropology. They include the World Atlas of Language Structures [35, WALS], the Atlas of Pidgin and Creole Language Structures [59, APiCS] and the World Loanword Database [34, WOLD]. WALS is a large database of typological features that includes phonological, grammatical and lexical properties for hundreds of languages. APiCS complements WALS by providing comparable data on grammatical and lexical structures for 77 pidgin and creole languages. WOLD is a lexical-semantic resource that provides mini-dictionaries for 41 languages, each containing 1,000–2,000 entries, including the loanword status of each. At the moment, these resources are available as RDF. Another data set currently being converted to RDF is provided by the Automated Similarity Judgment Program [10, ASJP] that comprises information about the basic phonological shape of 40 words for over 5,000 languages. Together with Langdoc/Glottolog and PHOIBLE, these resources are subject to currently on going efforts to integrate different typological data sets using a Linked Data approach.

### 12.4.4 Integration Efforts

Relational databases are the mainstay of typologists. Be it phoneme inventories [52], word order [25] or the number of genders [21], virtually every subfield of grammar has seen a listing of the different structures found in the languages of the world, often – but not always – adopting variants of a Language-Feature data model as shown in Fig. 12.3a. Unfortunately, these databases employ ad hoc formats, often not openly accessible or transparent, and almost never interoperable. A mobilization of these resources as Linked Data can help remedy this.

Interoperability efforts can build upon two major strands of existing research:

(a) The Typological Database System (TDS) is a multi-database query system designed before the advent of Linked Data.[19] It provides unified access to several distinct databases by means of an overarching ontology, and a similar architecture can be envisioned for our harmonization efforts.
(b) Alexis Dimitriadis and colleagues developed a 'Language-Construction-Feature' (LCF) schema for the construction of several linguistic databases, e.g., pertaining to anaphora [23]. The LCF schema represents the features of a language relative to a particular construction. In this way, it provides a greater level of granularity than other schemas, and thus represents a suitable generalization over these.

[19] http://languagelink.let.uu.nl/tds

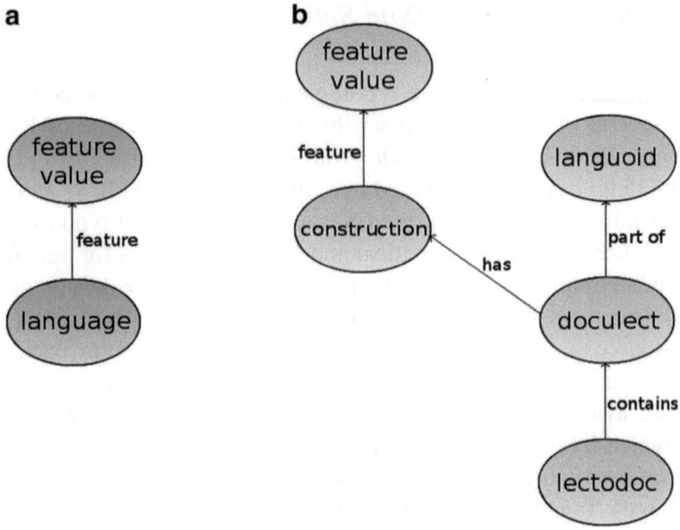


**Fig. 12.3** Data models for typological data bases. (**a**) Traditional language-feature schema. (**b**) Revised languoid-construction-feature schema

In March 2012, we conducted an exploratory workshop at the Max Planck Institute for Evolutionary Anthropology on the representation of typological databases as Linked Data. We found that the Language-Construction-Feature schema can be meaningfully applied to all domains of interest, but that the notion of 'Language' is better replaced with the notion of 'doculect' (i.e., a document-specific language variety that is an instance of a Glottolog 'languoid'). An ontology that models the relevant concepts has been posted online.[20]

Figure 12.3b shows the preliminary version of the data model that we are going to adapt. The central parts are the `languoid`, the `construction`, and the `featurevalue`. A traditional typological statement such as '*English is an SVO language*' does not have the element `construction`. Rephrasing this insight in the LCF-model yields: '*English has the English Main Clause Construction. The English Main Clause Construction has the value "SOV" for the feature "word order"*'. This allows us to model diverging information, such as '*English has the English Locative Inversion Construction. The English Locative Inversion Construction has the value "OVS" for the feature "word order"*'. The addition of the additional concept `construction` allows the representation of conflicting information. Without this addition, we would either have to state that English is rigid SVO (which is empirically wrong) or to state that it is 'mixed', without indication what extent and importance the respective constructions have.

[20]https://github.com/SebastianNordhoff/LingTyp.owl

**Table 12.1** Candidate typological databases for the conversion into Linked Data. The total number of data points is in the order of $10^7$

| Name | Domain | Features | Languages | Notes |
|---|---|---|---|---|
| WALS | Phonology, morphology, syntax, semantics, lexicon | 192 | 2,678 | Sparse |
| APICS | Phonology, morphology, syntax, semantics, lexicon | 253 | 114 | Only Creole languages |
| ASJP | Frequent words | 40 | 5,754 | Phonetically normalized and reduced |
| IDS | Words | 1,310 | 217 | See Sect. 12.4.2 |
| WOLD | Words | 1,460 | 395 | With indication of origin if borrowed |
| Numerals | Numbers | 40 | | 1–30, 40 ... 100, 200, 1,000, 2,000[21] |

Another source of conflicting information are disagreements among the sources used. Does the noun *police* require singular or plural agreement? Different grammars of English will give different answers to this (depending on whether they are using British English or American English). This raises the question of how this can be represented in RDF. We use the concept `doculect` [32, 64] (further elaborated by Good and Cysouw, 2011, Languoid, doculect & glossonym: specifying the notion 'language', unpublished manuscript) to model this. A `doculect` is the linguistic system described in a document of linguistic interest (a `lectodoc`). By linking the `construction` to `doculects` rather than to languages, we achieve two things: firstly, we are able to model disagreements among different sources. For instance, the number of vowels in Kabardian is given as 0, 2, or 4 by different authors. Assigning only one value to the language Kabardian here would be misleading. But assigning the three values to three `doculects`, and assigning the `doculects` in turn to Kabardian is possible. Secondly, we make the assertion more empirical by providing the exact source from where the information was drawn, and allow linking to the bibliographic domain of the Semantic Web.

The conversion of existing databases was found to be trivial from a conceptual point of view, but it represents a considerable effort because of the different file formats involved. We are planning to convert up to ten typological databases to RDF and make them available as Linked Data (Table 12.1).

The original databases did not have any collaborative editing in mind; the maintainer was the only person thought to contribute to many of the databases. However, the success of articles on small languages in Wikipedia and the increased connectivity of developing countries (where the majority of the world's languages are spoken) shows that there is crowdsourcing potential in this domain. We are

[21]Compiled by Eugene Chang, http://lingweb.eva.mpg.de/numeral/

planning to make our knowledge base available in OntoWiki,[22] which will allow users to contribute to the knowledge base in a controlled way. OntoWiki allows for collaborative editing with access control, versioning and rollback functionalities. The data is stored in RDF and users can draw on existing ontologies to structure their data or add new concepts to the ontology. We hope that this will be a good way to facilitate crowdsourcing for the aggregation of linguistic information.

Beyond this, representing both types of data, existing typological databases and crowd-sourced additions to these, in RDF, allows us to interface them more easily with other linguistic resources, including corpora, lexical-semantic resources, or – as discussed in the following section – terminology and metadata repositories.

## 12.5 Conceptual Interoperability of Language Resources

So far, we described interoperability and information integration from a *structural* point of view, i.e., how the same formalism (here, RDF) can be applied to represent annotated corpora, their linking with lexical-semantic resources and typological databases. This section illustrates how the same principles can be applied to enhance *conceptual interoperability* among and between linguistic resources; with metadata, annotations and other forms of linguistic resources represented in centralized terminology repositories, different linguistic resources can ground their concepts in these repositories just by linking to them. If these terminology repositories are linked with each other, or if the same repository is referred to, then we can verify that concepts, annotations or metadata categories are well-defined on the basis of a shared set of definitions.

### *12.5.1 Linguistic Terminology*

The last decades have seen numerous attempts to develop machine-readable taxonomies of linguistic terms for different purposes, including (but not limited to) linguistic glossing rules [3], standards for linguistic annotation and NLP [48], and ontologies for development and documentation of linguistic terminology [70].

In recent years, two resources have become particularly influential. The **General Ontology of Linguistic Description (GOLD)**[23] is an ontology for descriptive linguistics [27, 28]. It attempts to codify linguistic knowledge with the goal of facilitating automated reasoning over linguistic data. It provides a large number of URIs for linguistic resources to use for terminology resolution, in particular

[22]http://ontowiki.net

[23]http://linguistics-ontology.org

for morphosyntactic categories and their relations. GOLD is currently maintained by the Linguist List,[24] a multi-disciplinary community has grown around it, and a community process has been initiated that led to continuous improvement of the resource. GOLD directly employs the Semantic Web formalism OWL, and can thus be used as Linked Data, just by referring to concepts in GOLD. The ontology is freely accessible, but no explicit license information is provided at this time.

The **ISO TC37/SC4 Data Category Registry (ISOcat)**[25] serves as a host for community-defined, as well as user-defined terminologies [42, 46, also see Wright et al., this vol.]. Unlike GOLD, which formalizes linguistic terms in a single taxonomy, ISOcat is semi-structured in that it provides limited possibilities to express relations between linguistic concepts, which are, moreover, optional. By supporting user-defined data categories, it follows a grass-roots approach such that a broad range of possible applications can be covered, but no global coherence is enforced. Accordingly, ISOcat currently features different sub-profiles with partially redundant information. For example, the concept `noun` is represented by DC-1333 (introduced April 2004) and DC-2704 (introduced January 2010 as part of the ISOcat sub-profile for the Polish National Corpus). Recently, the GOLD data has been added as an ISOcat sub-profile, and introduced yet another data category, DC-3347, for nouns. To formalize the relations between these data categories, a Relation Category Registry (RELcat) has been developed that shall be used to define equivalence and other relations [71], but this integration process is still in its early stages. Similar to GOLD, ISOcat follows the vision of freely available data (Menzo Windhouwer, personal communication, March 2012), and it is accessible over the internet, but no explicit license information is provided. Technically, ISOcat data is available in a special-purpose XML format, but an RDF representation has been developed, as well, that can be used as a point of reference for Linked Data [78].

### *12.5.2 Application to Language Resources*

While GOLD and ISOcat provide higher-level information about linguistic terminology, concrete resources employ more specialized schemes, whose definitions do not only follow theoretical considerations, but also practical demands. If two categories overlap, for example, *his house* where *his* is both a pronoun (in a semantic sense) and a determiner (in its syntactic function), the author of an annotation scheme has to make a design decision how to annotate elements in the intersection, and these design decisions may differ in different annotation schemes and resources. Also, different terminological traditions may be involved, such that the same term is used in different ways. In order to avoid confusion between

[24] http://linguistlist.org/

[25] http://www.isocat.org

identical terms with different meanings in different language resources, but also to represent whether different terms have the same meaning (e.g., the word class 'adverbial participle' in Russian, which is also described as 'transgressive aspect' of verbs), resource-specific terminologies and reference terminologies should be physically separated and formalized independently. In this way, the definition of concepts (for either a resource or reference terminology) is clearly distinguished from its interpretation (which may be conducted by a layman not acquainted with the resource under consideration, say, an NLP engineer). With formal representations for both terminological systems, and explicit links between them, it becomes possible to trace back misinterpretations and erroneous equivalence statements to their origin. In this way, conceptual interoperability can be established in a sustainable and reproducible way that a direct link between a resource and a terminology repository would not have allowed.

This idea is implemented in the **Ontologies of Linguistic Annotation (OLiA)** architecture [15], where reference categories and annotation schemes for various linguistic phenomena and currently about 70 languages are formalized as independent ontologies. OLiA establishes interoperability between different annotation schemes by linking them to an overarching 'Reference Model'. Linked Data principles are applied to connect annotation schemes with the reference model, but also with external terminology repositories; interoperability with community-maintained data category registries can be achieved through the linking between the OLiA Reference Model and both GOLD and (an OWL/DL representation of) the morphosyntactic profile of ISOcat. Unlike GOLD and ISOcat, the OLiA ontologies are distributed under an open license (CC-BY/CC-BY-SA), and with the definitions provided in the Reference Model they thus can act as a central reference hub for linguistic annotations in the realm of Linked Open Data.

For a concrete example, we may consider the example from the MASC corpus again, where the phrase *Byzantine land* was described in POWLA as `?phrase has_cat "NP"`. Given the information that MASC syntax annotation follows the Penn Treebank [7], and a formal representation of this scheme,[26] the underlying data property can be rephrased as a URI reference to the annotation scheme, so that it becomes possible to query for `?phrase a penn-syntax:NounPhrase`. The concept `penn-syntax:NounPhrase` is defined as a subclass of `olia:NounPhrase`,[27] the corresponding concept in the OLiA Reference Model (which is further linked to GOLD and ISOcat). Correspondingly, we could query for `?phrase a olia:NounPhrase`. Unlike the original string-based query, this expression is no longer resource-specific, but could also be applied to a resource that uses another annotation scheme, e.g., that of the German TüBa-D/Z corpus [73], where a different tag for noun phrases is used (`NX`). Resources are thus conceptually interoperable.

[26] http://purl.org/olia/penn-syntax.owl

[27] http://purl.org/olia/penn-syntax-link.rdf

### 12.5.3 Language Metadata

The same principles can also be applied to represent linguistic metadata, i.e., information *about* linguistic resources, such as place and time of origin, the language, information about author, edition history and annotator, etc.

Different terminology repositories have been developed for this purpose. The Linked Open Data cloud currently contains two resources of this type that are also linked with each other, **Lexvo**[28] and **lingvoj**.[29] Both provide linguistic and non-linguistic information that can be used to formalize metadata about language resources. They capture information about languages, words, characters and other language-related entities as Linked Data, e.g., URIs that can be used in Linked Data resources to identify the languages represented in their data sets. Other Linked Open Data resources provide information that is not specific to linguistics, e.g., **GeoNames** [76] provides identifiers for geographic regions and places.

Section 12.4.1 described Langdoc/Glottolog, where Glottolog serves as a structuring device for the bibliography (Langdoc) in that it provides identifiers for languages, language families and dialects. But the languoids it provides can also be used to specify the language that a corpus is written in. Glottolog is linked with Lexvo, but whereas the language identifiers in Lexvo are just an implementation of ISO 639-3 codes, Glottolog provides a greater level of detail, because it covers more fine-grained differentiations between languages, dialects and language families that are necessary for typological research. For linguistic applications where this level of detail is necessary, Glottolog can also be used as a metadata repository.

### 12.5.4 Outlook: Building Bridges Between Linguistics and Natural Language Processing

Sections 12.2–12.4 described efforts to integrate different linguistic resources on the basis of common RDF specifications. For this task, it is, however, not sufficient to rely on interoperable representations of the data, but also, the harmonization of metadata is essential. As shown above, this can be accomplished if shared repositories are created, maintained and actively used by the participating communities. Efforts to establish such repositories are underway, but for the integration of typological databases, no standardized repository to identify dialects and languages has been established so far. ISO 639-3, as provided for instance by lexvo, provides more than 7,000 codes for languages, but does not include dialects. This means that certain types of resources cannot be annotated with the required degree of granularity. The multitree project contains codes for dialects and language families,

[28] http://www.lexvo.org

[29] http://www.lingvoj.org

but is not available in RDF. It furthermore assigns the same code to nodes in different authors' classifications. The code ALTC is for instance assigned to both 'Macro-Altaic' and 'Micro-Altaic', which are obviously different in scope. This represents a major conceptual problem and makes it difficult to use this repository for unambiguous reference in a Semantic Web context. Glottolog has as the goal to assign unique codes to every node of every tree, and represent the relations between the nodes as RDF. This will allow a more granular annotation of linguistic resources with information about the linguistic variety they are relevant for.

By building a user community around the set of resources and technologies described here, the necessary repositories can be developed and, for example, different legacy databases can be integrated with each other on the basis of shared data types, shared terminology repositories and Linked Data principles.

But not only databases can be harmonized with each other in this way. Also, bridges between disciplines such as typology and NLP can be established; if the same metadata repositories are used as reference for both annotated corpora and typological databases, it is actually possible to find resources provided by another community. For example, an increasingly growing branch of research in computational linguistics is dedicated to less-resourced languages, i.e., languages that are lacking fundamental NLP resources such as corpora, machine-readable lexicons, annotation schemes, part-of-speech taggers, etc. Although this line of research is not particularly focusing on *endangered* languages, the NLP community is increasingly aware of the band-width of typological variation, and interested in the challenges associated with it. Recent examples include the creation of annotated corpora, morphological analyzers or parsers for languages like Formosan [72], Wambaya [5], Lule Sámi [75], or Syriac (Neo-Aramaic, [53]), all of which are endangered (i.e., spoken by very few speakers).

One of the goals of NLP research in this respect is to develop technologies to create NLP resources through little effort, using techniques like annotation projection on parallel corpora [39], or where these are not available, to adapt existing resources from one language to another related language [75]. In the field of typology, these developments are hardly known, but they could be used to facilitate linguistic research. Even if projected tools achieve low performance, they can be applied to filter out sentences and constructions that are of particular interest. In fact, if a tool that is projected from a non-endangered language to a related minority language *fails* to analyze a sentence, this may indicate where the minority language structurally differs from its better documented sibling language. In this way, findings from NLP can actually trigger hypotheses for typological research.

On the other hand, typological resources may become of interest to NLP engineers in case novel tools have to be developed quickly. A well-known example is the rapid development of a machine translation system for Haitian Creole after the 2010 earthquake [50]. Before the catastrophe, Haitian Creole had mostly been studied out of academic interest, but in the aftermath resources for Haitian Creole (lexicons and corpora) were identified and used to facilitate the development of NLP tools to assist in tasks like machine translation.

These examples show how resources created in linguistic research and NLP can complement each other. Linking resources from typology (and other academic disciplines of linguistics), annotated corpora, machine-readable lexicons and linguistic bibliographies to the same set of identifiers for language and region can help to discover relevant resources for both disciplines. Beyond that, if the equivalent identifiers for linguistic phenomena are applied, then the information in these resources can be directly integrated with each other, e.g., glosses in a typological database can refer to the same set of linguistic categories as grammatical characterizations in a machine-readable lexicon developed for part-of-speech tagging in the NLP community.

The Linked Data paradigm provides a technological framework in which cross-references can be established; resources created by independent research groups working on different problems within different communities can be interlinked with each other and with a common specifications for metadata and linguistic terminology. If the resources are published under open licenses, the resulting 'cloud' of resources is then accessible to other groups, and a resource may be more easily reused for another application. In this way, a Linked Open Data (sub-)cloud of linguistic resources generates network effects as those sketched above, which may eventually lead to better (cross-disciplinary) visibility of existing resources, to a greater likelihood for this resource to be reused, to an improved flow of information between the participating researchers, and, of course, it will allow researchers to apply formalisms and technologies that have been developed in the context of the Semantic Web to novel applications, thereby anchoring linguistic research and NLP in a rich and vivid technological ecosystem. Initial steps towards the formation of such a Linguistic Linked Open Data cloud are described in the following section.

## 12.6 Towards a Linguistic Linked Open Data Cloud

This section summarizes the advantages and benefits of modeling linguistic resources as Linked Data, in particular, non-lexical-semantic resources that have previously and rarely been discussed in a Linked Data context. Further, it posits these efforts in the larger context of the **Open Linguistics Working Group (OWLG)**, a recent community effort whose goals include the creation of a Linked Open Data (sub-)cloud of linguistic resources. We call this network of resources the **Linguistic Linked Open Data (LLOD) cloud**.

### *12.6.1 Modeling Linguistic Resources as Linked Data: Advantages and Benefits*

One of the key advantages of publishing Linked Data is that resources are globally and uniquely identified and can be easily found through standard Web protocols. Moreover, resources can be easily linked to one another in a uniform fashion.

Chiarcos et al. [20] have recently summarized the benefits for the application of the Linked Data paradigm to linguistic resources. Aside from aspects of structural and conceptual interoperability that were already discussed in previous sections, they identified the following main benefits: (a) linking through URIs, (b) federation, (c) dynamic linking between resources, and (d) the availability of a rich ecosystem of formats and technologies.

**Linking through URIs:** One of the key ideas of Linked Data is that every single resource is identified by a Uniform Resource Identifier (URI) that figures both as a global identifier and as a Web address – i.e., a description of the resource is available if you request it from its URI on the Web. This can be as simple as providing a document for download at the given address. However, RDF allows for a standard description of such resources on the Web and hence for automatic processing of these resources. It is not necessarily the case that the data must be solely available as RDF, as the HTTP protocol supports *content negotiation*: as one example, the URL http://purl.org/vocabularies/princeton/wn30/synset-Byzantine-adjective-2 mentioned earlier in this chapter does resolve to a human-readable representation if opened in a browser, but to an RDF file if the HTTP header requests `application/rdf+xml`.

**Information integration at query runtime (federation):** As resources can be uniquely identified and easily referenced from any other resource on the Web through URIs, the connections between these resources can be navigated even during query runtime. In effect, this allows the creation of a linked web of data similar to the effect of hyperlinks in the HTML Web. Moreover, it is possible to use existing Semantic Web methods such as Semantic PingBack [74] to be informed of new incoming links to your resource. Semantic Pingback returns a location in the HTTP header whereby referencing resources that can be used to inform the user of possible connections to other resources. Along with HTTP-accessible repositories and resolvable URIs, it is possible to combine information from physically separated repositories in a single query at runtime. Information from different resources in the cloud can then be integrated freely.

**Dynamic import:** If cross-references between linguistic resources are represented by resolvable URIs instead of system-defined ID references or static copies of parts from another resource, it is not only possible to resolve them at runtime, but also to have access to the most recent version of a resource. For community-maintained terminology repositories like GOLD or ISOcat, for example, new categories, definitions or examples can be introduced occasionally, and this information is available immediately to anyone whose resources refer to GOLD or ISOcat URIs.

**Ecosystem:** RDF as a data exchange framework is maintained by an interdisciplinary, large and active community, and it comes with a developed infrastructure that provides APIs, database implementations, technical support and validators for various RDF-based languages, e.g., reasoners for OWL. For developers of linguistic resources, this ecosystem can provide technological support or off-the-shelf implementations for common problems, e.g., the development of a database that is capable of support flexible, graph-based data structures as necessary for multi-layer corpora (Sect. 12.2).

Beyond this, another advantage warrants a mention: the distributed approach of the Linked Data paradigm facilitates the distributed development of a web of resources and collaboration between researchers that provide and use this data and that employ a shared set of technologies. One consequence is the emergence of interdisciplinary efforts to create large and interconnected sets of resources in linguistics and beyond. One of these initiatives is described in the following section.

### *12.6.2 Ongoing Community Efforts*

Recent years have seen not only a number of approaches to provide linguistic data as Linked Data, but also the emergence of larger initiatives that aim at interconnecting these resources, culminating on the creation of a Linguistic Linked Open Data (LLOD) cloud, i.e., a Linked Open Data (sub-)cloud of linguistic resources.

One such example is the **Open Linguistics Working Group (OWLG)**,[30] a network open to anyone interested in linguistic resources and/or the publication of these under an open license. The OWLG is a working group of the Open Knowledge Foundation (OKFN),[31] a community-based non-profit organization promoting open knowledge (i.e., data and content that is free to use, re-use and to be distributed without restriction). The OWLG adopts the principles, definitions and infrastructure of the OKFN as far as they are relevant for linguistic data. The OKFN defines standards and develops tools that allow anyone to create, discover and share open data. The Open Definition of the OKFN states that "openness" refers to: "A piece of content or data [that] is open if anyone is free to use, reuse, and redistribute it – subject only, at most, to the requirement to attribute and share-alike."[32] The Open Definition is accompanied by a list of compliant licenses. One important aspect here is that openness should not constrain the use of data provided under these licenses, e.g., its commercial use, and that it should allow to complement these with proprietary resources. Adopting this understanding of openness for the LLOD cloud warantees that the resources contained in the cloud are available under *any circumstances*.

Since its formation in 2010, the Open Linguistics Working Group has grown steadily. One of OWLG's primary goals is to attain openness in linguistics through:

(a) Promoting the idea of open linguistic resources,
(b) Developing the means for the representation of open data, and
(c) Encouraging the exchange of ideas across different disciplines.

---

[30] http://linguistics.okfn.org

[31] http://okfn.org/

[32] http://opendefinition.org

Publishing linguistic data under open licenses is an important issue in academic research, as well as in the development of applications. We see increasing support for this in the linguistics community [65], and there are a growing number of resources published under open licenses [58]. There are many reasons for publishing resources under open licenses: for instance, freely available data can be more easily re-used, double investments can be avoided, and results can be replicated. Also, other researchers can build on this data, and subsequently refer to the publications associated with it. Nevertheless, a number of ethical, legal and sociological problems are associated with open data,[33] and the technologies that establish interoperability (and thus, re-usability) of linguistic resources are still under development. The OWLG represents an open forum for interested individuals to address these and related problems.

The OWLG maintains a home page, a mailing list, and a wiki. We conduct regular meetings, organize workshops (e.g., Linked Data in Linguistics, LDL-2012, held in conjunction with the 34th Annual Meeting of the German Linguistic Society, DGfS-2012, Frankfurt/M., Germany, March 2012, or Multilingual Linked Open Data for Enterprises, MLODE-2012, held in conjunction with the 3rd Conference on Software Agents and Services for Business, Research, and E-Sciences, SABRE-2012, Leipzig, Germany, September 2012) and document our efforts [19].

One central aspect in our work is the focus on *openness* and on the problems and benefits associated with using, maintaining, and distributing open linguistic resources. The OWLG provides a platform for sharing experiences and technology across discipline boundaries, as researchers work with field-specific technologies, but face similar issues. For instance, heterogeneous data, interoperability and legal questions arise in lexicography, corpus research, and linguistic typology alike.

At the time of writing, the OWLG consists of about 100 people from 20 different countries. Our group is relatively small, but continuously growing and sufficiently heterogeneous. It includes people from library science, typology, historical linguistics, cognitive science, computational linguistics, and information technology; the ground for fruitful interdisciplinary discussions has been laid out.

One of the activities that involve a large number of OWLG members is the development of the LLOD cloud. Independent research activities of many community members involve the application of RDF/OWL to represent linguistic corpora, lexical-semantic resources, terminology repositories and metadata collections about linguistic data collections and publications, and to many of them, the Linked Open Data paradigm represents a particularly appealing set of

[33] For example, complex copyright situations may arise if one resource (say, a lexicon) was developed on the basis of another resource (say, a newspaper archive), and researchers are uncertain whether the examples from the original newspaper contained in the lexicon violate the original copyright. Ethical problems may arise if a data base of quotations from a newspaper is linked to a data base of speakers, and this data base is further connected with, say, obituaries from the same newspaper. Even if this was done only in order to study generation-specific language variation, one may wonder whether such an accumulation of information violates the privacy of the people involved.

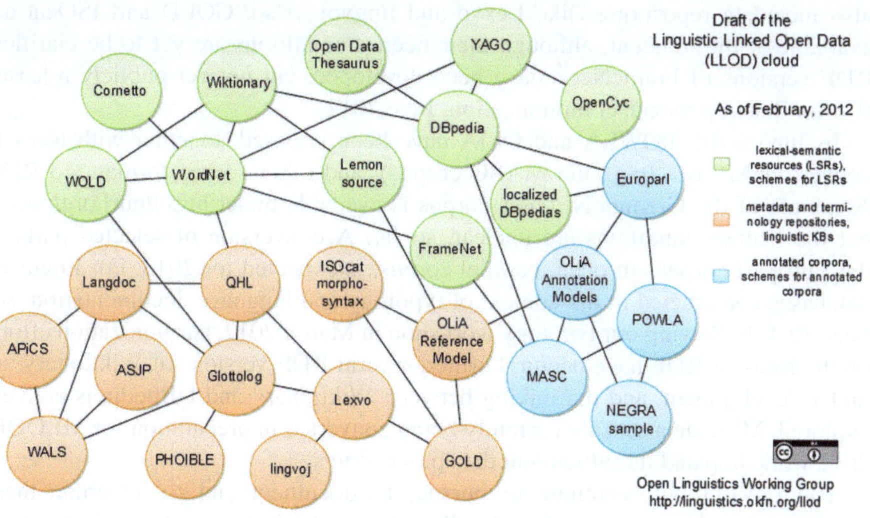


**Fig. 12.4** Draft version of the Linguistic Linked Open Data cloud

technologies. Consequently, a major goal in the recent past has been the creation of a LLOD cloud from this compilation. We selected 28 of these resources to investigate the possibility of establishing cross-links between them. A draft for the LLOD cloud diagram, inspired by the Linked Open Data diagram by Cyganiak and Jentzsch,[34] is shown in Fig. 12.4.

We apply the following criteria for a new linguistic resource to be included in the LLOD cloud diagram: (1) The data is resolvable through HTTP, (2) it is provided as RDF, (3) it contains links to another data set in the diagram, and (4) the entire data set must be available. In order to add a new data set, a contributor would have to create a Web or wiki page and announce the resource on the OWLG mailing list. The diagram itself is maintained in a repository and can be edited collaboratively.

As of September 2012, the LLOD cloud diagram has *draft* status. This means that resources and their linkings do not yet have to be provided (even though many of them are available already), but that their publication under LLOD conditions is promised by the data providers. To distinguish the draft from the final diagram, directionality of edges is not yet marked, because the linking is not necessarily already available. The shift from draft to official status will require that all resources shown in the diagram are published under LLOD conditions and is a process that we initiated with the MLODE-2012 workshop.

At the time of writing, the conversion of data sets to RDF and the creation of links between them is on going. Some of the resources are already available, including lexical-semantic resources like the DBpedia, different RDF versions of WordNet, Cornetto (Dutch WordNet), OpenCyc, and the Open Data Thesaurus, but

[34]http://lod-cloud.net

also metadata repositories like Lexvo and lingvoj. Also, GOLD and ISOcat are available at the moment, although their license conditions are yet to be clarified. RDF versions of FrameNet[35] have been developed, but not yet publicly released (Collin Baker, personal communication, June 2012).

In June 2012, POWLA and OLiA have been released, together with tools to convert GrAF data (e.g., the MASC corpus), and other source formats, to RDF. A sample of the German NEGRA corpus is available under http://purl.org/powla, but the license conditions are unclear, so far. A conversion of selected parts of the Europarl corpus, an open, parallel corpus, is expected for 2013. For linguistic databases constructed in the context of typology and language documentation, see Sect. 12.4. Following our typology workshop in March 2012, harmonization efforts on the basis of RDF have begun. Further, several RDF versions of Wiktionary are under development, and the linking between Wiktionary and DBpedia is actively explored. More data sets are currently being converted in preparation for MLODE-2012 workshop and its subsequent data proceedings.

The OWLG will continue to pursue, to document and to advertise these efforts, and invites other interested colleagues and/or initiatives to participate and collaborate with us.

## 12.7 Summary

This chapter described on going efforts to create a Linked Open Data Cloud of linguistic resources, the Linguistic Linked Open Data (LLOD) cloud.

Aside from lexical-semantic resources, that have long been standing in the center of interest of the Semantic Web community and to which the application of the Linked Data paradigm is an established technique, only few publications so far have addressed the representation of other types of linguistic resources specifically within the Linked Open Data cloud.

The primary objective of this chapter was to provide an overview and introduction with respect to the Linked Data paradigm in its application to specific types of linguistic resources. An implicit evaluation of the feasibility of the approach can be seen in its applicability to highly diverse resources and its (unique) potential to bring these together: Already the successful application of the Linked Data paradigm to diverse types of resources shows the genericity of the approach. From existing large-scale data bases with high access rates maintained by the Semantic Web community (e.g., DBpedia, see Sect. 12.3), we know that the technological infrastructure is capable to deal with large amounts of data at reasonable speed.

This chapter provides an overview with respect to the application of the Linked Open Data paradigm to resources that are *specific* to linguistic research, namely annotated corpora and linguistic data bases as Linked Data. We discussed benefits

[35]http://framenet.icsi.berkeley.edu

and advantages of this approach, mostly in terms of structural interoperability. We also showed how the Linked Data paradigm can be applied to facilitate the conceptual interoperability between and among corpora and linguistic data bases, and we gave a brief overview over the Open Linguistics Working Group of the Open Knowledge Foundation under whose umbrella the activities described here are conducted.

**Acknowledgements** In parts, this chapter is based on a number of earlier conference presentations, including [14, 18] and [38]. We would like to thank the contributors to these papers: Jonas Brekle, Philipp Cimiano, Judith Eckle-Kohler, Iryna Gurevych, Silvana Hartmann, Sebastian Hellmann, Jens Lehmann, Michael Matuschek, John McCrae, Christian M. Meyer, and Claus Stadler. We would also like to thank all other OWLG members, as well as the participants of LDL-2012. Further, we would like to express our gratitude towards the anonymous reviewers for feedback and comments. The research of the first author was partially supported by a DAAD postdoctoral fellowship at the Information Sciences Institute of the University of Southern California.

## References

1. Abney S, Bird S (2010) The Human Language Project: Building a universal corpus of the world's languages. In: 48th Annual Meeting of the Association for Computational Linguistics (ACL-2010), Uppsala, Sweden, pp 88–97
2. Baker C, Fellbaum C (2009) WordNet and FrameNet as complementary resources for annotation. In: 3rd Linguistic Annotation Workshop (LAW-2009), Suntec, Singapore, pp 125–129
3. Bakker D, Dahl O, Haspelmath M, Koptjevskaja-Tamm M, Lehmann C, Siewierska A (1993) EUROTYP guidelines. Technical report, European Science Foundation Programme in Language Typology
4. Baker C, Fillmore C, Lowe J (1998) The Berkeley FrameNet project. In: 36th Annual Meeting of the Association for Computational Linguistics (ACL-1998), Montréal, Canada, pp 86–90
5. Bender E (2008) Evaluating a crosslinguistic grammar resource: A case study of Wambaya. In: 46th Annual Meeting of the Association for Computational Linguistics: Human Language Technologies (ACL-2008: HLT), Columbus, Ohio, pp 977–985
6. Berners-Lee T (2006) Design issues: Linked data. http://www.w3.org/DesignIssues/LinkedData.html. Accessed 31 July 2012
7. Bies A, Ferguson M, Katz K, MacIntyre R (1995) Bracketing guidelines for Treebank II style Penn Treebank project. ftp://ftp.cis.upenn.edu/pub/treebank/doc/manual/root.ps.gz. Accessed 31 July 2012, version of January 1995
8. Bird S, Liberman M (2001) A formal framework for linguistic annotation. Speech Commun 33(1-2):23–60
9. Brandes U, Eiglsperger M, Herman I, Himsolt M, Marshall M (2002) GraphML progress report: Structural layer proposal. In: 9th International Symposium on Graph Drawing (GD-2001), Vienna, Austria, pp 501–512
10. Brown C, Holman E, Wichmann S, Velupillai V (2008) Automated classification of the world's languages. STUF Lang Typol Univers 61(4):286–308
11. Burchardt A, Padó S, Spohr D, Frank A, Heid U (2008) Formalising multi-layer corpora in OWL/DL – Lexicon modelling, querying and consistency control. In: 3rd International Joint Conference on NLP (IJCNLP-2008), Hyderabad, India
12. Carletta J, Evert S, Heid U et al (2003) The NITE XML toolkit: Flexible annotation for multi-modal language data. Behav Res Methods Instrum Comput 35(3):353–363
13. Cassidy S (2010) An RDF realisation of LAF in the DADA annotation server. In: 5th Joint ISO-ACL/SIGSEM Workshop on Interoperable Semantic Annotation (ISA-5), Hong Kong, China

14. Chiarcos C (2012) A generic formalism to represent linguistic corpora in RDF and OWL/DL. In: 8th International Conference on Language Resources and Evaluation (LREC-2012), Istanbul, Turkey, pp 3205–3212
15. Chiarcos C (2012) Ontologies of linguistic annotation: Survey and perspectives. In: 8th International Conference on Language Resources and Evaluation (LREC-2012), Istanbul, Turkey, pp 303–310
16. Chiarcos C (2012) POWLA: Modeling linguistic corpora in OWL/DL. In: 9th Extended Semantic Web Conference (ESWC-2012), Heraklion, Crete, pp 225–239
17. Chiarcos C, Dipper S, Götze M, Leser U, Lüdeling A, Ritz J, Stede M (2008) A flexible framework for integrating annotations from different tools and tagsets. TAL (Traitement automatique des langues) 49(2):217–246
18. Chiarcos C, Hellmann S, Nordhoff S, Moran S, Littauer R, Eckle-Kohler J, Gurevych I, Hartmann S, Matuschek M, Meyer C (2012) The Open Linguistics Working Group. In: 8th International Conference on Language Resources and Evaluation (LREC-2012), Istanbul, Turkey, pp 3603–3610
19. Chiarcos C, Nordhoff S, Hellmann S (eds) (2012) Linked data in linguistics. Representing and connecting language data and language metadata. Springer, Heidelberg
20. Chiarcos C, McCrae J, Cimiano P, Fellbaum C (2012) Towards open data for linguistics: Linguistic linked data. In: Oltramari A, Lu-Qin, Vossen P, Hovy E (eds) New Trends of Research in Ontologies and Lexical Resources. Springer, Heidelberg
21. Corbett G (2005) Number of genders. In: Haspelmath M, Dryer M, Gil D, Comrie B (eds) The World Atlas of Language Structures. Oxford University Press, Oxford
22. Declerck T (2006) SynAF: Towards a standard for syntactic annotation. In: 5th International Conference on Language Resources and Evaluation (LREC-2006), Genoa, Italy, pp 229–232
23. Dimitriadis A, Everaert M, Reinhart T, Reuland E (2005) Anaphora typology database. http://languagelink.let.uu.nl/anatyp. Accessed 31 July 2012
24. Dostert L (1955) The Georgetown-IBM experiment. In: Locke W, Booth A (eds) Machine translation of languages. Wiley, New York, pp 124–135
25. Dryer M (1997) On the six-way word order typology. Stud Lang 21(1):69–103
26. Eckart R (2008) Choosing an XML database for linguistically annotated corpora. Sprache und Datenverarbeitung 32(1):7–22
27. Farrar S, Langendoen DT (2003) A linguistic ontology for the Semantic Web. GLOT Int 7:97–100
28. Farrar S, Langendoen DT (2010) An OWL-DL implementation of GOLD: An ontology for the Semantic Web. In: Witt A, Metzing D (eds) Linguistic modeling of information and markup languages, Springer, Dordrecht
29. Francis WN, Kucera H (1964) Brown Corpus manual, revised edition. Technival report, Brown University, Providence, Rhode Island, 1979
30. Francopoulo G, George M, Calzolari N, Monachini M, Bel N, Pet M, Soria C (2006) Lexical Markup Framework (LMF). In: 5th International Conference on Language Resources and Evaluation (LREC-2006), Genoa, Italy, pp 233–236
31. Gangemi A, Navigli R, Velardi P (2003) The OntoWordNet project: Extension and axiomatization of conceptual relations in WordNet. In: Meersman R, Tari Z (eds) Proceedings of On the Move to Meaningful Internet Systems (OTM-2003), Catania, Italy, pp 820–838
32. Good J, Hendryx-Parker C (2006) Modeling contested categorization in linguistic databases. In: EMELD Workshop on Digital Language Documentation, East Lansing, MI
33. Greenberg J (1960) A quantitative approach to the morphological typology of languages. Int J Am Linguist 26:178–194
34. Haspelmath M, Tadmor U (eds) (2009) World Loanword Database. Max Planck Digital Library, Munich
35. Haspelmath M, Dryer M, Gil D, Comrie B (eds) (2008) The World Atlas of Language Structures online. Max Planck Digital Library, Munich
36. Hellmann S, Unbehauen J, Chiarcos C, Ngonga Ngomo AC (2010) The TIGER Corpus Navigator. In: 9th International Workshop on Treebanks and Linguistic Theories (TLT-9), Tartu, Estonia, pp 91–102

37. Hellmann S, Lehmann J, Auer S (2012) Linked-data aware URI schemes for referencing text fragments. In: 18th International Conference on Knowledge Engineering and Knowledge Management (EKAW-2012), Galway, Ireland
38. Hellmann S, Stadler C, Lehmann J (2012) The German DBpedia: A sense repository for linking entities. In: Chiarcos C, Nordhoff S, Hellmann S (eds) Linked data in linguistics, Springer, Heidelberg, pp 181–190
39. Hwa R, Resnik P, Weinberg A, Cabezas C, Kolak O (2005) Bootstrapping parsers via syntactic projection across parallel texts. Nat Lang Eng 11(3):311–325
40. Ide N, Pustejovsky J (2010) What does interoperability mean, anyway? Toward an operational definition of interoperability. In: 2nd International Conference on Global Interoperability for Language Resources (ICGL-2010), Hong Kong, China
41. Ide N, Romary L (2004) International standard for a linguistic annotation framework. Nat Lang Eng 10(3-4):211–225
42. Ide N, Romary L (2004) A registry of standard data categories for linguistic annotation. In: 4th International Conference on Language Resources and Evaluation (LREC-2004), Lisbon, Portugal, pp 135–139
43. Ide N, Romary L (2006) Representing linguistic corpora and their annotations. In: 5th International Conference on Language Resources and Evaluation (LREC-2006), Genoa, Italy, pp 225–228
44. Ide N, Suderman K (2007) GrAF: A graph-based format for linguistic annotations. In: 1st Linguistic Annotation Workshop (LAW-2007), Prague, Czech Republic, pp 1–8
45. Ide N, Baker CF, Fellbaum C, Fillmore CJ, Passonneau R (2008) MASC: The Manually Annotated Sub-Corpus of American English. In: 6th International Conference on Language Resources and Evaluation (LREC-2008), Marrakech, Morocco, pp 2455–2461
46. Kemps-Snijders M, Windhouwer M, Wittenburg P, Wright S (2008) ISOcat: Corralling data categories in the wild. In: 6th International Conference on Language Resources and Evaluation (LREC-2008), Marrakech, Morocco, pp 887–891
47. Kilgarriff A, Grefenstette G (2003) Introduction to the special issue on the Web as Corpus. Comput Linguist 29(3):333–347
48. Leech G, Wilson A (1996) EAGLES recommendations for the morphosyntactic annotation of corpora. http://www.ilc.cnr.it/EAGLES/annotate/annotate.html. Accessed 31 July 2012
49. Lehmann J, Bizer C, Kobilarov G et al (2009) DBpedia – A crystallization point for the Web of Data. J Web Semant 7(3):154–165
50. Lewis W (2010) Haitian Creole: How to build and ship an MT engine from scratch in 4 days, 17 hours, & 30 minutes. In: 14th Annual Conference of the European Association for Machine Translation (EAMT-2010), Saint-Raphaël, France
51. Lux M, Laußmann J, Mehler A, Menßen C (2011) An online platform for visualizing lexical networks. In: 2011 IEEE/WIC/ACM International Conference on Web Intelligence and Intelligent Agent Technology (WI-IAT 2011), Lyons, France, pp 495–496
52. Maddieson I (1984) Patterns of Sounds. Cambridge University Press, Cambridge/New York
53. McClanahan P, Busby G, Haertel R, Heal K, Lonsdale D, Seppi K, Ringger E (2010) A probabilistic morphological analyzer for Syriac. In: 14th Conference on Empirical Methods on Natural Language Processing (EMNLP-2010), Cambridge, MA, pp 810–820
54. McCrae J, Spohr D, Cimiano P (2011) Linking lexical resources and ontologies on the semantic web with Lemon. In: 8th Extended Semantic Web Conference (ESWC-2011), Heraklion, Crete. Springer, pp 245–259
55. Mendes P, Jakob M, García-Silva A, Bizer C (2011) DBpedia Spotlight: Shedding light on the Web of Documents. In: 7th International Conference on Semantic Systems (I-Semantics 2011), Graz, Austria
56. Mendes P, Daiber J, Rajapakse R et al (2012) Evaluating the impact of phrase recognition on concept tagging. In: 8th International Conference on Language Resources and Evaluation (LREC-2012), Istanbul, Turkey
57. Mendes P, Jakob M, Bizer C (2012) DBpedia for NLP: A multilingual cross-domain knowledge base. In: 8th International Conference on Language Resources and Evaluation (LREC-2012), Istanbul, Turkey

58. Meyers A, Ide N, Denoyer L, Shinyama Y (2007) The shared corpora working group report. In: 1st Linguistic Annotation Workshop (LAW-2007), Prague, Czech Republic, pp 184–190
59. Michaelis S, Maurer P, Haspelmath M, Huber M (eds) (to appear 2013) Atlas of Pidgin and Creole Language Structures. Oxford University Press, Oxford
60. Moran S (2012) Phonetics information base and lexicon. PhD thesis, University of Washington
61. Moran S (2012) Using linked data to create a typological knowledge base. In: Chiarcos C, Nordhoff S, Hellmann S (eds) Linked data in linguistics. Springer, Heidelberg, pp 129–138
62. Moran S, McCloy D, Wright R (2012) Revisiting the population vs phoneme inventory correlation. In: 86th Annual Meeting of the Linguistic Society of America (LSA-2012), Portland, OR
63. Morris W (ed) (1969) The American Heritage dictionary of the English language. Houghton Mifflin, New York
64. Nordhoff S, Hammarström H (2011) Glottolog/Langdoc: Defining dialects, languages, and language families as collections of resources. In: 1st International Workshop on Linked Science (LISC-2011), Bonn, Germany
65. Pederson T (2008) Empiricism is not a matter of faith. Comput Linguist 34(3):465–470
66. Ponzetto S, Navigli R (2009) Large-scale taxonomy mapping for restructuring and integrating Wikipedia. In: 21st International Joint Conference on Artificial Intelligence (IJCAI-2009), Pasadena, CA, pp 2083–2088
67. Prud'Hommeaux E, Seaborne A (2008) SPARQL query language for RDF. W3C Recommendation. http://www.w3.org/TR/rdf-sparql-query, Accessed Dec, 20th, 2013
68. Reed JW, Jiao Y, Potok TE, Klump BA, Elmore MT, Hurson AR (2006) TF-ICF: A new term weighting scheme for clustering dynamic data streams. In: 5th International Conference on Machine Learning and Applications (ICMLA-2006), Washington, DC, pp 258–263
69. Romary L, Zeldes A, Zipser F (2011) [Tiger2/] – serialising the ISO SynAF syntactic object model. Arxiv preprint arXiv:11080631
70. Schneider R (2007) A database-driven ontology for German grammar. In: Rehm G, Witt A, Lemnitzer L (eds) Data structures for linguistic resources and applications, Narr, Tübingen, pp 305–314
71. Schuurman I, Windhouwer M (2011) Explicit semantics for enriched documents. What do ISOcat, RELcat and SCHEMAcat have to offer?. In: 2nd Supporting Digital Humanities Conference, Copenhagen, Denmark
72. Su L, Sung L, Huang S, Hsieh F, Lin Z (2008) NTU corpus of Formosan languages: A state-of-the-art report. Corpus Linguist Lingust Theory 4(2):291–294
73. Telljohann H, Hinrichs E, Kübler S, Zinsmeister H, Beck K (2003) Stylebook for the Tübingen treebank of written German (TüBa-D/Z). Technical report, Seminar für Sprachwissenschaft, Universität Tübingen, Germany
74. Tramp S, Frischmuth P, Arndt N, Ermilov T, Auer S (2011) Weaving a distributed, semantic social network for mobile users. In: 8th Extended Semantic Web Conference (ESWC-2011), Heraklion, Crete, pp 200–214
75. Tyers F, Wiechetek L, Trosterud T (2009) Developing prototypes for machine translation between two Sámi languages. In: 13th Annual Conference of the European Association for Machine Translation (EAMT-2009), Barcelona, Spain, pp 120–127
76. Vatant B, Wick M (2012) GeoNames ontology. http://www.geonames.org/ontology. Accessed 31 July 2012, version 3.01
77. Weibel S, Kunze J, Lagoze C, Wolf M (1998) RFC 2413 – Dublin core metadata for resource discovery. http://www.ietf.org/rfc/rfc2413.txt. Accessed 31 July 2012, Network Working Group
78. Windhouwer M, Wright S (2012) Linking to linguistic data categories in ISOcat. In: Chiarcos C, Nordhoff S, Hellmann S (eds) Linked data in linguistics, Springer, Heidelberg, pp 99–107

# Chapter 13
# Community Efforts Around the ISOcat Data Category Registry

**Sue Ellen Wright, Menzo Windhouwer, Ineke Schuurman, and Marc Kemps-Snijders**

**Abstract** The ISOcat *Data Category Registry* provides a community computing environment for creating, storing, retrieving, harmonizing and standardizing *data category* specifications (DCs), used to register linguistic terms used in various fields. This chapter recounts the history of DC documentation in TC 37, beginning from paper-based lists created for lexicographers and terminologists and progressing to the development of a web-based resource for a much broader range of users. While describing the considerable strides that have been made to collect a very large comprehensive collection of DCs, it also outlines difficulties that have arisen in developing a fully operative web-based computing environment for achieving consensus on data category names, definitions, and selections and describes efforts to overcome some of the present shortcomings and to establish positive working procedures designed to engage a wide range of people involved in the creation of language resources.

S.E. Wright
Kent State University, Kent, OH, USA
e-mail: swright@kent.edu

M. Windhouwer (✉)
The Language Archive, DANS, The Hague, The Netherlands
e-mail: menzo.windhouwer@dans.knaw.nl

I. Schuurman
KU Leuven and Utrecht University, Leuven, Belgium and Utrecht, The Netherlands
e-mail: ineke.schuurman@ccl.kuleuven.be

M. Kemps-Snijders
Meertens Instituut Amsterdam, Amsterdam, The Netherlands
e-mail: marc.kemps.snijders@meertens.knaw.nl

I. Gurevych and J. Kim (eds.), *The People's Web Meets NLP*, Theory and Applications of Natural Language Processing, DOI 10.1007/978-3-642-35085-6_13,

## 13.1 Introduction

The ISOcat *Data Category Registry* (DCR) provides a community computing environment for creating, storing, retrieving, harmonizing and standardizing *data category* specifications (DCs) and *Data Category Selections* (DCSs) used to create a wide range of language resources.[1] According to formal definition, a DC is the 'result of the specification of a given data field' [5], which essentially implies that a DC comprises the concept (together with the name) of a data field, although in practice in the DCR, DCs include field names (complex DCs that can have content) and permissible instances (enumerated values that are listed for use with closed complex DCs). The intent of the DCR is to encourage the creators of language resources to use consistent DCs (and in some cases, consistent data models) in order to encourage the leveraging of semantic information across resource, application, and platform boundaries.

The DCR is hosted by The Language Archive, a unit of the Max Planck Institute for Psycholinguistics in Nijmegen (NL), and operated under the auspices of ISO TC 37, *Terminology and other language and content resources*, but it is nonetheless intended to function as an open service for DC creators and users working in a freely accessible web context. The 'community' served by the DCR can only be defined by breaking down a broader sense of community into a set of constituents. At the broadest level, this community is made up of a number of *Communities of Practice* (CoP), each of which comprises a group of people who share a professional or scholarly commitment to a particular domain. CoPs can frequently be broken down into user groups associated with the use of a resource, application, or platform. In practice, some user groups are involved in creating the DCR, but others may simply be desirous of consulting the resource in order to retrieve and apply data found there. Roles in this context include Guests, who may access information in the DCR, and Experts, who may freely register as such in order to create their own DCs and DCSs.

On a more formal basis, *Thematic Domain Groups* (TDGs) have been established inside the DCR to perform the more structured task of standardizing and harmonizing specific DCs for use in more controlled environments where a high level of interoperability is desirable. In addition to ISO-appointed TDGs, there are also semi-official infrastructure groups that have taken on responsibility for maintaining some of the DCSs inside the DCR, such as CLARIN-NL/VL and Athens Core (see Sect. 13.5).

This chapter recounts the history of DC documentation in TC 37, beginning from paper-based lists created for lexicographers and terminologists and progressing to the development of a web-based resource for a broad range of CoPs and user groups. While describing the considerable strides that have been made to collect a very large comprehensive collection of DCs, it also outlines difficulties that have arisen

[1] See http://www.isocat.org.

in developing a fully operative web-based computing environment for achieving consensus on data category names, definitions, and selections. The final sections of the paper describe efforts to overcome some of the present shortcomings and to establish positive working procedures designed to engage a wide range of CoPs involved in the creation of language resources.

## 13.2 Historical Perspective

### *13.2.1 Communities of Practice and User Groups*

Each CoP making up the stakeholders of the DCR has its own 'cultural history', defined best practices, and terminological usage. These groups are often aware that they are for the most part defining the same or very similar sets of language-related concepts, but individual researchers are often oblivious to the fact that they are almost always generating different faceted views of those concepts within the framework of the broad environment of language resources. In principle at least, it would be highly desirable to be able to leverage these knowledge units across language resources, regardless of theoretical approach, application environment, or computing platform, especially in situations where definitions have been standardized or otherwise represent the consensus of known experts. Nevertheless, the ability to access and process such information depends on identifiable models and mappable element identifiers (DC names and persistent identifiers) in order to ensure common understanding, both at the human and the machine-processing level. Even with intelligent mapping, it is important to bear in mind that incommensurability and the indeterminacy of language that inevitably occurs between Communities of Practice is likely to lead to anomalies and semantic inconsistencies [10].

These language resources include:

(a) Lexicography and machine-readable lexicons, such as machine translation glossaries
(b) Discourse-purposed terminology, i.e., termbanks, terminology management for writers and translators
(c) Subject-purposed controlled vocabularies, such as thesauri and SKOS resources
(d) Language-related and linguistic databases of various types
(e) Annotation frameworks for use in corpus collection and markup
(f) Metadata used in language-related and even other database environments (CMDI[2] and ISO 11179 communities[3])

[2]http://www.clarin.eu/cmdi/

[3]http://metadata-standards.org/11179/

(g) Ontologies, especially formats for leveraging 'term'-related objects across platforms and varying formats (OntoIOp community,[4] OWL,[5] RDF, Common Logic[6])
(h) Archives for endangered languages
(i) Documentation of sign (gestural) languages used by the deaf and hearing impaired
(j) Assessment schemes for translation or annotation schemes used for parallel bi- and multilingual corpora
(k) Annotation schemes for recorded interpreting sessions

### *13.2.2 Starting Out on Paper*

It should come as no surprise that early language resources created by some CoPs were paper-based. The most well-known early example of a structured community-based effort to create a major language resource was perhaps the Oxford English Dictionary (OED): begun in 1857, the 'community' in question grew from a relatively small group of dictionary-aficionados to include hundreds of men and women scholars scattered across the English-speaking world documenting words and word forms using quasi-uniform 'slips' designed primarily to document usage and provenance as identified in significant works of English literature. Here the designation of types of information (main forms, part of speech, etymology, etc.) in word-oriented lexical entries is achieved by the now-famous Oxford entry layout, which uses font variation to represent the different kinds of information contained in a lexicographical entry (see Fig. 13.1).

The terminology management community also began on paper before the computer era, creating the tradition of the pre-printed paper fiche, ISO A5 (or other smaller sized) cards or slips of light or heavier paper stock segmented into sections for the various data categories associated with concept-oriented, frequently multilingual terminological entries [14]. With the advent of the wide-spread use of computers for word-processing and then glossary production by technical writers and translators in the 1980s and 1990s, previously visionary efforts led to real projects to standardize data category names, and for a while, the abbreviations used to cope with the space limitations first of paper fiche and then of DOS-era term entries.

By the late 1990s ISO TC 37/SC 3/WG 1 elaborated ISO 12620:1999, Data Categories [4] as a companion standard to the SGML-based ISO 12200, Computer applications in terminology – Machine-readable terminology interchange format

[4]http://ontoiop.org/

[5]http://www.w3.org/standards/semanticweb/

[6]http://iso-commonlogic.org

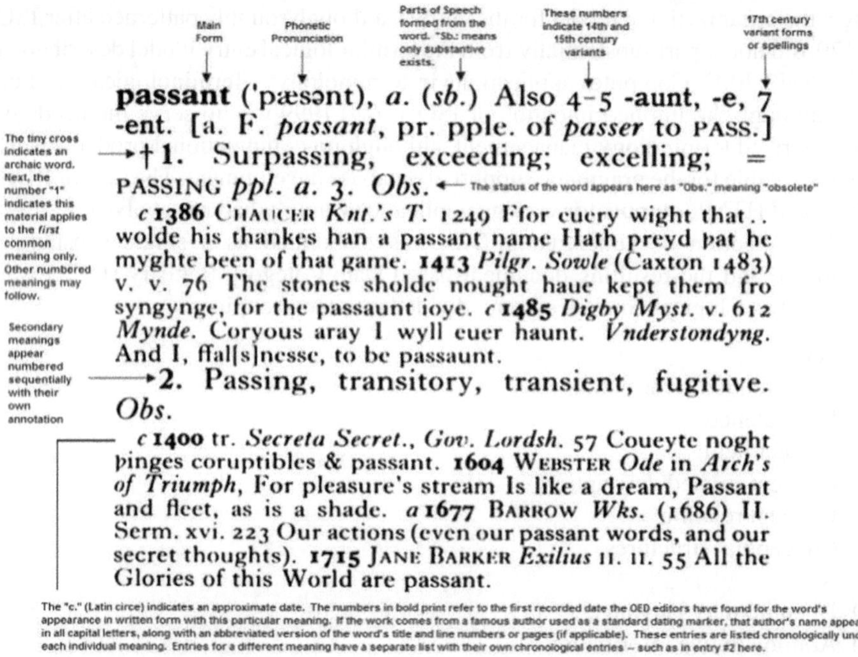


**passant** ('pæsənt), *a.* (*sb.*) Also 4–5 **-aunt, -e,** 7 **-ent.** [a. F. *passant*, pr. pple. of *passer* to PASS.] †**1.** Surpassing, exceeding; excelling; = PASSING *ppl. a.* 3. *Obs.*

*c* **1386** CHAUCER *Knt.'s T.* 1249 Ffor euery wight that .. wolde his thankes han a passant name Hath preyd þat he myghte been of that game. **1413** *Pilgr. Sowle* (Caxton 1483) v. v. 76 The stones sholde nought haue kept them fro syngynge, for the passaunt ioye. *c* **1485** *Digby Myst.* v. 612 *Mynde.* Coryous aray I wyll euer haunt. *Vnderstondyng.* And I, ffal[s]nesse, to be passaunt.

**2.** Passing, transitory, transient, fugitive. *Obs.*

*c* **1400** tr. *Secreta Secret., Gov. Lordsh.* 57 Coueyte noght þinges coruptibles & passant. **1604** WEBSTER *Ode* in *Arch's of Triumph*, For pleasure's stream Is like a dream, Passant and fleet, as is a shade. *a* **1677** BARROW *Wks.* (1686) II. Serm. xvi. 223 Our actions (even our passant words, and our secret thoughts). **1715** JANE BARKER *Exilius* II. II. 55 All the Glories of this World are passant.

**Fig. 13.1** Sample OED entry[7]

(MARTIF) – Negotiated interchange [3]. Although intended as a support standard for the machine-processing of terminological data, the standard itself was created as a paper document modeled roughly on the approach specified by the ISO 11179, Metadata Registries (MSR) family of standards, which describes the standardization of data element concepts for use as metadata. The six-part ISO 11179 standard was also under active development at the time, which made keeping pace between the working groups extremely difficult, coupled with the fact that language experts working in TC 37, although they were increasingly involved in computing environments, became uncomfortable with aspects of ISO 11179 as it moved further and further away from familiar lexicographical and terminological practice.

In its paper incarnation, ISO 12620:1999 was developed primarily as a structured list of categories under the control of a single project leader, with group input via the still-customary ISO comments process. (This somewhat archaic process uses an MS Word™ template, in which experts enter comments identified by document paragraph number. Although this process is cumbersome during data entry, an effective macro then combines comments from individuals to produce a master comment document that provides an effective guide for group discussions.) At this

[7] See http://www.oxford-royale.co.uk/news/2010/12/04/new-online-edition-of-oxford-english-dictionary.html

stage, the organizational model for the entries, although roughly patterned after ISO 11179, did not depart substantially from the terminological entry model described in ISO 16642:2003, Computer applications in terminology – Terminological markup [6]. Although the implied intention of ISO 12620:1999 was to serve the needs of computerized terminology management, although no computational products were made available for the pragmatic support of database development. The idea of RDF, XML, and HTML outputs for varying applications simply had not evolved.

The collection of hundreds of DCs was elaborated as a single substantial document, and the resulting document-based Data Category Registry (DCR) was presented as a logical concept system, divided into ten groupings:

(a) Term
(b) Term-related information
(c) Equivalence
(d) Subject field
(e) Concept-related description
(f) Concept relation
(g) Conceptual structures
(h) Note
(i) Documentary language
(j) Administrative information

This classification of the DCs reflected TC 37-recommended policy to create concept systems when elaborating terminology collections, but even though all the experts involved in the process were terminologists, differences among individual user groups were great enough that more time was spent in the end discussing ordering systems than defining data category concepts. A further drawback of the paper-based hierarchical system was the creation of mnemonic DC identifiers (e.g., A.2.2.1 in Fig. 13.2), which hindered the addition of new concept at any point in the ordering system because any new item was bound to break the continuity of the existing system.

This experience led to the insight that the assertion of relations between DCs can be highly individualistic, and even within closely related CoPs, may vary widely not only from application to application, but also in the selection of required DCs and in the relations that function within a given computing environment. As a consequence, further elaboration of the data categories within TC 37, both in paper and in computerized forms, has dispensed with the notion of any single ordering system, giving rise to the implementation of external Relation Registries that should be allowed to proliferate to meet individual needs.

As noted, the elaborators of ISO 12620:1999 were terminologists, although many probably also had experience in lexicography. The working group represented administrators of large national term banks, well-trained translation-oriented terminologists managing terminology in large-scale commercial environments, and translator trainers, mostly working at the university level. There was no formal link between these experts and a broader range of working terminologists, and no mechanism for 'outsiders' to contribute data categories or to comment on those

A.2.2.1 part of speech
NONADMITTED NAME 1: grammatical category
NONADMITTED NAME 2: word class
DESCRIPTION: A category assigned to a word based on its grammatical and semantic properties.
PERMISSIBLE INSTANCES: Examples of parts of speech commonly documented in terminology databases can include:
a) noun
b) verb
c) adjective

**Fig. 13.2** Data Category specification from ISO 12620:1999; note the short list, which reflects the tendency of Terminology to limit the set of permissible values used for */part of speech/*

that were included. Furthermore, the breadth of the collection, although powerful, proved daunting to those unfamiliar with the elaboration process. In addition to these factors, TC 37 was rapidly growing to include additional CoPs, in particular corpus linguists elaborating annotation frameworks for the markup of a wide variety of text corpora, as well as computational linguists involved in the creation of linguistic metadata, again for a wide variety of applications. Consequently, as is so often the case with a standard, ISO 12620:1999 was outdated and limited in range even at the moment it went into effect as an international standard.

### *13.2.3 Moving from Paper to the Web*

As TC 37 language experts prepared to move forward with the specification and standardization of data categories for a broader range of language resources, it was obvious that reworking ISO 12620 as a traditional paper document did not make sense. In order for data categories to truly function as standardized units they first and foremost needed to be readily accessible in processable form as an open resource – ideally in the form of an ISO 11179-style metadata registry. The transition from paper to a database environment evolved through two major stages under the support of multiple funding environments. During the HLT-funded SALT project (Standards-based Access to multilingual Lexical and Terminological resources), the working group remained focused on lexicography (including machine translation lexicons) and terminology management and developed an XML-based format and tools for storing, exchanging, and processing these data based on the DCs defined originally in ISO 12620:1999 [1].

The LIRICS-related,[8] SYNTAX project was initially planned to be the DCR, but in the end constituted a significant pilot project that made progress toward a global resource by incorporating a wider range of CoPs, most prominently Morphosyntax,

[8]http://lirics.loria.fr retrieved 2012-8-30

and laid the groundwork for some of the features that became standard in ISOcat during the course of the CLARIN project [9, 12].

By the development of ISOcat, it was clear that any new Data Category Registry needed to feature:

- Its own data model designed to reflect the needs and functions of a DCR apart from the terminology data model specified in ISO 16642
- Open involvement of the broader linguistics community, encompassing a wide range of Communities of Practice
- Within the open-access framework, the ability to establish a core of CoP-specific experts with the authority and tools to standardize DCs and in some cases, DCSs
- Non-mnemonic, persistent DC identifiers that could be referenced from anywhere at any time in web environments, which enables
  - Referenceability to reliable definitions from any language resource anywhere on the web
  - The anchoring of assertions in Relation Registries to individual data category specifications
- The ability to share data categories (such as */part of speech/*) across CoPs, which builds efficiency into the system, but also lays the ground work for leveraging semantics between language resources
- The ability to subset data categories for individual or group applications
- The ability to create groups of expert users, to share responsibilities with these users, and to discuss issues involving shared DCs
- The ability to keep DCs or DCSs private or to make them public, as needed
- The ability to represent or output DCs and DCSs in a variety of modes, e.g., as HTML tables (which can be imported to a number of other formats), as RDF, in a native XML mode representing entire data sets, etc.

The creation of the ISOcat DCR paralleled the elaboration of ISO 12620:2009. Unlike its predecessor, the new standard did not list data categories – the catalog of DC statistics in Sect. 13.3 bear witness to the impossibility of continuing a paper document that would reflect the wealth of items needed in today's computing environments. Instead, the standard specifies the data model for the new DCR and outlines the roles to be played by a set of administrative groups who would propose, elaborate, and eventually pass judgment on the standardization of data category specifications proposed for standardization. The entire DC submission and approval process as defined in the new standard is then physically implemented within the functional structure of the ISOcat database.

For better or worse, great effort went into conforming the ISOcat system to a then-valid ISO document called Annex ST of the ISO directives, a standardization management system designed to enable the creation of so-called 'standards as databases' [2]. The vision of Annex ST was to mirror ISO balloting procedures during the approval process for individual data categories. As a consequence, the ISO Committee Draft (CD) and Draft International Standard (DIS) are reflected in the proposed work of the Thematic Domain Groups (TDGs) followed up by final approval by the DCR Board. This procedure has given rise to a complex balloting system implemented programmatically inside ISOcat. Unfortunately, after considerable effort went into planning, describing and actually implementing this system, ISO Central Secretariat withdrew Annex ST along with its plans to provide a central home for standardized data element descriptions. Although this turn of events may be viewed on the one hand with chagrin, on the other hand it affords the ISOcat team the opportunity to explore a simplified voting procedure in the future. It could even be that the system described for the CLARIN projects (see Sect. 13.5) will point the way to a more workable approach.

## 13.3 Community Support in ISOcat

Based on the needs of the various CoPs, ISO 12620:2009 specifies the foundation of an open DCR, i.e., where anyone can add new data categories, but with the aim to build a standardized core of data categories. Thus the vision underlying ISOcat is the one of openness: everyone can add the data categories needed by him or her. These individual users can be members of one or more user groups which can correspond with various CoPs. To allow all the group and community members in general to be involved in the process of selecting and possibly defining suitable data categories, ISOcat offers various means to support cooperation. The corner stone is the ability for everyone to become a registered ISOcat user. A registered user (expert) can create a group and invite other users to become a member of this group. Individual or selections of data categories can be shared by a user with a group on various levels: (a) for reading, (b) for extending the selection or (c) for editing data categories. User groups can thus work together towards a stable and usable selection of data categories. Once they are satisfied with the selection they have assembled, they can make it and the data categories that they have created public.

Once data categories and selections are public, anyone can inspect these data category specifications. Sometimes a data category created for one community is very close to what is needed by another. The owner of the data category might not be known by the interested community. To foster contact and hopefully harmonization of data categories, ISOcat offers the means to send a mediated email to any registered user, e.g., a data category owner or a group member. The mediation of ISOcat allows users to get in contact without exposing entrusted email addresses on the Internet.[9]

Discussions involving more community members or more than one community can be supported by ISOcat fora, which can be created on demand. These fora can be private and only accessible by the members of a group, or they can be public and accessible by anyone, even guests, i.e., non-registered ISOcat users. Groups can

---

[9]The email address of the sender is exposed to the recipient so only the first introduction email is mediated.

have fora of both kinds. Although fora have been created and have been briefly used, none of them has seen continuous active use by a user group or CoP (see Sect. 13.2.3 for discussion of a pilot test).

Additional functionality is provided to the Thematic Domain Groups (TDGs) and the DCR Board to support the standardization of data categories. Each TDG can have a public and private forum to discuss the needs of the thematic domain. When data categories are submitted for standardization, a public forum for discussions during the standardization process is created. These submission-specific fora are open to TDGs, the DCR Board and everyone during the whole standardization process. The various phases are also supported by timed ballots, which allow TDG or DCR Board members to accept or reject individual candidate data categories. Although several TDGs have worked extensively on DCs and DCSs, at the time of writing, no submission for standardization has been processed.

The statistics in Table 13.1 illustrate the current status of individual and community efforts in the ISOcat DCR. Based on these statistics some trends in the CoP and TDG efforts around ISOcat can be seen. More than half of the data categories are shared and more than half of the users cooperate with other users. Communication happens mainly outside of ISOcat, as the fora are underused. A quarter of the data categories are not assigned to any thematic domain, so they cannot be standardized. The cause could be that ISOcat users do not feel or understand the need to assign DCs to profiles or the current set of (linguistic) thematic domains is not broad enough. Some groups within TC 37 have also been slow to approve TDGs, or having done so, to activate their collections.

## 13.4 Standardization Community Efforts

### *13.4.1 Standards as Databases*

The extensive catalog of DCs described statistically in Sect. 13.3 actually poses a huge barrier to the effective use of the DCR by any but the most experienced and dedicated users. Unless users have a fairly firm notion of what their needs may be, the contemplation of many hundreds of DCs is not an encouraging prospect. To make matters worse, there are many duplicates throughout the collection, which may give rise to questions concerning which of several possible options may be useful in a given situation. The relative quality of the entries varies as well. As described above, the original set of terminological DCs was elaborated in a series of carefully controlled projects by trained terminologists. There are no significant doublettes within this DCS, but since the profiles for the different TDGs were created separately, especially during the SYNTAX phase of the process, there are duplicates between this set and Morphosyntax in particular.

With regard to the Metadata DCS, duplicates tend to be of a different sort: here metadata collections from several different sources were batch loaded into the DCR with little effort to harmonize duplicates. Definition style varies drastically, and it can be very difficult to guess whether a given item covers the same data category concept as another, or whether it might be a subordinate or related data category instead – or possibly some unrelated item that only appears to be relevant. There are also spelling variants, which ideally should be consolidated using the option to list variable data category names. Furthermore, some DCSs are associated with existing or soon-to-be completed TC 37 standards, such as ISO 30042 (TBX) [8] and ISO 24611 (MAF) [7]. As a consequence, there have been complaints that the collection is unreliable. In the end this results in a high pressure to standardize important DCs within the DCR and to be able to designate the associated sets as standardized DCSs.

### *13.4.2 First Trials*

As noted in Sect. 13.3, the ISOcat model includes balloting procedures and the option to create both public and private fora for discussing the proper specification of data categories during the standardization process. As a trial run in early summer 2011, the system designers planned a pilot submission of a short set of data categories taken from the Metadata TDG, with the intention of 'walking through' the complete process of balloting and approval.

The submission subset: The system administrator selected a sub-set of 42 DCs from the Metadata TDG, 27 of which were complex DCs, and the remainder simple DCs making up value domains for some of the complex DCs. The selection was not especially motivated, in that it did not reflect a coherent subset reflecting any given application or approach, but some items were related (e.g., */audio/, /video/, /capture method/, /media type/*), and there was a variety of data category types. Two items presented a problem for discussion because they were not clearly differentiated either by data category name or definition:

*/environment/*

*Description of the environmental conditions under which the recording was created.*

*/running environment/*

*Specification of the running environment that is required to execute the tool/service.*

The definitions pose a number of problems. First of all, they are both tautological in that the name of the data category is repeated in the definition. The concept of 'environment' might imply that there is some sort of link between the two categories, but what that relation might be is unclear. What sort of 'conditions' are referenced in the first definitions – are they audio-related conditions, or more ambient conditions? This environment involves a 'recording' – potentially a field recording? – while the second requirement appears to involve the 'execution of a tool/service'. Closer examination of the DCR reveals yet another data category specification:

*/recording environment/*

*The environment where the recording took place.*

One might certainly assume that environment and recording environment are unnecessary duplicates. It is precisely this kind of conundrum that the standardization process could at least theoretically be intended to resolve. If, for instance, we assume that *environment* and *recording environment* are identical, we could conceivably give preference to one and suppress the other, but what if in the parent database from which one or the other came there are relations that are incompatible with the data environment in the other database? The question clouds prospects for seamless harmonization, but there is no easy solution within ISOcat itself. Making use of the RELcat Relation Registry (cf. Sect. 13.6), however, we can state that *environment* is in a same-as relation with *recording environment.*

The DCR Chair reviewed the selected entries and updated them to conform to the requirements of ISO 12620:2009, which are more stringent than the requirements originally implemented during the SYNTAX project. This step alone is slightly problematic from a community computing standpoint because the reasons for some of the requirements, although valuable within a global computing context, are not always clear to the average ISOcat user, so even new DC specifications sometimes lack the requisite items to clear the automated quality check before being submitted for standardization. The chair also added comments in the */explanation/* and */note/* sections, and proposed proper form for definitions where necessary. She also tried to introduce questions designed to sort out the 'environment' issues raised in the previous paragraph.

Although the DCs included in the Metadata DCS were indeed discussed during the various meetings of the so-called Athens Core[10] group, the totality of the collection represents an aggregation of a number of separate DCSs already in use throughout the Thematic Domain. No effort had been taken during this phase of the collection process to (1) harmonize definitions and express hierarchical relations among major DCs or to (2) weed out duplicate and to explicitate similar, but different DCs. The significant difference between this approach and the rigorous sorting and redefining that took place with the Terminology DCs reflects a clear theoretical and methodological difference in approach – the kind of difference that embodies the Kuhnian incommensurability inherent in the interfaces between the various CoPs working within the broader scope of language resources in TC 37. Operating as a strict terminologist, the DCR Chair was proceeding on the premise that the major foci of the standardization process included:

- Refining definitions to conform to TC 37 rules for concept-oriented definitions
- Weeding out and harmonizing duplicate or closely related DCs
- Coordinating subsets with the Metadata subset for coherent, application-oriented presentation

**The pilot assessment group:** The pilot assessment group was made up of Metadata TDG members, from which various were also active members of the Athens

[10]The Athens Core group was named after the first meeting of large number of metadata modellers for language resources, which took place in Athens in 2009. A series of (online) meetings resulted in a set of more than 200 metadata elements, which were made publically available in the ISOcat DCR.

Core group. They may or may not have actually been familiar with the mechanics of the ISOcat system itself, but they should have been more or less oriented to strategies for on-line discussion.

**The pilot procedure:** An initial teleconference was designed to familiarize everyone with the procedures being tested. The intention was to spend roughly 2 weeks discussing the issues raised in the existing comments and to introduce any new ones that the assessment group encountered in looking at the list, to reach consensus on any problematic issues, and then to run a test ballot on the subset presented for this test run. Unfortunately, the actual 'test run' did not play out anything like this initial intention. Some of the group failed to realize that there was a stated time frame for comments and were distracted by other projects. Some useful comments came in almost 2 months after the beginning of the test. Two of the group became involved in fairly detailed discussions, but were unable to reach consensus for lack of input from other members. Finally everyone did some commenting, but comments were coming in at random times and no coherent thread of discussion arose. One source of frustration was that no one seemed to be able to answer critical questions involving the conflicts regarding the 'environment' items and no one seemed capable of taking the initiative to suggest resolutions. All told, the experiment was basically unsuccessful and did not result in any balloting phase. Efforts to resurrect the discussion failed because key members were drawn off onto other projects.

**Conclusions:** The system designers (primarily the system administrator and the DCR Chair) have attempted to analyze the difficulties encountered during this dry run.

- It is difficult to judge whether aspects of the interface and functionality played a role because there is really a need for an outside evaluation – the designers are too close to the design and too familiar with it to render a really unbiased judgment. Unfortunately, no one really spent enough time with the system to give that kind of constructive feedback.
- Certainly the participants were all competent in terms of computer experience and subject area expertise, but what was probably lacking was any compelling motivation to invest time and effort into the project – this might be caused by the fact that several had evaluated these DCs several times already in the Athens Core group.
- The ad hoc nature of the subset – that is, the fact that they did not represent any coherent set that anyone might want to use in a given application, significantly impaired buy-in.
- Although the chair tried to take a leading role in suggesting consensus positions, no one really 'owned' the set or had any experience working with it. She inappropriately expected the experts in the group to chime in with explanations and history concerning the conflicting DCs, but since this 'history' was imbedded in the ad hoc way in which they had been collected, there was no source of information that might have resolved issues. This factor poses a serious consideration in any crowd computing environment if there is any desire to

eventually extract some sort of order out of the chaos of group work – it may be challenging to reconstruct any coherent relationship among the DCs. This condition points to the potential need for relation registries of some sort. Additional justification information or the ability to reconstitute original subsets would be extremely useful, but this kind of subordinate information is precisely what many people consider a nuisance to record when creating new DCs or batch-loading existing DCSs.

## *13.4.3 New Standardization Efforts*

### 13.4.3.1 ISO 30042: Systems to Manage Terminology, Knowledge, and Content: TermBase eXchange (TBX)

As a consequence of the issues cited above, the Metadata TDG decided to lower the priority of standardization and first get user experience with the current set of public DCs, following the procedures outlined in Sect. 13.5. Two new projects are scheduled for fall 2012 that are more directly related to standardization. The first is to begin the re-standardization of the terminology DCs for ISO 30042 (TBX), which represents a substantial subset of the old ISO 12620:1999 described in Sect. 13.2. The existing DCs, plus a fair number of new ones that have been specified in the DCR since 2009 (in all 619 DCs according to Table 13.1) pose such a large set that no one anticipates that any TDG will want to tackle them all in a single pass, so the current plan is to deal with meaningful subsets and to refine working methods (and possibly interface options) as experience is gained.

The submission subset: The first subset to be discussed is a highly popular, widely used set of DCs called the 'TBX-Basic Data Category Selection', which is designed for use as a member of the so-called TBX family of terminology interchange formats.[11] Together these DCs comprise those data fields and permissible values that might be used in a relatively simple terminological entry. Not only does this project provide a coherent set of DCs; it also is a familiar one to the TDG that will be evaluating the DCs. The definitions are for the most part ones that have previously been standardized, although there are instances where suggestions for editorial changes have been made in recent years. There are some issues involving variant names, but there are no doublettes in the group.

One challenge will be that some items (for instance */part of speech/* comprise a subset of the larger Morphosyntax DC. In theory, such a DC would be 'owned by' Morphosyntax and shared with other TDGs, such as Terminology or Lexicography. Figure 13.3 illustrates the ISOcat solution for ensuring reusability in such cases. Here, Morphosyntax has around 125 values, reflecting the rich options provided by part of speech taggers for corpus markup. Terminology shows four values, and

[11] See http://www.ttt.org/oscarstandards/tbx/tbx-basic.html

**Table 13.1** Statistics of the content of the ISOcat DCR and its fora[12]

Users

- 506 registered users

Groups

- 47 groups of users
- 17 of these groups are public, i.e., members have made selections public
- 6 is the average group size, i.e., number of members of a group
- 239 users are members of one or more groups

Data categories

- 4,128 candidate data categories
- 623 deprecated or superseded data categories
- 3,446 public data categories
- 2,734 shared data categories (other group members can also edit these data categories)

Data category selections

- 392 selections
- 100 selections are shared by groups (other group members can also edit these selections)

Standardization

- 11 established TDGs
- 3 proposed TDGs
- 71 registered TDG and DCR Board members (not all members officially assigned by ISO members have registered)
- 3,279 candidate data categories, which include
  - 907 Metadata candidate data categories
  - 838 Morphosyntax candidate data categories
  - 619 Terminology candidate data categories
  - 128 Syntax candidate data categories
  - 121 Lexicography candidate data categories
  - 117 Sign Language candidate data categories
  - 103 Semantic Content Representation candidate data categories
  - 92 Translation candidate data categories
  - 72 Language Codes candidate data categories[13]
  - 35 Lexical Semantics candidate data categories
  - 33 Multilingual Information Management candidate data categories
  - 21 Lexical Resources candidate data categories
  - 13 Language Resource Ontology candidate categories
- 254 candidate data categories belong to multiple TDG related profiles
- 1,071 data categories don't belong to any TDG related profile

Fora

- 4 group related fora
- 1 public group related forum
- 4 TDG related fora
- 3 public TDG related fora
- 13 topics
- 41 posts

[12] The language codes from the ISO 639 family of standards are not represented by data categories in ISOcat. ISO TC 37 is currently working on a new code registry to make them easily accessible. The experiences of the ISOcat DCR will be taken into account during that process.

[13] These numbers are based on the status of the ISOcat DCR on July 25th 2012.

**Data Category: part of speech**

**... skipped Administration Information...**

**2. Description Section**

| | |
|---|---|
| **Profile** | Morphosyntax |
| **Profile** | Terminology |
| **2.1Data Element Name Section** | |
| **Data Element Name** | part of speech |
| **Source** | GP, ISO 12620 |
| **[+] 2.2 English Language Section** | |
| **[+] 2.3 Czech Language Section** | |
| **[+] 2.4 French Language Section** | |

**3. Conceptual Domain**

| | |
|---|---|
| **Data Type** | string |
| **Profile** | Morphosyntax |
| **Value** | /adjective / |
| **Value** | /adposition / |
| **Value** | /adverb/ |
| **Value** | /adverbialPronoun/ (adverbial pronoun) |
| **Value** | /affirmativeParticle/ (affirmative particle) |
| **Value** | /affixedPersonalPronoun/ (affixed personal pronoun) |
| **Value** | /allusive Pronoun / (allusive pronoun) |
| **Value** | /article/ |
| **... skipped 125 values...** | |
| **Value** | /voiceNoun / (voice noun) |
| **Value** | /weakPersonalPronoun/ (weak personal pronoun) |

**4. Conceptual Domain**

| | |
|---|---|
| **Data Type** | string |
| **Profile** | Terminology |
| **Value** | /adjective/ |
| **Value** | /adverb/ |
| **Value** | /noun/ |
| **Value** | /verb/ |

**Fig. 13.3** Profile-specific conceptual domain

Lexicography, if there were a special set for this, might have less than a dozen. Coordinating consistency with sets like this will entail interaction between the TDGs in order to achieve consensus on definitions in particular.

Another discontinuity that occurs is more difficult to resolve. */Part of speech/* is classified in both the Morphosyntax and Terminology profiles as a complex closed data category with declared values, where the values in the smaller set are a subset of the larger. If, however, one domain required a DC to be, say, complex open, and the other required it to be a simple DC (a value of another DC), the difference between the two instantiations would be unresolvable. For instance, in Fig. 13.3, */noun/* is a simple DC for both profiles, but suppose someone configured */noun/* as complex closed, with the values */strong/, /weak/* and */irregular/*. The only solution in such a case is to create two data category specifications for each instantiation. One could use a RELcat relation to indicate that there are shared semantics between the two DCs.

In addition to the hope that there will be greater motivation on the part of the TDG members during this process, efforts will be made to set up strict deadlines for everyone to comment on specified DCs, and there will be periodic discussions of comments, after which a formal ballot will take place for subsets of the DCS. Since this TDG is accustomed to using the ISO paper document and comments system, the DCS will be circulated as an HTML document and people will have the option of commenting in the forum or using the ISO comments template. Ideally, a set of Relation Registry relations should be created for the subset as well in order to provide additional support for definition harmonization and the general understanding of the DCs. Once the TBX-Basic subset has been completed, it can be declared a standardized subset. Then the Terminology TDG can move on to other subsets of the so-called TBX-Default DCS, always working with meaningful subsets, such as term-related DCs, administrative DCs, etc.

One of the tasks that must be addressed in standardizing DCs and DCSs is to remove all extraneous notes and explanations, along with personal references. Many sources, for instance, are identified with the initials of the TDG chairs who elaborated the DCs in question, and other comments refer to other members of discussion groups. In these cases, since the finished product will be a standardized ISO text, sources could be changed simply to 'ISO 30042', for instance.

## 13.5 Infrastructure Community Efforts

### *13.5.1 CLARIN(-NL/VL)*

The European CLARIN (Common Language Resources and Infrastructure Technology)[14] project intends to make available an integrated and interoperable research infrastructure of language resources and technology. This calls for clear, well defined metadata and other content-related linguistic concepts: both people and machines should, for example, be able to conclude what a concept A used in resource X means, and whether it means the same as in resources Y and Z.

CLARIN adopted ISOcat mainly as an integral part of the Component MetaData Infrastructure (CMDI) ISOcat contains all definitions and CMDI the metadata schemata to put the definitions in a specific context. But soon it was realized that ISOcat could also be used as vehicle for definitions of all content related concepts used within the CLARIN community. However, it turned out that some issues had to be solved, mainly due to the current lack of stable, standardized DCs in various domains. Furthermore, there was a demand for a way to relate DCs with each other,

[14]http://www.clarin.eu

as well as the possibility to associate DCs by their persistent identifiers with a specific (version of a) schema or standard.

As ISOcat also allows for the construction of DCSs, it is also possible to bring together all DCs relevant for such a project.[15] This way a user can determine which instantiation of a concept is used in a specific project, and whether it is the same as that used by another one, be it in the same language or another one. This, of course, presumes that DCs are available for many (all?) concepts, as well as DCSs for many resources.

Therefore, initiated by CLARIN-NL, the CLARIN communities in the Netherlands and Flanders[16] require that project-specific DCSs be constructed for all their projects. Currently, mainly existing resources are being made CLARIN-compatible. All metadata and content-descriptive linguistic concepts used are to be entered into ISOcat, and of course, no new DCs are to be created unless a DC with the proper definition is not yet available. But in all cases a project-specific DCS is to be constructed.

One of the challenges for ISOcat is to prohibit the proliferation of DCs, the basic building blocks of ISOcat, as everybody is allowed to add his/her contributions. As a result, there can be several DCs for apparently the same concept, while it is often unclear what the crucial differences are between the various instantiations.

Also taking into account that there are not yet any ISOcat DCs which are marked as standard, i.e., DCs that are marked for preferred usage, the CLARIN NL/VL community has decided to create a (larger) user group, CLARIN-NL/VL, of which at least everybody taking part in one of the various CLARIN-NL and/or CLARIN-VL projects (some 40 projects) is to be a member, while other expert-users of ISOcat working on Dutch are regularly invited to join as well. Currently, there are 69 members, covering a rather broad series of domains, from Metadata to Morphosyntax, Syntax and Semantic Content Representation to Sign Language and Audio.

Guidelines have been developed in order to ensure well defined, technically correct and reusable DCs, which are on the one hand both as general as possible and as specific as necessary, while, on the other hand, they are formulated in a way that also makes them useful for other projects and even other languages as well. While these guidelines were developed with the CLARIN, and especially the CLARIN NL/VL communities, in mind, they can be used by others as well. Indeed, as TC 37 considers revising ISO 12620:2009, these guidelines may serve as a model for moving away from the now cumbersome standardization process described in the standard.

Note that in principle an existing DC should be reused (see Fig. 13.4), unless this is impossible.

---

[15]Use of a Schema Registry (SCHEMAcat), will allow the storage of resource schemata persistently, each with a persistent identifier (PID) of its own. SCHEMAcat also allows the storage of different versions of a schema. ISOcat is related to SCHEMAcat, while there are also direct links between SCHEMAcat and RELcat, see also cf. [11]

[16]The northern part of Belgium with Dutch as its official language.

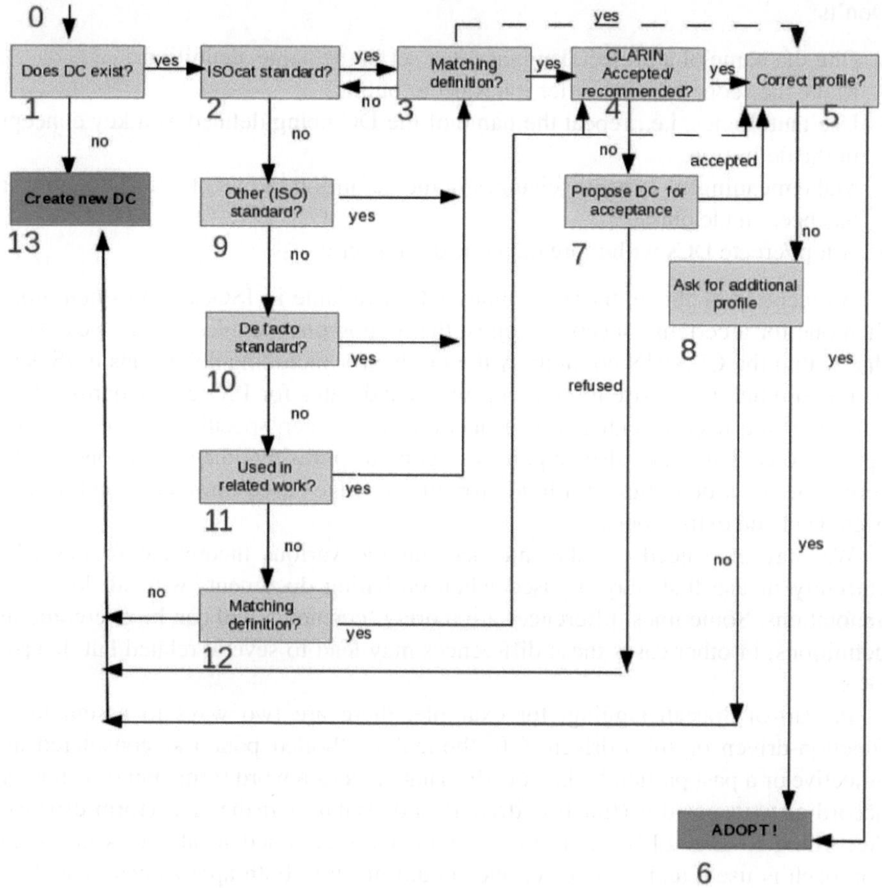


**Fig. 13.4** How to decide whether an existing DC can be adopted

Dos:

- Make your definition clear and concise
- Make your definition as general as possible
- Disambiguate your definition, i.e., when you use a concept such as ‘noun’ in it, it should be made clear which instantiation is meant by referring to it in the note section
- Mention in the justification why an existing DC cannot be used, when it belongs to a (de facto) standard[17]
- Assign the correct profile from the start
- Provide examples

[17] Only for DCs reflecting standards or contained in CLARIN-accepted/recommended.

Don'ts:

- Cite the name of a project, language, tag set in the name, definition, ...
- Make the scope 'private' after it has been 'public'
- Use tautologies, i.e., repeat the name of the DC being defined as a key concept in the definition
- Make meaningful semantic changes in the definition of one of your DCs after it has been made public[18]
- Adopt/create DCs with more than one definition

As mentioned above, there are many DCs available in ISOcat, and often more than one for a certain concept, amongst them good ones and less good ones. Note that within the CLARIN community the motive for including definitions in ISOcat is not so much the desire to come up with candidates for ISOcat standardization, but rather the need to enhance interoperability between specific resources and/or applications. This may also require less general phrasings than when one might formulate for a definition that holds for all languages used nowadays, and maybe even dead and extinct ones.

We may also need to take into account the various theoretical frameworks currently in use that may be used when enriching documents with all kinds of annotations. Sometimes differences are merely 'cosmetic' and can be overcome in definitions; in other cases these differences may lead to several related but diverse DCs.

In Part-of-Speech tagging, for example, there are two ways to assign tags: function-driven or form-driven.[19] Is 'boiled' in 'boiled potatoes' considered an adjective or a past participle, i.e. a verb? That is, does a word form change category according to its position (function-driven), or does it remain the same (form-driven)? According to EAGLES[20] it is to be explicitly mentioned in all tag sets which approach is used, and, of course, one should not mix both approaches. It will be clear that the choice made influences the definition of the word classes involved, such as adverb, adjective, and verb.

Apart from this, there are some groups of words that are considered pronouns in some tag sets but determiners in other ones, even when the same language is involved.

A third, and last, example concerns another domain, that of Semantic Content Representation. In SemAF Part1 (Time and events), an 'event' is defined as 'something that can be said to obtain or hold true, to happen or to occur'. In the document in which the standard is described, this definition comes with a note saying "This is a very broad notion of event, also known in the literature as

[18]In such a case a new DC should be constructed (with the same name), the old one should get the status 'superseded' and be linked with the new one.

[19]A third, assigning two tags, a function-driven plus a form-driven, is rarely used.

[20]EAGLES: Expert Advisory Group on Language Engineering Standards, cf. especially http://www.ilc.cnr.it/EAGLES96/annotate/node24.html#SECTION00065000000000000

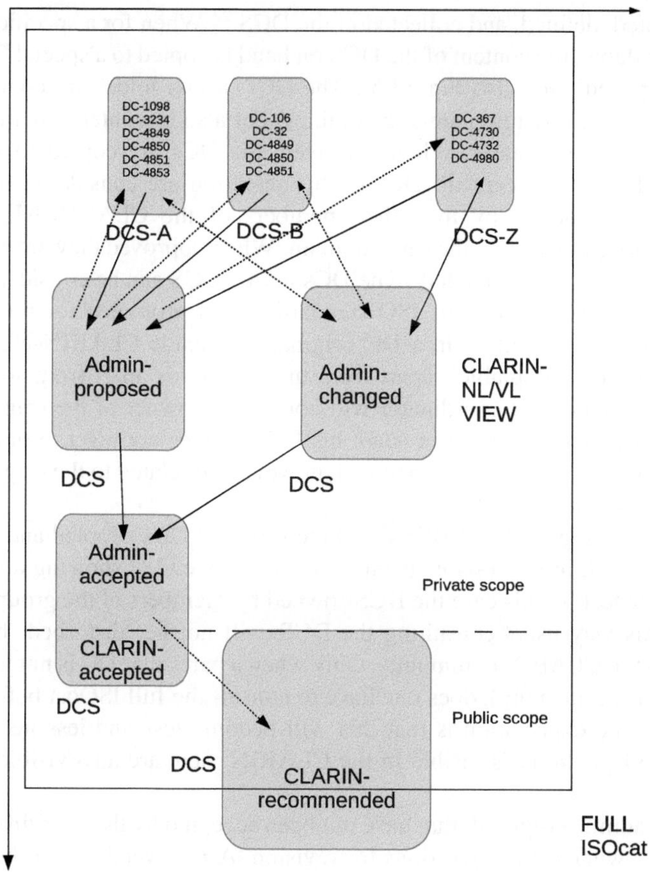


**Fig. 13.5** The organization of CLARIN NL/VL

'eventuality' which includes all kinds of actions, states, processes, etc. It is not to be confused with the more narrow notion of event as something that happens at a certain point in time [. . .]".[21]

It will be clear that in this case as well two DCs will be created for a concept, in which the narrow one is the daughter of the broader one (to be expressed using RELcat).

The idea in CLARIN-NL/VL is that for all CLARIN projects a DCS is constructed containing all DCs adopted and created for that project (see Fig. 13.5). Note that also that linguistic concepts used in the definitions of DCs are also to be

[21]Cf. http://semantic-annotation.uvt.nl/ISO-TimeML-08-13-2008-vankiyong.pdf, Sect. 3.1 (retrieved 2012-08-31)

disambiguated, defined, and collected in the DCS.[22] When for a specific project all DCs are available, the content of the DCS on hand is copied to a special DCS, called 'admin-proposed' (see also Fig. 13.5). The DCs in this folder are controlled with respect to technical requirements (guidelines) but also to content, by the CLARIN ISOcat content coordinator. When approved, the DCs are copied to yet another DCS, called 'admin-accepted'. DCs in this selection are considered for approval with regard to content by the other members of the CLARIN-NL/VL group, especially those working in the same domain. When approved they are promoted to a DCS 'CLARIN-recommended'. The DCs in this DCS can be considered de facto standardized (but are not (yet) ISO standards!) and promoted for reuse by third parties.[23] When a definition in a DC originating outside CLARIN-NL/VL seems to be useful, i.e., 'adoptable', apart from the profile, or apart from some smaller adaptations, the content coordinator will contact the owner of the original DC. In case the owner does not want to adapt his DC (or does not react to our request) a new DC should be created. This one will, however, be related to the original one via the RELcat Relation Registry.

As shown in Fig. 13.5, all DCSs, and therefore all DCs adopted and created, by CLARIN-NL/VL are accessible using a so-called 'view',[24] showing a limited part of the full ISOcat, in this case the DCSs owned by members of the group CLARIN-NL/VL. This way users consulting the DCR will not be 'disturbed' by DCs not relevant for the CLARIN community. Only when a particular DC is not yet available (in the intended meaning), does one have to consult the full ISOcat before creating a new DC. The expectation is that this will become less and less necessary. But of course, all public DCs visible in the CLARIN view are also visible in the full ISOcat.

DCs in 'admin-proposed' that have not been accepted by the coordinator are sent back to the owner with suggestions for revision. A new version is to be submitted to a DCS 'admin-changed', and after approval of the coordinator copied to the DCS 'admin-accepted' for approval with respect to content by the group.

In principle, DCs are only made 'public' when they are CLARIN-recommended (or, sometimes, admin-accepted: some DCs are created for historical reasons, but these concepts can no longer be recommended, so they will be copied to a DCS CLARIN-accepted). The reasons for this belated 'public'-status is that a DC should be settled before making it available to the world outside. Note that the very fact that non-standardized DCs, which holds for all DCs at this very moment, are still subject to semantic changes, discourages potential users from taking advantage of ISOcat. The CLARIN public DCs, however, are quite stable.

---

[22]For the time being, links with DCs for such concepts are mentioned in the note section.

[23]In the future such a DC will be explicitly recognizable as such.

[24]http://www.isocat.org/interface/index.html?view=CLARIN-NL/VL

```
@prefix relcat: <http://www.isocat.org/relcat/set/> .
@prefix rel: <http://www.isocat.org/relcat/relations#> .
@prefix dc: <http://purl.org/dc/elements/1.1/> .
@prefix isocat: <http://www.isocat.org/datcat/> .
relcat:cmdi {
   isocat:DC-2573 rel:sameAs dc:identifier .
   isocat:DC-2482 rel:sameAs dc:language .
   ...
   isocat:DC-2556 rel:subClassOf dc:contributor .
   isocat:DC-2502 rel:subClassOf dc:coverage .
}
```

**Fig. 13.6** Equivalence relationships

## 13.6 RELcat a Relation Registry

As indicated in the previous sections it has become clear that there are various needs to relate data categories to each other, e.g., because they cover the same underlying concept or are related in another way. To accommodate this the development of a Relation Registry named RELcat has started, cf. [13]. RELcat allows the storage of sets of relations containing the view of one user, a user group or even a CoP. A set of relations consists of typed individual relationships between two data categories. The type system starts out from a core taxonomy of relationship types:

(a) Related
  (a) Same as
  (b) Almost same as
  (c) Broader than (the inverse of the 'narrower than' relationship)
    (i) Superclass of (the inverse of the 'subclass of' relationship)
    (ii) Has part (the inverse of the 'part of' relationship)
  (d) Narrower than (the inverse of the 'broader than' relationship)
    (i) Subclass of (the inverse of the super class of' relationship)
    (ii) Part of (the inverse of the 'has part' relation- ship)

This taxonomy is just a start as it can easily be extended by types occurring in existing vocabularies, e.g., OWL and SKOS. Using this mechanism, equivalence relationships between ISOcat metadata (CMDI) DCs and Dublin Core elements have been established as shown in Fig. 13.6.

This example also illustrates the fact that the alpha implementation of RELcat is based on an RDF quad store. This implementation already stores several relation sets and a read only interface including query services based on SPARQL. At the time of writing it mainly lacks a convenient user interface allowing users to create and manage these sets.

## 13.7 Conclusions and Future Work

In this chapter we described the various efforts of CoPs and TDGs around the ISOcat DCR, which have resulted in the availability of a wide variety of DCs, but also uncovered problems with the envisioned ISO standardization procedure. The lack of standardized DCs sometimes leaves users of ISOcat on their own in making a choice between very similar DCs. Both ISO TC37 and larger user communities, i.e., CLARIN-NL/VL, have rekindled or started initiatives to assist the user. These efforts are supported by new functionality in ISOcat to make recommendations of these user communities more explicit, which creates an intermediate level of community approved DCs between the solely private DCs and the official ISO standardized DCs.

Thus it can be concluded that within the web of collaborative linguistic resources ISOcat is still struggling to find good practices to live up to its aim of defining widely accepted linguistic concepts. However, due to the open nature of ISOcat, CoPs can already establish their own practices and create well defined DCs and DCSs. With the appearance of Relation Registries like RELcat it will always be possible, when these DCs don't become a part of the standardized subset, to create crosswalks from these DCs to future standardized DCs.

## References


1. Budin G, Melby A (2000) Accessibility of multilingual terminological resources – current problems and prospects for the future. In: Proceedings of the second international conference on language resources and evaluation (LREC'00), Athens, Greece. ELRA
2. ISO (2008) Annex ST (normative) procedure for the development and maintenance of standards in database format. Technical report, International Organization of Standardization, Geneva, Switzerland
3. ISO:12200 (1999) Computer applications in terminology – machine-readable terminology interchange format (MARTIF). Technical report, International Organization of Standardization, Geneva, Switzerland
4. ISO:12620 (1999) Computer applications in terminology – Data categories. Technical report, International Organization of Standardization, Geneva, Switzerland
5. ISO:12620 (2009) Terminology and other language and content resources – specification of data categories and management of a Data Category Registry for language resources. Technical report, International Organization of Standardization, Geneva, Switzerland
6. ISO:16642 (2003) Computer applications in terminology – terminological markup framework. Technical report, International Organization of Standardization, Geneva, Switzerland
7. ISO:24611 (2012) Language resource management – morpho-syntactic annotation framework (MAF). Technical report, International Organization of Standardization, Geneva, Switzerland
8. ISO:30042 (2008) Systems to manage terminology, knowledge and content – TermBase eXchange (tbx). Technical report, International Organization of Standardization, Geneva, Switzerland
9. Kemps-Snijders M, Ducret J, Romary L, Wittenburg P (2006) An API for accessing the Data Category Registry. In: Proceedings of the fifth international conference on language resources and evaluation (LREC'06), Genoa, Italy. ELRA

10. Kuhn T (2000) The road since structure. University of Chicago Press, Chicago. Chap Commensurability, Comparability, Communicability
11. Schuurman I, Windhouwer M (2011) Explicit semantics for enriched documents. What do ISOcat, RELcat and SCHEMAcat have to offer? In: Proceedings of the second supporting digital humanities conference, Copenhagen, Denmark
12. Váradi T, Krauwer S, Wittenburg P, Wynne M, Koskenniemi K (2008) CLARIN: Common language resources and technology infrastructure. In: Proceedings of the sixth international conference on language resources and evaluation (LREC'08), Marrakech, Morocco. ELRA
13. Windhouwer M (2012) RELcat: a Relation registry for ISOcat data categories. In: Proceedings of the eighth international conference on language resources and evaluation (LREC'12), Istanbul, Turkey. ELRA
14. Wright SE, Budin G (2001) Handbook of terminology management: application-oriented terminology management. John Benjamins Publishing, Amsterdam

# Index



I. Gurevych and J. Kim (eds.), *The People's Web Meets NLP*, Theory and Applications of Natural Language Processing, DOI 10.1007/978-3-642-35085-6,

MIX
Papier aus verantwortungsvollen Quellen
Paper from responsible sources
FSC® C105338

If you have any concerns about our products,
you can contact us on
**ProductSafety@springernature.com**

In case Publisher is established outside the EU,
the EU authorized representative is:
**Springer Nature Customer Service Center GmbH**
**Europaplatz 3, 69115 Heidelberg, Germany**

Printed by Libri Plureos GmbH
in Hamburg, Germany